T0405915

Optical Fluorescence Microscopy

Alberto Diaspro
Editor

Optical Fluorescence Microscopy

From the Spectral to the Nano Dimension

Editor
Prof. Alberto Diaspro
Head, Nanophysics Unit
Senior Scientist
The Italian Institute of Technology -IIT
Via Morego, 30
16163 Genova
Italy
alberto.diaspro@iit.it

Department of Physics
University of Genova
Via Dodecaneso, 33
16146 Genova
Italy
diaspro@fisica.unige.it

ISBN 978-3-642-15174-3 e-ISBN 978-3-642-15175-0
DOI 10.1007/978-3-642-15175-0
© Springer Heidelberg Dordrecht London New York

Cover illustration: High resolution two-photon 3D immunofluorescence image of MAP-2 localised to neauronal-like cells in 11 days old embryoid body; see Fig.13.7 in "Near Infrared 3-Dimensional Nonlinear Optical Monitoring of Stem Cell Differentiation", Uday. K. Tirlapur and Clarence Yapp

Cover design: deblik Berlin, Germany

Printed on acid-free paper

Springer is part of Springer Science+Business Media (www.springer.com)

Preface

It was morning, and the new sun sparkled gold across the ripples of a gentle sea. . . . It was another busy day beginning. . . . But way off alone, out by himself beyond boat and shore, Jonathan Livingston Seagull was practising. . .No limits, Jonathan? he thought, and he smiled. His race to learn had begun.

– AD free excerption from "Jonathan Livingston Seagull – a story" by Richard Bach-First published in Great Britain by Turnstone Press 1972. This edition published by Element 2003

The intention behind this book is to provide a sort of starting point for those who are interested in the hard task of fully exploiting the capabilities offered by an optical microscopy fluorescence approach. It is fundamental to have some basic concepts about optical microscopy and fluorescence to understand their enormously high potential in basic and applied research. The classical spatial domain of optical microscopy is the submicron level dictated by the physics laws. Notwithstanding this we are moving to the nano-dimension, i.e., a spatial domain from 5 to 100 nm, exploiting the photophysics of the fluorescent markers being used and the technological abilities of detecting low signals and manipulating light in phase and amplitude including the utilization on nonlinear phenomena related to light–matter interactions.

When thinking about optical microscopy and the images coming from different contrast mechanisms, we mainly reason about intensity. In the last years, thanks to the rapid growth of fluorescent markers, we started thinking about spectral properties in terms of both emission and excitation. The advent of spectral microscopes and of white light laser sources enabled to improve the information content of microscopic images. At the very same time, there was an increasing demand for collecting information in terms of dynamics and of molecular interactions. To this end, there are two important methods such as fluorescence recovery after photobleaching (FRAP) that is able to provide information related to molecular diffusion and forster-fluorescence resonance energy transfer (FRET) that allows monitoring events at the nano-level. Within this scenario, spectroscopic techniques and single particle tracking approaches were developed along with an increasing knowledge about scattering phenomena affecting or "helping" image formation processes including those nonlinear phenomena known as second and third harmonic generation processes.

Modern optical microscopy cannot prescind from integration with other data sources like those coming from the IR interactions, X-rays, and electron microscopy. For this reason, this book contains important contributions as complement to the fluorescence microscopy methods mentioned above. I personally think that this is a good collection of contributions for having the appropriate control on optical methods in fluorescence microscopy toward super-resolution methods that have their own bases on most of the topics treated here from photophysics of fluorescent molecules to point spread function engineering.

I hope the reading of this book can stimulate new projects and ideas. I have so many people to thank including you, reader, that I would like to mention my wife Teresa, my daughter Claudia, and our puppy Sissi, a fantastic Cavalier king Charles dog, for sharing myself and their time with this project that reached the end thanks to the patience of all the authors and the professionality of Sabine Schwarz and Jutta Lindenborn. A special mention is for Prof. Antonino Zichichi, President of the Ettore Majorana Foundation and Centre for Scientific Culture in Erice, Sicily, where this book was born following the 36th Course of the Antonio Borsellino School of Biophysics. A warm thanks to all the students of the school. The IIT-Italian Institute of Technology is acknowledged for giving me an important opportunity for a new adventure in the optical microscopy field.

Genova, Italy Alberto Diaspro

Contents

Contributors

Naveen K. Balla Division of Bioengineering, National University of Singapore, 9 Engineering Drive, Singapore 117576, Singapore; Computation and Systems Biology, Singapore-MIT Alliance, National University of Singapore, Singapore 117576, Singapore

Rolf Borlinghaus Leica Microsystems CMS, Am Friedensplatz 3, 68165 Mannheim, Germany, Rolf.Borlinghaus@leica-microsystems.com

Kevin Braeckmans Laboratory of General Biochemistry and Physical Pharmacy, Ghent University, Harelbekestraat 72, 9000 Ghent, Belgium, Kevin.Braeckmans@UGent.be

E. Brunello Università degli Studi di Firenze, Florence, Italy

Gertrude Bunt Molecular and Cellular Systems, Department of Neuro and Sensory Physiology, University Medicine Goettingen, Humboldtallee 23, 37073 Goettingen, Germany, gbunt@gwdg.de

M. Caccia Dipartimento di Fisica G.Occhialini, Università di Milano Bicocca, Piazza della Scienza 3, 20126 Milan, Italy

Patrizio Candeloro Lab. BIONEM, Dipartimento di Medicina Sperimentale e Clinica, Università "Magna Graecia" di Catanzaro, Catanzaro, Italy

G. Chirico Dipartimento di Fisica G.Occhialini, Università di Milano Bicocca, Piazza della Scienza 3, 20126 Milan, Italy, giuseppe.chirico@mib.infn.it

G. Cojoc Lab. BIONEM, Dipartimento di Medicina Sperimentale e Clinica, Università "Magna Graecia" di Catanzaro, Catanzaro, Italy

M. Collini Dipartimento di Fisica G.Occhialini, Università di Milano Bicocca, Piazza della Scienza 3, 20126 Milan, Italy

L. D'Alfonso Dipartimento di Fisica G.Occhialini, Università di Milano Bicocca, Piazza della Scienza 3, 20126 Milan, Italy

Gobind Das Lab. BIONEM, Dipartimento di Medicina Sperimentale e Clinica, Università "Magna Graecia" di Catanzaro, Catanzaro, Italy; CalMED s.r.l., Catanzaro C.da Mula Loc. Germaneto, Campus Universitario Edificio Area Medicina, 88100 Catanzaro, Italy

Francesco DeAngelis Lab. BIONEM, Dipartimento di Medicina Sperimentale e Clinica, Università "Magna Graecia" di Catanzaro, Catanzaro, Italy; CalMED s. r.l., Catanzaro C.da Mula Loc. Germaneto, Campus Universitario Edificio Area Medicina, 88100 Catanzaro, Italy

Stefaan C. De Smedt Laboratory of General Biochemistry and Physical Pharmacy, Ghent University, Harelbekestraat 72, 9000 Ghent, Belgium

Enzo Di Fabrizio Lab. BIONEM, Dipartimento di Medicina Sperimentale e Clinica, Università "Magna Graecia" di Catanzaro, Catanzaro, Italy, Enzo.Difabrizio@iit.it; CalMED s.r.l., Catanzaro C.da Mula Loc. Germaneto, Campus Universitario Edificio Area Medicina, 88100 Catanzaro, Italy; INFM–TASC–S.S., 163.5 Area Science Park, Basovizza, Trieste, Italy

Jo Demeester Laboratory of General Biochemistry and Physical Pharmacy, Ghent University, Harelbekestraat 72, 9000 Ghent, Belgium

Hendrik Deschout Laboratory of General Biochemistry and Physical Pharmacy, Ghent University, Harelbekestraat 72, 9000 Ghent, Belgium

Alberto Diaspro LAMBS-IFOM, MicroScoBio, Department of Physics, University of Genoa, Via Dodecaneso 33, 16145, Genoa, Italy; Nanophysics, Italian Institute of Technology, Via Morego 30, 16163 Genova, Italy, alberto.diaspro@iit.it

R. Elangovan Università degli Studi di Firenze, Florence, Italy

Isabel Escobar Optics Department, Universitat de València, 46100 Burjassot, Spain

Umberto Fascio CIMA, Interdepartmental Center for Advanced Microscopy, University of Milan, Via Celoria, 26 – 20133 Milano, Italy, umberto.fascio@unimi.it

L Fusi Università degli Studi di Firenze, Florence, Italy, luca.fusi@unifi.it

Francesco Gentile Lab. BIONEM, Dipartimento di Medicina Sperimentale e Clinica, Università "Magna Graecia" di Catanzaro, Catanzaro, Italy; CalMED s.r.l., Catanzaro C.da Mula Loc. Germaneto, Campus Universitario Edificio Area Medicina, 88100 Catanzaro, Italy

M. Irving King's College, Cambridge, UK

Maria Laura Coluccio Lab. BIONEM, Dipartimento di Medicina Sperimentale e Clinica, Università "Magna Graecia" di Catanzaro, Catanzaro, Italy; CalMED s.r.l., Catanzaro C.da Mula Loc. Germaneto, Campus Universitario Edificio Area Medicina, 88100 Catanzaro, Italy

Zeno Lavagnino Department of Physics, LAMBS, University of Genova, Genova, Italy, Italian Institute of Technology, Genova, Italy

B. Lettiero Dipartimento di Medicina Sperimentale, Università di Milano Bicocca, Via Cadore 48, 20052 Monza, Italy

Carlo Liberale Lab. BIONEM, Dipartimento di Medicina Sperimentale e Clinica, Università "Magna Graecia" di Catanzaro, Catanzaro, Italy; CalMED s.r.l., Catanzaro C.da Mula Loc. Germaneto, Campus Universitario Edificio Area Medicina, 88100 Catanzaro, Italy

M. Linari Università degli Studi di Firenze, Florence, Italy

V. Lombardi Università degli Studi di Firenze, Florence, Italy

Manuel Martínez-Corral Optics Department, Universitat de València, 46100 Burjassot, Spain, manuel.martinez@uv.es

Federico Mecarini Lab. BIONEM, Dipartimento di Medicina Sperimentale e Clinica, Università "Magna Graecia" di Catanzaro, Catanzaro, Italy; CalMED s.r.l., Catanzaro C.da Mula Loc. Germaneto, Campus Universitario Edificio Area Medicina, 88100 Catanzaro, Italy

G. Miserocchi Dipartimento di Medicina Sperimentale, Università di Milano Bicocca, Via Cadore 48, 20052 Monza, Italy

T. Narayanan ESRF, Grenoble, France

P. Panine ESRF, Grenoble, France

M. Panzica Dipartimento di Fisica G.Occhialini, Università di Milano Bicocca, Piazza della Scienza 3, 20126 Milan, Italy

G. Piazzesi Università degli Studi di Firenze, Florence, Italy

Franco Quercioli Consiglio Nazionale delle Ricerche, Istituto Nazionale di Ottica, Largo E. Fermi 6, 50125, Firenze, Italy, franco.quercioli@ino.it

M. Reconditi Università degli Studi di Firenze, Florence, Italy

Shakil Rehman Division of Bioengineering, National University of Singapore, 9 Engineering Drive, Singapore, 117576, Singapore; Singapore Eye Research Institute, 11, Third Hospital Avenue, #05-00, Singapore, 168751 Singapore, shakil@nus.edu.sg

I. Rivolta Dipartimento di Medicina Sperimentale, Università di Milano Bicocca, Via Cadore 48, 20052 Monza, Italy

Emiliano Ronzitti LAMBS-IFOM, MicroScoBio, Department of Physics, University of Genoa, Via Dodecaneso 33, 16145 Genoa, Italy; SEMM-IFOM-IEO, University of Milan, Via Adamello 16, 20139 Milan, Italy

Genaro Saavedra Optics Department, Universitat de València, 46100 Burjassot, Spain

Emilio Sánchez-Ortiga Optics Department, Universitat de València, 46100 Burjassot, Spain

Anna Sartori-Rupp Institut Pasteur – Imagopole, 25 rue du Dr Roux, F – 75015 Paris, France, anna.sartori-upp@pasteur.fr

Elijah Y.Y. Seng Division of Bioengineering, National University of Singapore, 9 Engineering Drive, Singapore 117576, Singapore

Colin J.R. Sheppard Division of Bioengineering, National University of Singapore, 9 Engineering Drive, Singapore 117576, Singapore; Department of Diagnostic Radiology, National University of Singapore, 5 Lower Kent Ridge Road, Singapore 119074, Singapore

L. Sironi Dipartimento di Fisica G.Occhialini, Università di Milano Bicocca, Piazza della Scienza 3, 20126 Milan, Italy

Y.-B. Sun King's College, Cambridge, UK

Uday K. Tirlapur Stem Cell Epigenetics and Biomolecular Imaging Group, The Botnar Research Centre, Institute of Musculoskeletal Sciences, Nuffield Orthopaedic Centre, University of Oxford, Oxford 3 7LD, UK, uday.tirlapur@ndorms.ox.ac.uk

Giuseppe Vicidomini LAMBS-IFOM, MicroScoBio, Department of Physics, University of Genoa, Via Dodecaneso 33, 16145 Genoa, Italy; Department of NanoBiophotonics, Max Planck Institute for Biophysical Chemistry, Am Fassberg, Göttingen, Germany

Fred S. Wouters Laboratory for Molecular and Cellular Systems, Department of Neuro- and Sensory Physiology, Centre for Physiology and Pathophysiology, University Medicine Göttingen, Humboldtallee 23, 37075 Göttingen, Germany; DFG Research Center for Molecular Physiology of the Brain and Excellence Cluster EXC171 for Microscopy on the Nanometer Scale, Göttingen, Germany, fred.wouters@gwdg.de

Clarence Yapp Stem Cell Epigenetics and Biomolecular Imaging Group, The Botnar Research Centre, Institute of Musculoskeletal Sciences, Nuffield Orthopaedic Centre, University of Oxford, Oxford 3 7LD, UK

F. Cella Zanacchi LAMBS-IFOM, MicroScoBio, Department of Physics, University of Genoa, Via Dodecaneso 33, 16145 Genoa, Italy; Nanophysics, Italian Institute of Technology, Via Morego 30, 16163 Genova, Italy

Chapter 1
Fundamentals of Optical Microscopy

Franco Quercioli

1.1 Introduction

1.1.1 The Human Visual System

It would not be correct to start a chapter on the fundamentals of optical microscopy without even a brief mention of the essential characteristics of the human visual system for a better understanding of the underlying principle of microscope architectures. The human visual system is a stereo, auto-focus, auto-exposure imaging system. In "color mode" (photopic vision), the eye's retina disposes of a 7-Mega pixel detector (cones) and more than 100 Mega pixels (rods) in monochrome vision (scotopic). Its spatial resolution is about 7 line pairs per millimeter and its temporal resolution about ten frames per second. However, the most amazing characteristic is perhaps its dynamic range of about 90 dB and sensibility: the retina can, in fact, detect as few as 50–100 photons per second.

The eye can accommodate its focal length by changing the optical power of the crystalline lens. The near point of accommodation is the minimum distance at which the eye can focus clearly; it changes with age and from person to person. For a normal eye, the generally agreed value is, by convention, 25 cm. At this minimum distance, the retinal image of the naked eye is at its maximum. To go beyond this limit, some sort of magnifier is needed. The magnifying glass (or simple microscope) is such a device in its simplest form.

F. Quercioli
Consiglio Nazionale delle Ricerche, Istituto Nazionale di Ottica, Largo E. Fermi 6, 50125, Firenze, Italy
e-mail: franco.quercioli@ino.it

A. Diaspro (ed.), *Optical Fluorescence Microscopy*,
DOI 10.1007/978-3-642-15175-0_1,

1.1.2 History

Who knows when and where the first magnifying device was unveiled. Certainly, a number of ancient works of art could not have been accomplished without the use of some kind of magnifier, perhaps a simple drop of water or some other kind of transparent material.

The early history of lenses is in fact controversial. Lens-shaped objects (made of rock crystal or glass) had been in fact produced in the eastern Mediterranean since the Bronze Age, but most of these ancient "lenses" must have only been decorative objects and not intended as magnifying devices (Plantzos 1997).

To date, the earliest lenses identified are from the Fourth and Fifth Dynasties of Egypt (roughly from 2620–2400 BC and 1750–1700 BC). They were parts of eye structures for insertion into funerary statues, and not actually used as magnifiers (Enoch 1998, 2000).

In 1853, Sir David Brewster presented a rock crystal lens to the British Association for the Advancement of Science that had been found in excavations at Nineveh, the ancient capital of Assyria. This famous "Nineveh lens," now at the British Museum, had been found in deposits dated around 600 BC, but in this case also, doubts were raised about its function (Brewster 1855; Barker 1930).

The earliest written records of lenses can be traced to ancient Greece; a burning glass is mentioned in Aristophanes' comedy of "the Clouds" (around 420 BC). Burning glasses were also known to the Romans as found in the writings of Pliny the Elder (23–79 AD), who also describes Emperor Nero as viewing the gladiatorial games through an emerald. Again, both Pliny (Hist. Nat.) and Seneca the Younger (Quest. Nat.) described the magnifying effect of a glass globe filled with water. Lenses have also been found in excavations in Ercolano e Pompei; moreover, the name lens itself comes from the Latin word for lentil.

It took about 1,000 years for the presentation of a theoretical explanation of the magnifying properties of the lens, which is attributed to the Arabians Ibn Sahl (ca 940–ca 1000) and Ibn al-Haytham (965–1038). Known in the West as Alhazen, Ibn al-Haytham wrote the first major optical treatise. Roger Bacon in his *Opus Majus* (1267) described the magnification of small objects using convex lenses.

During this same period, magnification devices known as reading stones were of common use in Europe. They were hemispherical lenses which, once placed on top of a text, magnified the letters (Schmidt et al. 1999; Graydon 1998).

The early intensive use of a simple lens as a true microscopic device is due to Antony van Leeuwenhoek at the end of the seventeenth century. He is commonly considered to be the first microbiologist.

Although more recent and documented, even the history of the compound microscope is, in some respects, controversial. The first compound microscopes were produced around the year 1600, quite the same time as the telescope.

Some scholars attribute the invention to Hans and Zacharias Janssen; however, others contend that the invention is to be credited to Galileo. In *Il Saggiatore* (1623), the Pisan scientist mentioned a "telescope modified to see objects very close."

What is not at all a matter of debate is the origin of its name: in 1625 a member of the Accademia dei Lincei and friend of Galileo, Johannes Faber conferred on the instrument, until then called "occhialino," "cannoncino," "perspicillo," and "occhiale," the name of "microscopio."

The name comes from the Greek words μικρον (micron) meaning "small" and σκοπειν (skopein) meaning "to look at" (IMSS 2007).

From a substantial point of view, nothing really new happened for about two centuries until the theoretical work of Ernst Abbe (1840–1905), who revolutionized the design of optical instruments with his fundamental contribution to the wave theory of image formation and his renowned formula of the limit of resolution of an objective.

Only nowadays the Abbe limit has been circumvented in some very special experimental conditions, and this will open the way to an entirely new and fascinating chapter of microscopy: optical nanoscopy (Hell 2007).

1.1.3 The Basic Structure

As we have previously said, the retinal image of an object located at the near point of accommodation, at a distance of 25 cm from the eye, has maximum dimensions.

To go beyond this limit, a magnifier is needed. The simple microscope or magnifying glass allows the observed object to be brought closer, to a distance equal to the focal length f of lens. The magnification factor with respect to the naked eye is $M = 25/f$ (Fig. 1.1).

In the compound microscope, a second lens is added: the objective (Fig. 1.2). It produces a real intermediate image of the object at a fixed distance of 16 cm which is, by standard, the length of the microscope tube. This image is then further magnified by a second lens, the eyepiece, which behaves like the simple microscope previously described.

This kind of setup is very effective to obtain maximum performances. The objective is designed to best match the object characteristics, mainly to set the

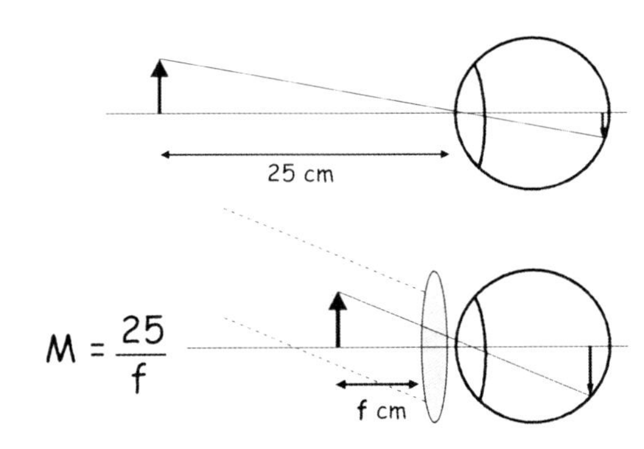

Fig. 1.1 The simple microscope or magnifying glass

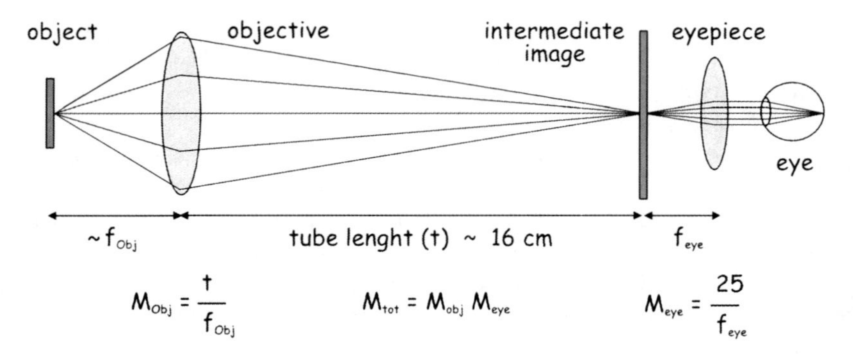

Fig. 1.2 The compound microscope

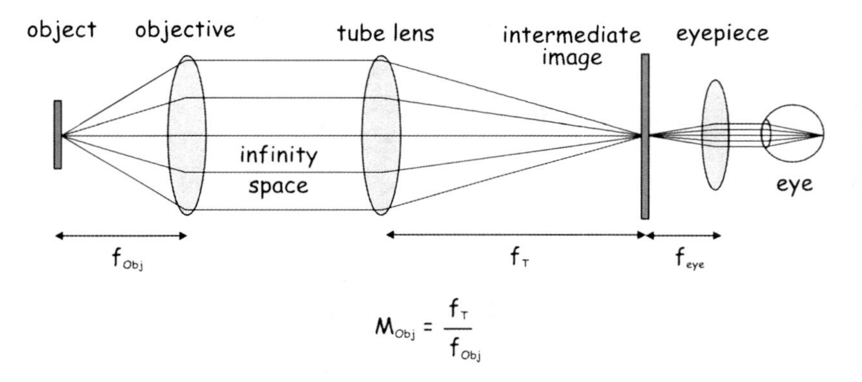

Fig. 1.3 The compound microscope with infinity-corrected objectives

system maximum spatial resolution, and the eyepiece, on the other hand, has to match the detector characteristics (retina).

This basic design of the compound microscope did not change since its invention for about four centuries, until the late eighties, when manufacturers began to address infinity-corrected objectives, and a third optics, the tube lens, was added (Fig. 1.3). This new design is not very different from the original one in principle , but it has great practical advantages. In the infinity space, in fact, many different kinds of optical devices can be located without affecting the optical performances of the instrument with the introduction of aberrations.

After four centuries, optical microscopy is still the observation technique of choice for biological systems. Despite the electron microscopy era and the scanning probe ones, optical microscopy is still alive and well because it allows a three-dimensional investigation of living specimens in their physiological environment.

Basically optical microscopy consists in recording a light signal in some points in space and time.

Light, space, and time are the physical quantities involved in optical microscopy.

1.2 Light

In optical microscopy, light is used to probe a sample. To do this with maximum efficiency, causing minimum damage to the sample itself, suitable imaging configurations are exploited to get maximum signal-to-noise ratio (*S*/*N*). All these methods are named contrast techniques and are the subject of this section.

1.2.1 Ray Optics

Light is a form of energy; according to the specific experimental condition, it is easier to describe light using one or the other of these pictures: rays, waves, or particles.

Rays are used in the realm of geometrical optics. Reflection, refraction, and absorption (at least from an empirical point of view) phenomena are well described using this model.

Ray tracing is the technique used to design almost all the optical elements of a microscope and to draw the light path inside the instrument.

1.2.2 Bright Field Microscopy

Bright field microscopy (Fig. 1.4) represents the oldest, most common, and simplest configuration. The spatial distribution of the object absorption characteristics or its reflectivity, according to whether transmission or reflection observation geometry is used, is converted into image intensity modulations.

Living objects are made up almost of water, which is invisible because it does not absorb in the visible wavelength range. The earliest solution used to achieve some contrast in the image was the staining of the sample, a technique which is still in wide use for histological preparations.

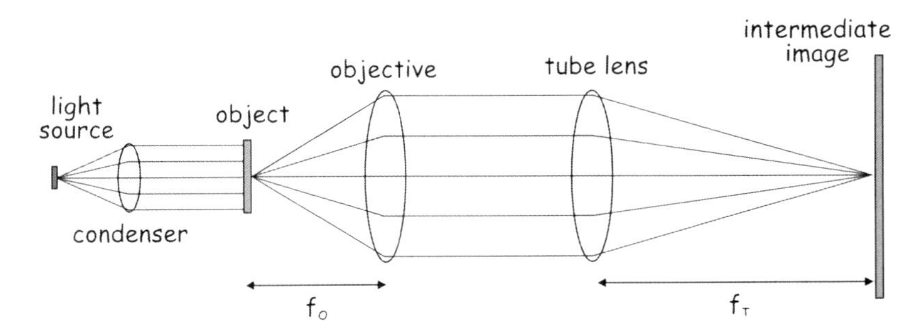

Fig. 1.4 Bright field microscopy

However, staining, at least in the past, was not compatible with living specimens, and less invasive contrast techniques had been looked for.

1.2.3 Köhler Illumination

To understand how all these contrast enhancement techniques work, a closer look must be taken at the light path inside the microscope (Fig. 1.5).

The figure shows, in a very simplified form, the commonly used Köhler illumination setup. A light source is placed in the back focal plane of a condenser and illuminates the specimen. The objective forms an image onto the intermediate image plane. This image can be directly acquired by a CCD camera or can be observed through an eyepiece, and in this case, a real image is formed on the eye's retina. The object plane, the intermediate image plane, and the retina are all conjugate planes, and the same relationship holds for the source plane, the objective back focal plane, and the eye's pupil plane.

1.2.4 The Spatial Frequency Plane

An object that does not absorb light is transparent and thus invisible; however, it can be capable of deviating light rays due to refraction phenomena if it possesses thickness or refractive index modulations. The light rays propagating at different directions are then focused by the objective onto its back focal plane at different points (Fig. 1.6).

Each point on this plane collects the light coming from a single propagation direction which in turn corresponds to a single scattering direction generated by the object, more physically speaking: a single object spatial frequency. This plane takes the name of spatial frequency plane or Fourier plane (Goodman 1968).

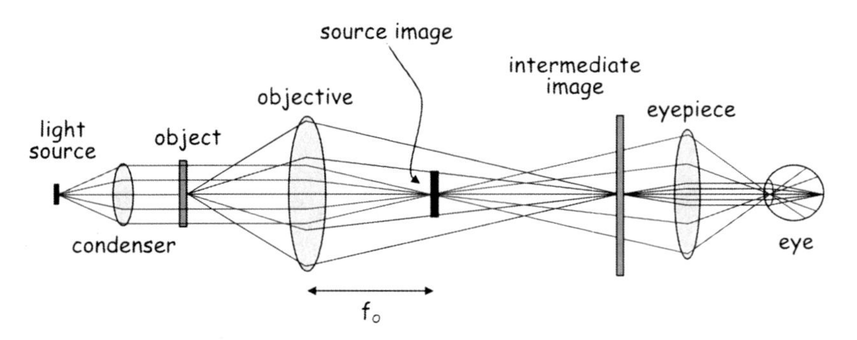

Fig. 1.5 Köhler illumination

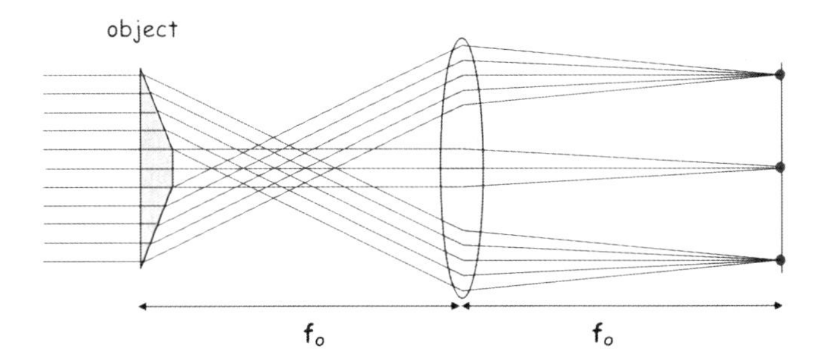

Fig. 1.6 The spatial frequency plane

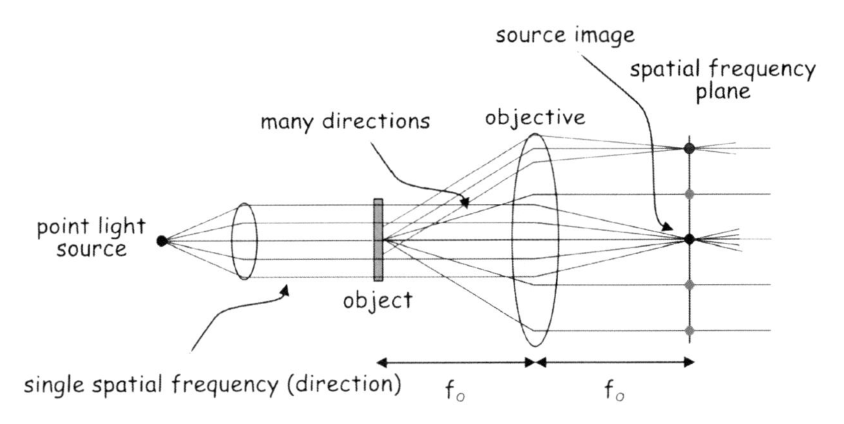

Fig. 1.7 The spatial frequency filtering

More physically speaking, the amplitude distribution in this plane is proportional to the Fourier transform of the object transmittance function.

1.2.5 The Spatial Frequency Filtering

The undeviated light, which contributes to the image background, goes to focus on axis. If we stop this light, we obtain an image with a dark background.

This technique, known as dark field microscopy (Fig. 1.7), is the simplest example of contrast enhancement based on spatial frequency filtering (Goodman 1968).

A point source, though useful for modeling purposes, is not a good realistic representation.

A real source has finite dimensions, and for an extended source, each point on the spatial frequency plane is no more related to a single object spatial frequency (Fig. 1.8).

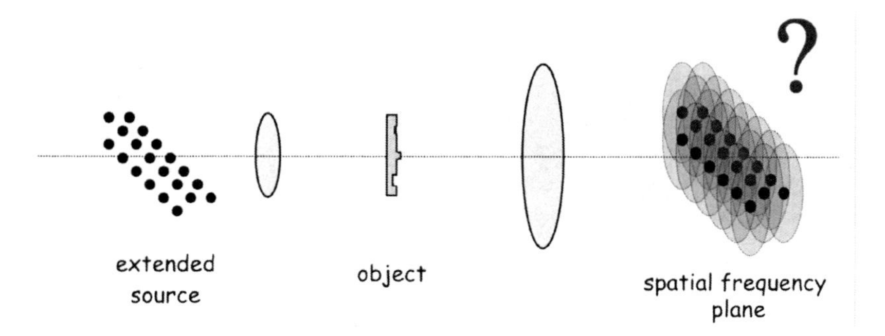

Fig. 1.8 The spatial frequency plane with an extended source

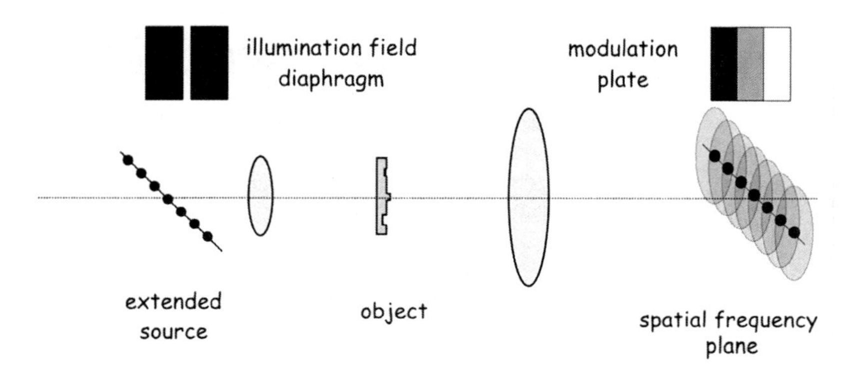

Fig. 1.9 The spatial frequency plane with a linear source

The trade-off between a light-efficient two-dimensional source and a point source with spatial filtering feasibility is one dimensional.

In practice, a diaphragm is used to shield the source into a linear shape so that a one-dimensional symmetry is realized (Fig. 1.9). In the Fourier plane, perpendicular to the linear source image, different spatial frequencies are separated and can be filtered. This is, for example, the case of Hoffmann modulation or oblique illumination techniques (Hoffman and Gross 1975).

Alternatively, a circular symmetry (Fig. 1.10) can be utilized as for the annular shaped source in phase contrast microscopy (Goodman 1968).

1.2.6 Digital Processing

The remarkable characteristic of an optical system to generate the Fourier transform of an object transmittance in the back focal plane of an objective was

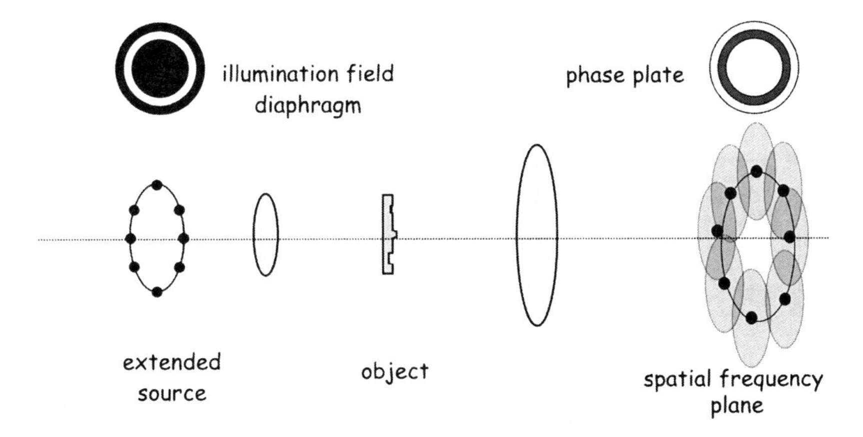

Fig. 1.10 The spatial frequency plane with a circular source

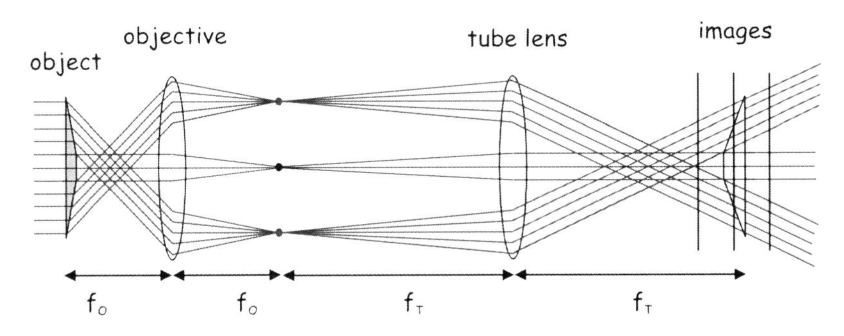

Fig. 1.11 Quantitative phase imaging (QPI)

intensively exploited in the years 1970–1980 for spatial frequency filtering, or simply optical filtering, in the field of optical signal processing techniques. An optical system can, in fact, be regarded as a sort of analog computer (Goodman 1968). In more recent years, due to the widespread availability of digital computers, new signal processing techniques have become possible, and microscopy also has benefitted.

In quantitative phase imaging (QPI) (Curl et al. 2004), for example, the in-focus image and the light distribution at two other nearby planes along the optical axis are acquired (Fig. 1.11). Different ray directions impinge upon different areas in the three image planes and, using suitable computing algorithms, the information about the refractive characteristics of the object are retrieved (that is, its thickness and refractive index modulation). Classical contrast enhancement techniques like dark field or phase contrast can also be obtained by computer simulation. At least two commercial devices based upon this principle are available on the market (Bossard 2007).

1.2.7 Wave Optics

To take a step further toward more complex contrast techniques, we must abandon the geometrical optics approach for the wave optics one, and a little bit of mathematics is also needed.

Light is an electromagnetic wave. The simplest waveform is a sinusoidal one, which mathematically describes a pure monochromatic light (Fig. 1.12). In scalar approximation (Born and Wolf 1999), the electric field F can be described as follows:

$$F = A\sin(2\pi nz/\lambda + 2\pi\nu t). \tag{1.1}$$

It represents a monochromatic plane wave propagating along the z-axis, let us say, the microscope axis. The characteristics of the wave are: the field amplitude A, which represents the strength of the field; the refractive index n of the propagation medium (air, glass, specimen); the wavelength λ, which is the distance in space between two consecutive maxima along the propagation axis z; the analogous distance in the time domain is called period, while its inverse is the frequency ν; the whole argument of the sine function is called phase and locates the position along the optical cycle.

To be a real electromagnetic wave, the relationship $\lambda\nu = c$ must hold, where c indicates the velocity of light.

When light interacts with matter, as the sample on the x, y-stage of a microscope, A can be changed due to absorption mechanisms. Spatial absorption modulations of the object are converted into field amplitude modulations $A(x, y)$. On the other hand, thickness $t(x, y)$ or refractive index $n(x, y, z)$ modulations modify the phase of the illuminating wave:

$$\Phi(x, y) = (2\pi/\lambda)\int_0^t n(x, y, z)\mathrm{d}z. \tag{1.2}$$

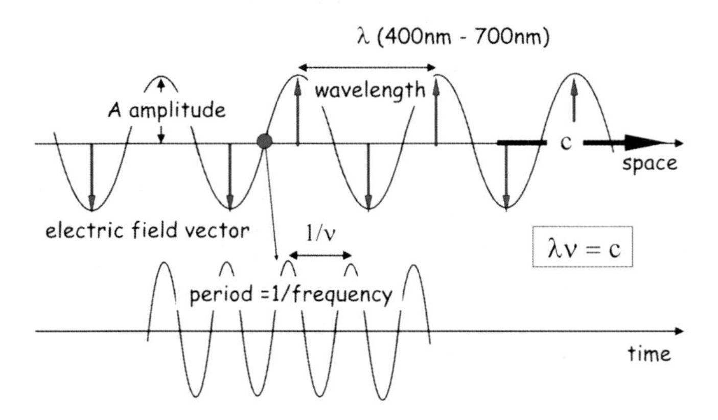

Fig. 1.12 The scalar approximation of a sinusoidal (monochromatic) light wave

This is because a light beam inside a medium is slowed down by a factor n ($v = c/n$) and thus accumulates a phase delay (Fig. 1.13). The integral in (1.2) is named optical path.

The transmitted field can then be written as follows:

$$F = A(x, y)\sin[\Phi(x, y) + 2\pi\nu t]. \tag{1.3}$$

A detector, be it the eye, a CCD camera, a photomultiplier, or whatever, senses energy, the temporal mean of the energy which, in the end, is proportional to the squared value of the field:

$$I \propto \langle F^2 \rangle = A^2(x, y)\langle \sin^2[\Phi(x, y) + 2\pi\nu t]\rangle = A^2(x, y)/2, \tag{1.4}$$

where I is the intensity and the symbol $\propto$ stands for "proportional to", while the brackets $< >$ mean temporal average. The amplitude A is the only measurable quantity. A nonabsorbing object does not produce any modulation of the field amplitude A, but only phase Φ modulations, unfortunately, these cannot be detected and the sample is invisible.

1.2.8 *Interference*

Biological specimens are almost nonabsorbing objects. One of the main efforts of optical microscopy had been to make them visible without the use of staining techniques but exploiting only the properties of light.

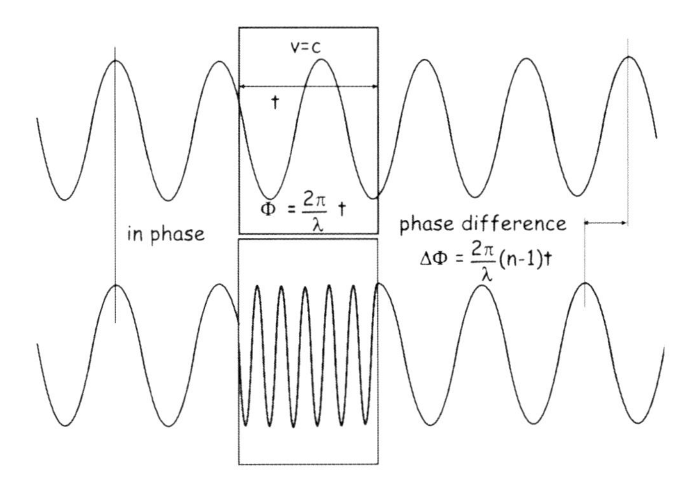

Fig. 1.13 The phase delay due to thickness t or refractive index n modulations

Light can be added. When two waves have the same phase, they are said to be in phase ($\Phi_1 = \Phi_2 = 0$ for simplicity), and their amplitudes sum up, and, if they have the same amplitude ($A_1 = A_2 = A$), the total energy is four times that of a single component.

$$I \propto <[A\sin(2\pi\nu t) + A\sin(2\pi\nu t)]^2> = (2A)^2 <\sin^2(2\pi\nu t)> = 4(A^2/2). \quad (1.5)$$

On the other hand, when two waves are in phase opposition ($\Phi_1 = 0$; $\Phi_2 = \pi$), a negative effect holds and the total energy is zero.

$$\begin{aligned} I \propto &<[A\sin(2\pi\nu t) + A\sin(\pi + 2\pi\nu t)]^2> \\ &= <[A\sin(2\pi\nu t) - A\sin(2\pi\nu t)]^2> = 0. \end{aligned} \quad (1.6)$$

Light can be subtracted! How is it then that switching on two lamps never darkens a room; neither does it give four times the light of a single lamp? In fact, at each instant of time, the space is full of tiny regions, with size of the order of a wavelength, where the intensity is zero or 4I, however these fluctuations last only for few femtoseconds. This is because natural light does not behave so well as the previous equations would show.

Natural light is a real mess. The ideal monochromatic wave of (1.3) does not exist. This lack of monochromaticity can be modeled by considering the spatial phase term Φ as not constant but varying very rapidly in time; this is naïvely due to the not well-defined value of λ in (1.2). The general equation is as follows:

$$\begin{aligned} I \propto &<\{A_1 \sin[\Phi_1(t) + 2\pi\nu t] + A_2 \sin[\Phi_2(t) + 2\pi\nu t]\}^2> \\ &= {A_1}^2/2 + {A_2}^2/2 + A_1A_2< \cos[\Phi_1(t) - \Phi_2(t)]>. \end{aligned} \quad (1.7)$$

Two distinct cases can then occur:

$$I \propto {A_1}^2/2 + {A_2}^2/2 \quad (1.8)$$

when $\Phi_1(t)$ and $\Phi_2(t)$ are completely uncorrelated (Fig. 1.14)

$$I \propto {A_1}^2/2 + {A_2}^2/2 + A_1A_2 \cos(\Delta\Phi) \quad (1.9)$$

when $\Phi_1(t)$ and $\Phi_2(t)$ are completely correlated, that is, they singularly change in a random manner but their difference remains constant in time: $\Phi_1(t) - \Phi_2(t) = \Delta\Phi$ (Fig. 1.15).

In the first case [see (1.8)], the total intensity is the sum of the single intensities, as common experience shows. The two light components are said to be incoherent. The second case, on the other hand, exhibits the same "strange" behavior described previously and (1.9) reduces to (1.5) and (1.6) for $A_1 = A_2 = A$, and $\Delta\Phi = 0, \pi$, respectively. This time, the two light components are said to be coherent. This

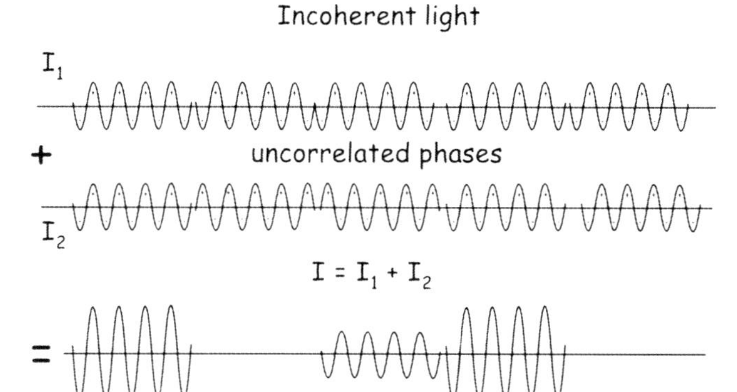

Fig. 1.14 Incoherent light

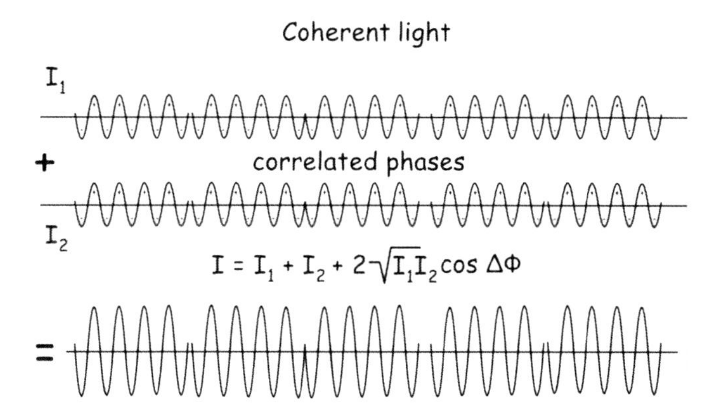

Fig. 1.15 Coherent light

"strange" phenomenon is named interference, and, as shown in (1.9), it depends on the phase difference $\Delta\varphi$. Interference can then be profitably exploited to measure phase. However, to fulfill the condition for interference [see (1.9)], the two light components must be correlated, which necessitates that they be originated from the same point source at the same time (Fig. 1.16). Fluorescence emission is an incoherent process and interference contrast methods cannot be used.

1.2.9 Phase Contrast

Phase contrast microscopy was introduced by Frits Zernike in 1930. For his invention, he was awarded the Nobel prize in physics in 1953 (Zernike 1955).

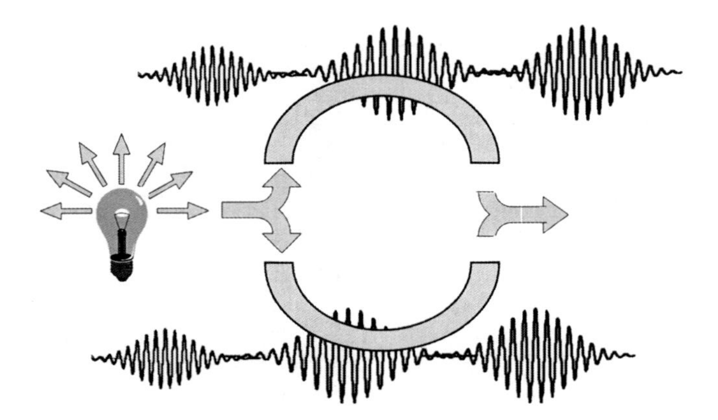

Fig. 1.16 Interference occurs when two light beams originate from the same point source at the same time

When a monochromatic plane wave illuminates a transparent sample located in the x, y-plane of the microscope stage, only the phase $\Phi(x, y)$ of the transmitted wave field is spatially modulated [see (1.3)].

$$\begin{aligned} F &= A\sin[\Phi(x,y) + 2\pi\nu t] \\ &= A\sin(2\pi\nu t)\cos[\Phi(x,y)] + A\cos(2\pi\nu t)\sin[\Phi(x,y)]. \end{aligned} \tag{1.10}$$

For a low phase modulation depth: $\Phi(x, y) \ll 1$, the following approximations hold: $\cos[\Phi(x, y)] \cong 1$; $\sin[\Phi(x, y)] \cong \Phi(x, y)$ and (1.10) becomes:

$$F \cong A\sin(2\pi\nu t) + A\Phi(x,y)\cos(2\pi\nu t). \tag{1.11}$$

These two components of the transmitted wave field propagate, pass through the microscope objective, and sum up (interfere) in the image plane. For an ideal optical system, the image field is identical to the object one [see (1.11)], and the detected intensity I is a constant ($A^2/2$), as previously said [see (1.4)] with no contrast.

Along the way, in the spatial frequency plane, the first component of (1.11), which represents a constant term not dependent on x- and y-coordinates, proceeds undeviated and goes to focus on axis, while the second one, whose amplitude $A\,\Phi(x, y)$ is spatially modulated, spreads out in the objective back focal plane (Fourier plane).

Being spatially separated in the Fourier plane, these two components can be independently processed.

In dark field filtering, the first term is eliminated:

$$F_{\mathrm{DF}} \cong +\ A\Phi(x,y)\cos(2\pi\nu t), \tag{1.12}$$

and the image intensity will be:

$$I_{\mathrm{DF}} \propto \langle F_{\mathrm{DF}}{}^2 \rangle = A^2\Phi^2(x,y)\langle\cos^2(2\pi\nu t)\rangle = A^2\Phi^2(x,y)/2. \tag{1.13}$$

A bright intensity distribution is proportional to the squared value of the sample phase on a dark background.

Dealing in a similar way, for phase contrast filtering, instead of shielding the on axis light, a $\pi/2$ phase shift is introduced ($a - \pi/2$ phase shift could also be used), and the image wave field will be:

$$F_{FC} \cong A\sin(2\pi\nu t + \pi/2) + A\Phi(x,y)\cos(2\pi\nu t) \\ = A\cos(2\pi\nu t)[1 + \Phi(x,y)], \tag{1.14}$$

and the detected energy:

$$I_{FC} \propto \langle F_{FC}{}^2\rangle = A^2\langle\cos^2(2\pi\nu t)\rangle[1 + \Phi(x,y)]^2 \\ = (A^2/2)[1 + \Phi^2(x,y) + 2\Phi(x,y)] \cong A^2/2 + A^2\Phi(x,y); \tag{1.15}$$

where the $\Phi^2(x, y)$ has been neglected according to the initial assumption: $\Phi(x, y) \ll 1$.

This time the phase term $\Phi(x, y)$ modulates the image intensity.

In writing (1.14) and (1.15), the coherence characteristics of the two interfering components have been taken into account.

The two components are in fact coherent because all the light comes from a single point source.

For an annular source, each point source along the circumference produces the same image intensity contribution given by (1.15); all these contributions then sum up incoherently in the image plane because, this time, the light comes from different point sources.

The $\pi/2$ phase shift in the back focal plane of the phase contrast objective is realized with an annular shaped dielectric deposition, clearly visible looking through the optics, whose thickness t and refractive index n satisfy the relation: $nt = \lambda/4$.

1.2.10 *Differential Interference Contrast*

Differential interference contrast (DIC) technique was introduced by Nomarsky in the mid-1950s (Fig. 1.17) (Nomarski 1955). It consists of two lateral shearing interferometers in cascade. The first shear is applied to the source beam to produce two identical copies of the illuminating field shifted by a distance Δx one with respect to the other. The second, and opposite, shift of $-\Delta x$ is applied to the transmitted fields. The resulting interfering components are coherent because each illuminating point interferes with the copy of itself.

The light from each of the twin illuminating couples passes through the sample at two laterally displaced points shifted by the distance Δx, and the transmitted field and intensity can then be written as follows:

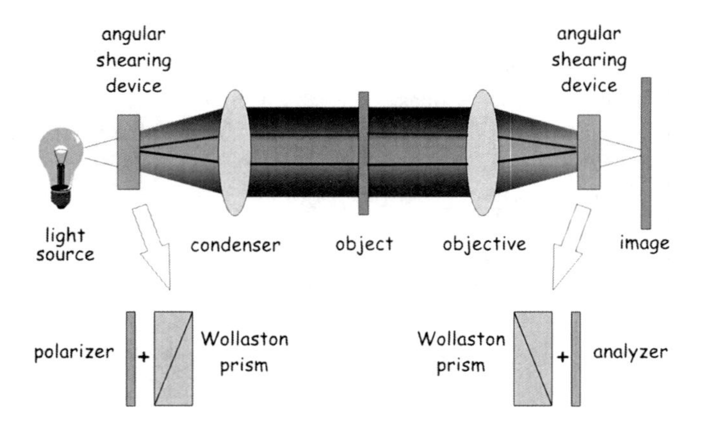

Fig. 1.17 Differential interference contrast (DIC) technique

$$\begin{aligned} F_{\mathrm{DIC}} &\cong [A\sin(2\pi\nu t) + A\Phi(x+\Delta x, y)\cos(2\pi\nu t)] - [A\sin(2\pi\nu t) \\ &\quad + A\Phi(x,y)\cos(2\pi\nu t)] \\ &\cong A\Delta x\Phi'(x,y)\cos(2\pi\nu t)\,, \end{aligned} \tag{1.16}$$

$$I_{\mathrm{DIC}} \propto \langle F_{\mathrm{DIC}}{}^2 \rangle = A^2\Delta x^2\Phi'^2(x,y)/2\,, \tag{1.17}$$

where the approximation for a low phase modulation depth has been taken into account [see (1.11)].

The minus sign of the second field component can be obtained by introducing a π phase shift in one arm of the interferometer path. Φ' is the sample phase derivative along the shift direction (x), and from this derives the name of the technique.

In this classical configuration, the lateral shears are generated by a polarizer and a Wollaston prism placed in the front focal plane of the condenser and another prism followed by an analyzer in the objective back focal plane.

The use of polarized light can be sometime unsuitable when birifringent objects are placed in the optical path as, for example, plastic Petri dishes. Zeiss has introduced a modified configuration, named PlasDIC (Wehner 2003; Danz et al. 2004), in which the light in the illuminating beam path is not polarized. To produce a coherent source field, a device named "coherent diaphragm" (some sort of diffracting diaphragm) substitutes the polarizer+Wollaston prism device in the illuminating path, while a polarizer+prism+analyzer is placed in the imaging light path.

1.2.11 Digital Holographic Microscopy

The use of a highly coherent light source, such as a laser, allows more degrees of freedom in the choice of the interferometric setup. In digital holographic microscopy (DHM), a laser beam is split into a reference beam F_R, which proceeds directly toward the imaging detector, and an object beam F_O, which illuminates the sample and then interferes with the other (Cuche et al. 1999a, b) (Fig. 1.18):

$$F_R = R\sin[(2\pi/\lambda)x\sin\beta + 2\pi\nu t], \tag{1.18}$$

$$F_O = A(x,y)\sin[\Phi(x,y) + 2\pi\nu t]. \tag{1.19}$$

Equation (1.19) represents the image field that carries the information of amplitude $A(x, y)$ and phase $\Phi(x, y)$ of the sample transmission, while (1.18) represents the reference plane wave that propagates off-axis along a direction making an angle β with the microscope z-axis. The hologram intensity will be [see (1.9)]:

$$\begin{aligned} I_H &\propto \langle (F_R + F_O)^2 \rangle \\ &= R^2/2 + A^2(x,y)/2 + RA(x,y)\cos[\Phi(x,y) - (2\pi/\lambda)x\sin\beta]. \end{aligned} \tag{1.20}$$

From the third term of (1.20), the two functions $A(x, y)$ and $\Phi(x, y)$ can be extracted and the whole sample information recovered. In a classical holographic setup, (1.20) represents the hologram plate transmittance. The plate, once illuminated by a copy of the reference plane wave, reconstructs the object field (Goodman 1968). This kind of analog processing can be replaced with a digital one. Instead of using a photographic plate, a CCD camera acquires the image which is then stored and processed by a computer using Fourier transform routines. To obtain a faithful recovery either analogically or digitally, the modulating spatial frequency $\sin\beta/\lambda$ must be higher than three times the maximum frequency of the object field (Goodman 1968). This condition imposes high spatial resolution requirements for the CCD detector (Ferraro et al. 2006).

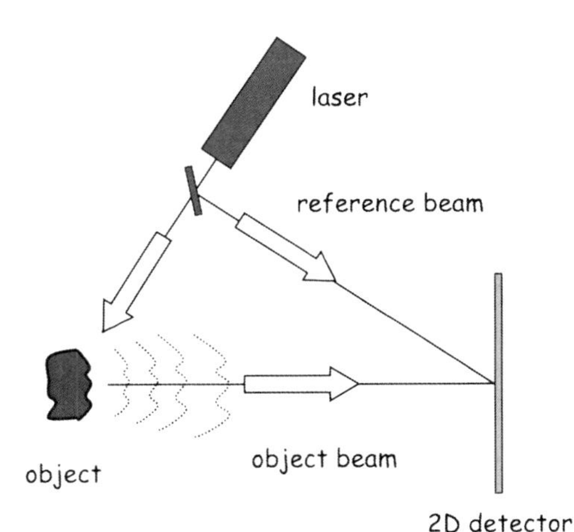

Fig. 1.18 Principle of holography

1.2.12 Polarization Contrast

Light is indeed a vector wave. Inside a homogeneous and isotropic medium, light propagates as a transversal wave where the electric and magnetic vectors are perpendicular to each other and to the direction of propagation. For an ideal monochromatic plane wave, the electric vector (and consequently the magnetic one) oscillates in time, at the optical frequency ν, tracing out a line, a circle, or an ellipse for linear, circular, or elliptical state of polarization, respectively.

As previously mentioned, however, natural light does not behave so well. The polarization state changes very rapidly in time and it does not exist in a well-defined state. Natural light is said to be nonpolarized.

A linear polarizer is an optical device that filters out all but the light whose electric vector oscillates along a single direction (the polarizer axis). When this linearly polarized light illuminates an anisotropic material, the polarization state of the transmitted light changes; this phenomenon is called birefringence. A second polarizer (analyzer) converts the modulated polarization into intensity variations (Fig. 1.19). The simplest birefringent materials are the uniaxial ones, where a single axis of anisotropy (optical axis) exists. Light with linear polarizations perpendicular (ordinary) and parallel (extraordinary) to the optical axis experiences different refractive indices: n_{O} and n_{E}, respectively. Using again the general equation for interference [see (1.9)]:

$$I_{\mathrm{P}} \propto P_{\mathrm{O}}^2/2 + P_{\mathrm{E}}^2/2 + P_{\mathrm{O}}P_{\mathrm{E}}\cos(\Delta\Phi). \tag{1.21}$$

$$\Delta\Phi = \Phi_{\mathrm{O}}(x,y) - \Phi_{\mathrm{E}}(x,y) = (2\pi/\lambda)\int_0^t \Delta\, n\mathrm{d}z. \tag{1.22}$$

P_{O} and P_{E} are the amplitudes of the output field components polarized along the ordinary and extraordinary axis of the birefringent specimen, respectively. Their magnitudes depend on the orientations of the input polarizer and output analyzer with respect to the specimen optical axis.

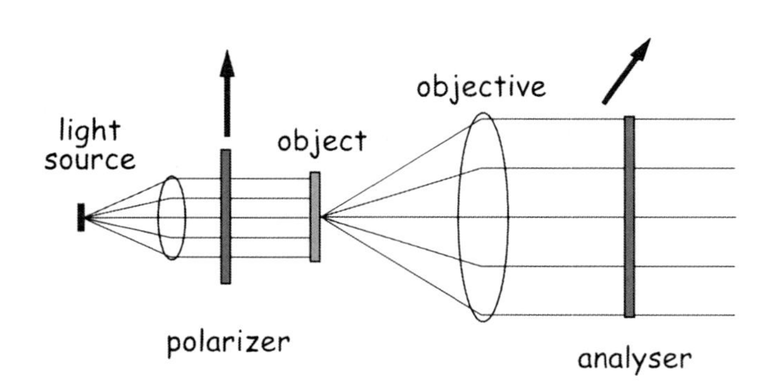

Fig. 1.19 Polarized light microscopy

$\Phi_O(x, y)$ and $\Phi_E(x, y)$ are the phase delays experienced by the ordinary and extraordinary rays, respectively. The difference in refractive index, $\Delta n = n_O(x, y, z) - n_E(x, y, z)$, is the birefringence value. As the phase difference $\Delta\Phi$ strongly depends on the wavelength, (1.21) will assume different values for different λ, and under white light illumination, the sample will assume a widely colored aspect (interference colors).

Polarized light microscopy has been extensively used in the field of mineralogy where many crystals exhibit birefringent behaviors. The technique, however, has proved its utility in the field of biology to observe macromolecular assemblies such as chromosomes, muscle fibers, microtubules, fibrous protein structures, collagen, and amyloid deposits (Massoumian et al. 2003; Inoué 2008; Rieppo et al. 2008; Kaminksy et al. 2006).

1.2.13 Wavelength Contrast

Fluorescence microscopy, confocal microscopy, multiphoton microscopy, higher harmonic generation microscopy, Raman microscopy, fluorescence lifetime imaging microscopy are widespread observation techniques where a sample is excited with a spectrally band-limited light centered at a wavelength λ_{EX} and then observed at another spectral band around an emission wavelength λ_{EM} (Diaspro 2001; Lakowicz 2006). All these techniques work due to complex quantum mechanical phenomenon involving the interaction of light with matter which are beyond the scope of this introductory chapter and will be treated in more details in the following chapters of the book.

What is of concern here is the optical devices that are utilized in these techniques. Figure 1.20 shows a simple scheme of the basic configuration. A broadband light source, be it mercury or xenon arc lamps, a tungsten-halogen, a multi-wavelength laser, or even the novel kind of supercontinuum source (Wadsworth et al. 2002; Dudley et al. 2006), is spectrally filtered with an excitation band-pass interference filter. A dichroic beam splitter reflects the excitation wavelengths toward the objective and then the sample. The fluorescent emission, at longer wavelengths, is gathered by the objective and passes through the barrier filter, which selects the output spectral band of interest. Simple intensity measurements or more complex statistics, for example, lifetime measurements (FLIM), can then be performed (Becker 2005).

The key components of a fluorescence microscope setup are the three optical devices: excitation, dichroic and barrier filters which are usually combined into a single block.

The underlying physical principle of operation of these devices is again interference.

Figure 1.21 shows a light beam that propagates through a dielectric film of thickness t and refractive index n. At the boundary with the outside medium, with a different refractive index, a transmitted and a reflected beam are generated. This

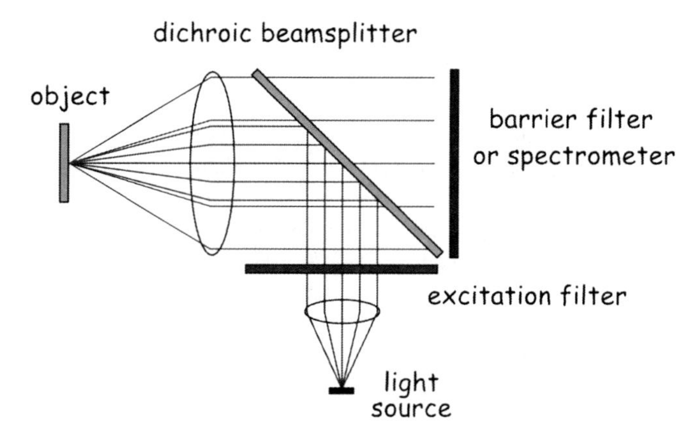

Fig. 1.20 Fluorescence microscopy

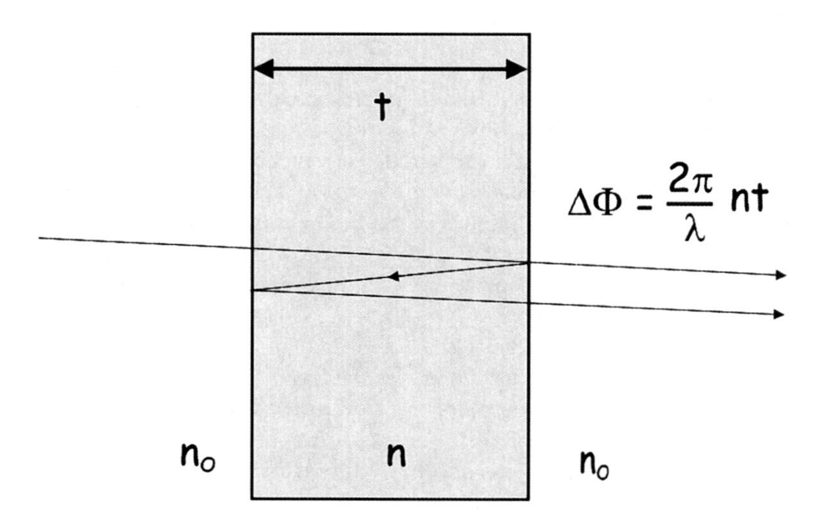

Fig. 1.21 The principle of dielectric films and interference filters

last beam is again reflected by the other boundary and, finally transmitted, it sums up to the first one. The transmitted intensity is:

$$I_T \propto {A_1}^2/2 + {A_2}^2/2 + A_1 A_2 \cos[(4\pi/\lambda)nt], \tag{1.23}$$

where, from (1.2), the phase difference, $\Delta\Phi = (2\pi/\lambda)2nt$, has been substituted into (1.9).

A_1 and A_2 are the amplitudes of the transmitted and the twice-reflected beams, respectively; they are connected by the relation $A_2 = r^2 A_1$, r being the amplitude reflection coefficient (Born and Wolf 1999).

The product nt is the optical thickness of the dielectric film. For a half-wave thick film ($nt = \lambda_F/2$), a constructive interference occurs and the output intensity assumes the highest value at the chosen filter wavelength λ_F. A quarter-wave thick film ($nt = \lambda_F/4$), on the other hand, produces, at the same wavelength, the lowest transmission output, which means the highest reflectivity.

Combining together stacks of quarter-wave and half-wave films using vacuum deposition techniques, interference filters with well-defined spectral characteristics can be designed.

Another spectral selecting device is the Lyot filter (Lyot 1933). Its operation is based on the wavelength-dependent transmission characteristics of birefringence as previously described. It consists of a sequence of birefringent crystalline plates (e.g., of quartz) and polarizers. If electrically variable birefringent elements are used (liquid crystal or electro optic), then a filter with a tunable spectral range is obtained [liquid crystal tunable filter (LCTF)]. Other increasingly used tunable filters are the acousto-optic tunable filter (AOTF) or acousto-optic beam splitter (AOBS). They exploit the scattering effect of a refractive index periodic pattern generated into a crystal (e.g., of quartz) by a high-frequency acoustic wave. They exploit the diffraction phenomenon.

1.2.14 Diffraction

Diffraction is the spreading out of light from its linear propagation, not caused by reflection or refraction, when an advancing wavefront is partially blocked by an aperture. "Lumen propagatur seu diffunditur non solum directe, refracte ac reflexe, sed etiam alio quodam quarto modo, DIFFRACTE"

This is the sentence that Francesco Maria Grimaldi used to define the phenomenon he had discovered in his book *De Lumine* (1665).

The correct explanation of the phenomenon in the framework of wave optics is given by Huygens who stated (Born and Wolf 1999) that each point of a wavefront is the source of a new spherical wave, and all these wavelets sum up to produce the wavefront at a later time. Interference again! Fresnel and after him Kirchhoff put the Huygens intuition on a sound mathematical basis.

A diffraction grating is an optical device whose absorption, reflection, or optical thickness is a periodic function of one spatial coordinate.

The simplest case is that of an on–off periodic pattern (Fig. 1.22). Following Huygens' principle, from each of the apertures, secondary spherical waves are generated. There is only a discrete number of angles θ_k at which all the wavelets are in phase and thus interfere constructively:

$$\sin\theta_k = k\lambda/p; \quad k = 0,\ \pm 1,\ \pm 2, \ldots, \tag{1.24}$$

where p is the grating period and k is called the diffraction order. The diffracted light propagates only along these directions.

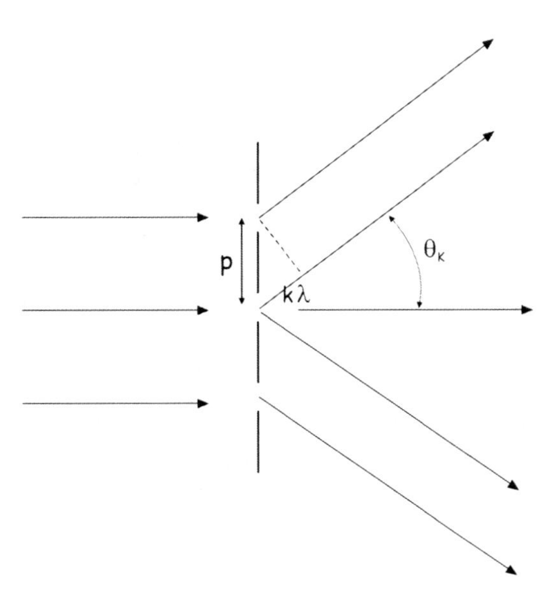

Fig. 1.22 The diffraction grating

Apart from the zeroth diffracted order, all other directions are highly wavelength dependent.

Diffraction gratings are widely used as spectral selective devices, besides dispersive prisms, either for illuminating sources (monochromators) or for spectral analysis in spectroscopes instead of simple barrier filters, allowing better identification of dye emissions (spectral fingerprinting) by means of deconvolution algorithms.

1.3 Space

1.3.1 Field and Resolution

The field of view, or view field or even simply field (not to be confused with the field amplitude of a light wave), is the observable area in the sample plane of the microscope (this classical definition can be extended to the observable volume).

The resolution is the ability of an imaging system to reproduce details in the object as separate entities in the image. Its exact value, in mm^{-1} or, as commonly written, line pairs per millimeter (lp/mm), depends on the adopted measuring criterion (e.g., the Rayleigh criterion) (Goodman 1968; Born and Wolf 1999); however, in principle, it corresponds to the maximum object spatial frequency that the optical system is able to image. It is the inverse of the minimum distance at which two points can still be distinctly imaged.

An ideal lens images each point that belongs to a semi-infinite object volume into an equally semi-infinite image volume. Moreover, the image of a point is still a point with zero dimensions. Both field and resolution are then infinite (Fig. 1.23).

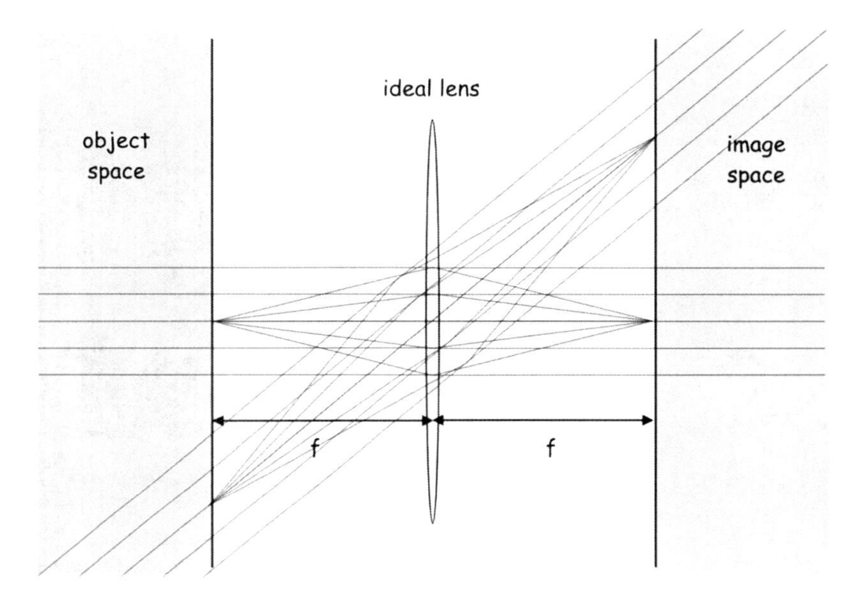

Fig. 1.23 The ideal lens, object field and resolution are both infinite

Ideal lenses unfortunately do not exist! Real lenses suffer from aberrations (Born and Wolf 1999). Chromatic aberration is due to dispersion phenomena, that is, different wavelengths are refracted at different angles and go to focus at different points. Spherical aberration is caused by the surface profile of the lens: the spherical shape is not, in fact, the ideal one and paraxial rays are focalized farther away than marginal rays. Off-axis aberrations are then those that depend on the distance from the optical axis (field): coma, astigmatism, distortion, and field curvature. To minimize all these aberrations, a very complex optical design is needed. A real objective is a device made up of tens of lenses with different thicknesses, refractive indexes, surface curvatures, air spacing, and so on, and all these efforts will help only to minimize aberrations but not to completely eliminate them! Moreover, the objective must be used following precise experimental requirements.

All these prescriptions and device performances are written onto the objective barrel (Fig. 1.24). The first piece of information refers to the objective quality in terms of aberration correction (from the least to the highest: achromat, plan achromat, fluorite, plan fluorite, plan apochromat); then the magnification power, the numerical aperture, the liquid matching fluid, the image position (e.g., infinity corrected), the cover glass thickness, and the working distance are written. Other special characteristics, if any, are also reported, for example its use for DIC, phase contrast, dark field, polarization techniques.

Among the constraints to which a real objective is submitted, there is the sample plane position which is no longer freely selectable, but is set at a fixed distance (the

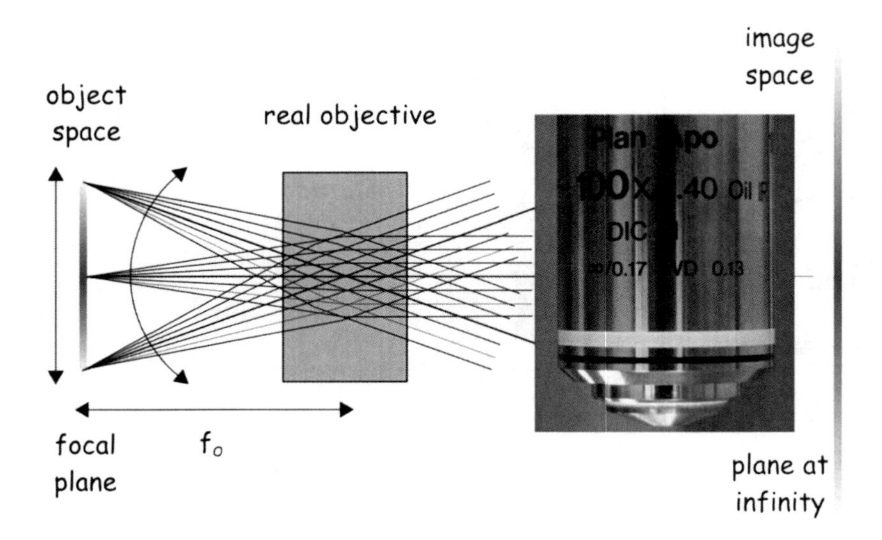

Fig. 1.24 The real objective, object field and resolution are both finite

working distance, WD) which is, for modern objectives, the front focal plane; consequently, the image will be at infinity.

In conclusion, the field is no more the semi-infinite space volume of the ideal lens, but it is restricted to a single plane and, furthermore, to only a tiny area near the optical axis.

And what about resolution? Even if it could be possible to eliminate all the aberrations, resolution could not grow indefinitely because light cannot concentrate into a point due to its wave-like nature. Diffraction from the clear aperture (pupil) of the objective gives rise to a light distribution, with finite dimensions in the vicinity of the geometric focus, which is named point spread function (PSF) (Fig. 1.25). In the focal plane, the amplitude PSF is the Airy function (Born and Wolf 1999); the radius of its first zero is:

$$d = 0.61\lambda/\mathrm{NA} \cong \lambda/2\mathrm{NA}, \tag{1.25}$$

where the numerical aperture NA $= n\sin\alpha$, and α is the objective semi-angular aperture. Equation (1.25) is the well-known Abbe formula; its inverse is the spatial resolution along the x- and y- (lateral) directions.

The PSF distribution along the optical axis has a similar shape, and the distance between the maximum and its first zero is given by:

$$r_Z = 2\lambda n/\mathrm{NA}^2. \tag{1.26}$$

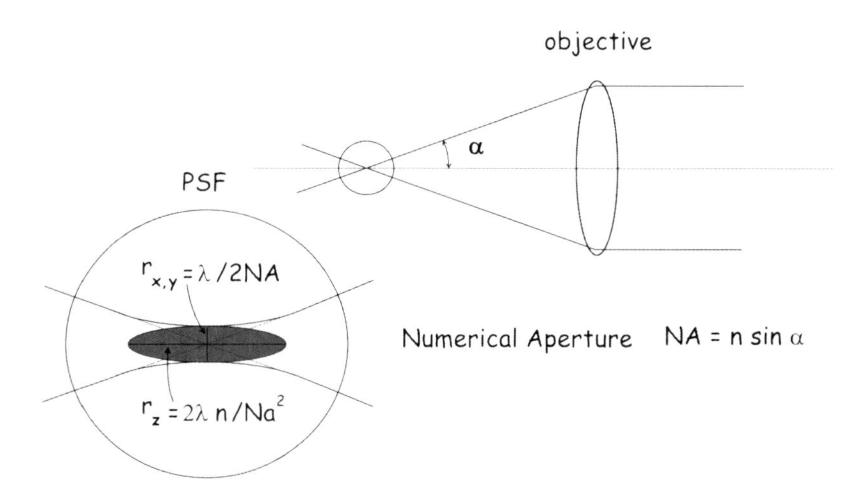

Fig. 1.25 The point spread function (PSF)

The ratio of the longitudinal and transversal amplitude PSF dimensions is:

$$r_Z/r_A = 4n/\mathrm{NA} = 4/\sin\alpha > 4. \tag{1.27}$$

The shape of the PSF is a sort of ellipsoid, whose major axis, which is at least four times the minor one, is oriented along the optical axis.

The images of two points displaced along z by a distance r_Z can still be distinguishable. The inverse of (1.26) can thus be assumed as the axial resolution for two points. But what happens instead for a uniformly (incoherently) illuminated plane? Is it possible to locate its position along the z-axis? The answer is no, and for a very fundamental reason: the conservation of energy. At every distance from the sample plane, the objective collects the same energy:

$$\int\int \mathrm{PSF}^2 \mathrm{d}x\,\mathrm{d}y = \text{constant}, \tag{1.28}$$

where PSF^2 is the intensity profile. The resolution, this time, is zero. The result is consistent with the "missing cone" of the optical transfer function (OTF) in the spatial frequency domain (Frieden 1967; Sheppard 1986a, b). This result would be the same even for the perfect point-like image of an ideal lens.

1.3.2 Field Extension

To explore a bigger object space, scanning strategies must be performed (Fig. 1.26). Z-scanning is commonly used for acquiring three-dimensional (3D) images, while xy-scanning is not as popular. This last approach is mainly used in object scanning

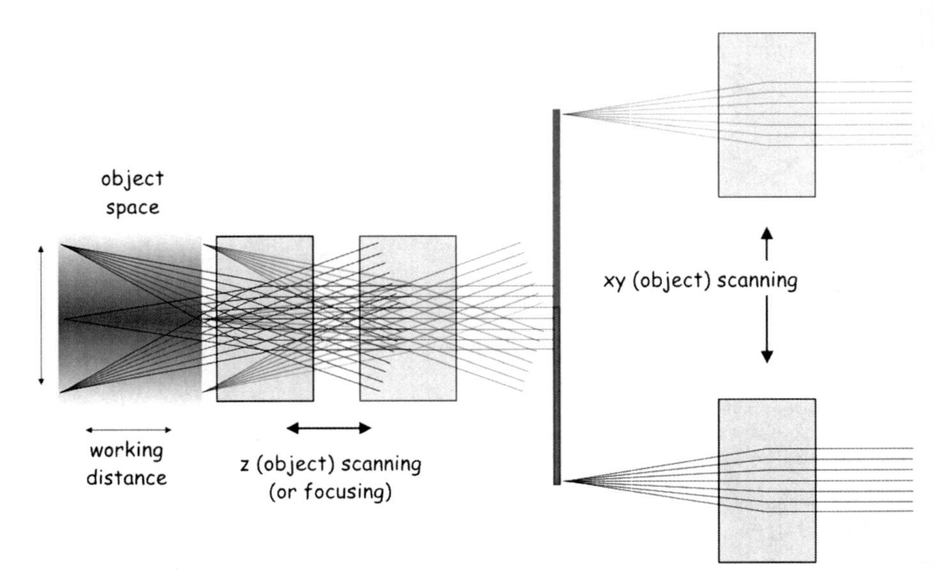

Fig. 1.26 Scanning strategies to extend the object field

technique, which is normally too slow for biological applications. On the other hand, it would produce quite perfect images as the objective can operate on-axis, where aberrations are at their minimum.

In 3D microscopy, z-scanning is commonly performed mechanizing the vertical stage with a computer-controlled motor at the focusing knob of the instrument or, for very fine movements, by means of piezoelectric actuated objectives.

A stack of two-dimensional (2D) images are then acquired and computer routines reconstruct the full 3D image afterward. Due to the slow axial decay behavior of the intensity distribution PSF^2, each image is blurred by out-of-focus light. Deconvolution routines are often used for contrast enhancement procedures (Pawley 2006); it should be made clear, however, that this kind of digital processing does not improve resolution at all. To increase resolution, a faster PSF^2 axial decay is necessary, that is, the OTF missing cone must be eliminated.

DHM can also be considered a 3D imaging technique, a scanningless approach. In fact, after recording both the amplitude and the phase of the field, it is possible to reconstruct the original wavefront and thus the whole 3D image.

1.3.3 Resolution Enhancement

Confocal microscopy is by far the most widespread technique for axial resolution improvement, and throughout this book the topic is treated extensively. Instead of the whole object field (wide-field microscopy), only a point at a time is illuminated, and moreover, a point-like detector is used.

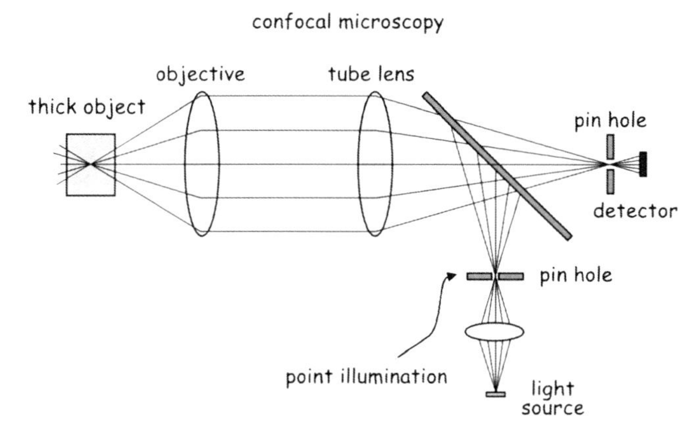

Fig. 1.27 Confocal microscopy

Looking at the confocal setup (Fig. 1.27), it is apparent that here the energy conservation principle, which is responsible for the lack of axial resolution in wide-field microscopy, is no more fulfilled. The out-of-focus light is, in fact, stopped by the pin-hole and does not reach the detector, departing from the right focal position, and the collected energy fades away. As both the excitation and detection events are ruled by the same function PSF^2, this time the integral of (1.28) becomes (Diaspro 2001; Pawley 2006; Wilson and Sheppard 1984):

$$\int\int \mathrm{PSF}^4 \mathrm{d}x\,\mathrm{d}y \sim 1/z^2 \quad (\text{for } z >> \lambda). \tag{1.29}$$

The integrated energy falls off, moving away from the focal plane. In other words, the image of a uniformly illuminated plane object becomes progressively weaker at the increase of defocus. This is a well-known optical sectioning capability of confocal microscopy. For example, the integrated intensity FWHM (sort of optical section thickness) for a unitary NA objective is 1.4λ (Wilson and Sheppard 1984).

Nonlinear microscopies such as two-photon microscopy and second harmonic generation microscopy are also ruled by (1.29) and have similar optical sectioning capabilities.

This time, the quadratic effect is an intrinsic characteristic of the physical phenomenon itself, and the pin-hole filtering is no longer necessary, so a large area detector can be used.

For both confocal and nonlinear microscopies, raster scanning strategies are necessary to inspect the whole object field. Laser beam scanning in the x- and y-directions and object scanning in the z-axis are the normal procedures.

Structured illumination microscopy (Fig. 1.28) is a wide-field alternative to confocal microscopy. The object is illuminated with a grid pattern, and three images are acquired for three different grid positions, stepped one third of the

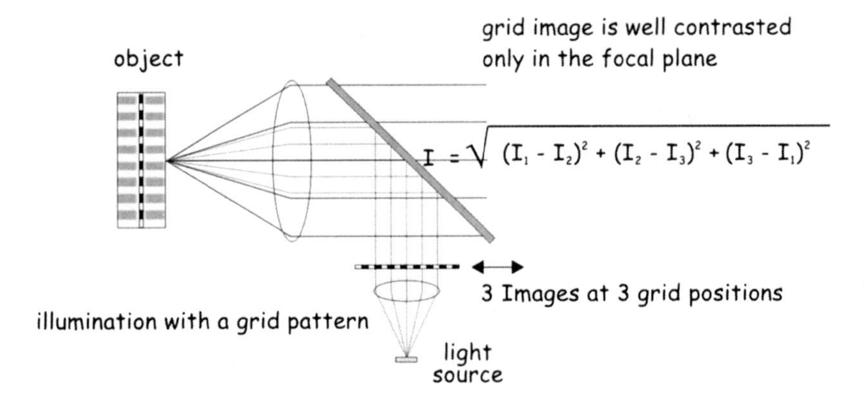

Fig. 1.28 Structured illumination microscopy

grid period apart. A computing algorithm reconstructs the in-focus image filtering out defocused light (Neil et al. 1997; Bauch and Schaffer 2006).

Resolution depends on numerical aperture; the PSF asymmetry between the axial and lateral dimensions is due to the uneven illumination and detection geometries. One semi-space is in fact missing. To overcome this lack of symmetry in some configurations, two opposing objectives are used to obtain a more uniform PSF shape. These techniques are: 4Pi microscopy, so named because a full 4π solid angle is used (Hell and Stelzer 1992); I^2M, which stands for image interference microscopy because the sample is observed from both sides and the two image fields are made to interfere; I^5M, incoherent interference illumination (I^3)+image interference microscopy (I^2M), where the sample is illuminated and observed from both sides (Gustafsson et al. 1999).

The limit of optics has been reached. To go further, some new physics is needed.

1.3.4 Resolution Enhancement Using Knowledge

A possible way to circumvent the diffraction limit is to provide the missing information in another nonoptic way. This kind of technique does not improve the PSF at all, but, with the addition of some new knowledge, it could be possble to locate the position of a sample point with better accuracy within its diffracted image spot. Förster resonance energy transfer (FRET) is a significant example of this ensemble of imaging procedures (Lakowicz 2006). FRET (Fig. 1.29) occurs when a donor chromophore and an acceptor are close enough to each other; the typical distance is the Förster radius and it is of the order of some nanometers. FRET can be considered a bottom-up approach.

Other ingenuous bottom-up approaches are three techniques named FPALM (Hess et al. 2006), PALM (Betzig et al. 2006), and STORM (Rust et al. 2006). The principle of operation is quite similar for all of them: if an image spot is certainly

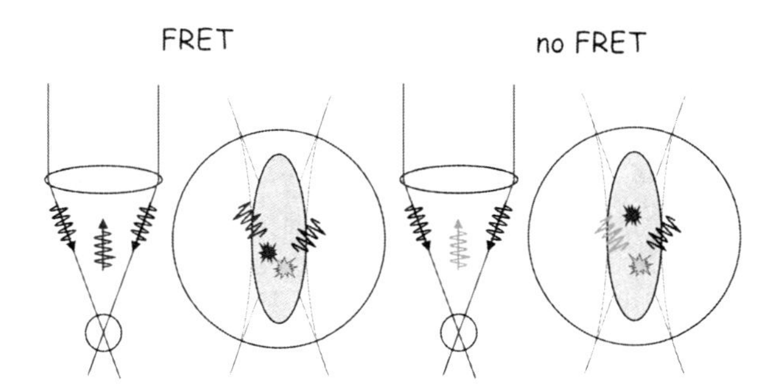

Fig. 1.29 Förster resonance energy transfer (FRET)

due to only one single dye molecule, then the spot center can be singled out with high accuracy and only that point is recorded as the image of that molecule, instead of the whole intensity PSF distribution. The final image is developed using only those center points (Fig. 1.30).

To fulfill the required experimental conditions, the image spot density must be very low to avoid any spatial superposition of two or more spots. This can be obtained by switching on only few fluorophores at the time in the sample and build up the complete image step by step by integration of hundreds of partial images. However, this procedure is also the weak point of the methods because of the very long measuring time needed to accumulate so many images.

1.3.5 Resolution Enhancement Using Matter

The top-down approach to overcome the diffraction limit is to effectively reduce the PSF dimensions. This cannot be done with light alone, and the only way is to exploit the interaction with matter.

Total internal reflection fluorescence (TIRF) microscopy (Born and Wolf 1999; Axelrod et al. 1983; Toomre and Manstein 2001) exploits the total internal reflection at the boundary between a high (glass) and a low (air or water) refractive index medium to generate an evanescent wave, which then illuminates and excites the specimen (Fig. 1.31). The evanescent wave decays exponentially, leaving the interface, and its vertical extent can be less than a wavelength, typically of the order of 100 nm. The surface of the sample, which is in contact with the interface, can then be imaged with that higher vertical resolution.

Scanning near field optical microscopy (SNOM) (Pohl 1991) extends the resolution improvement to three dimensions, this being the lateral resolution set by the physical dimensions of the probe (Fig. 1.32).

SNOM and TIRF microscopies are both surface imaging techniques. They avoid the loss of information due to far-field propagation by a near-field detection in the close vicinity of a boundary. The interaction with matter is mandatory. But why not exploit the material characteristics of the sample itself?

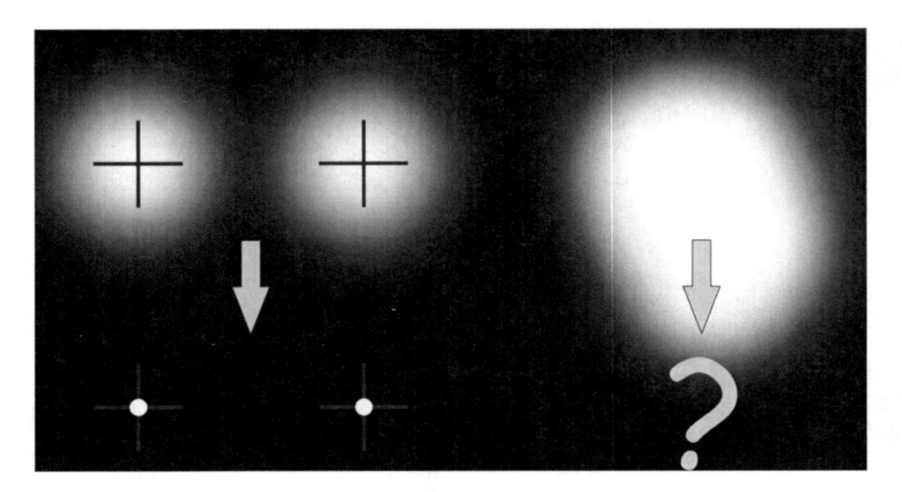

Fig. 1.30 Principles of FPALM, PALM, and STORM techniques

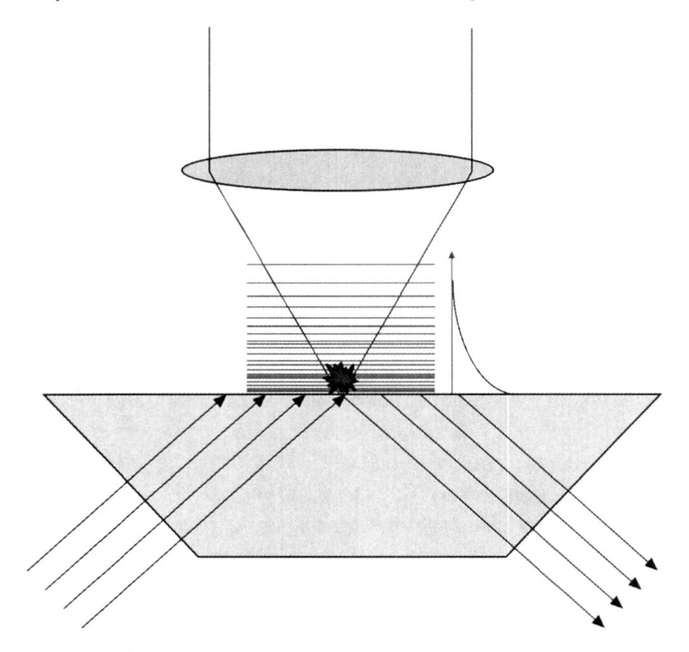

Fig. 1.31 Total internal reflection fluorescence microscopy (TIRFm)

Stimulated emission depletion (STED) (Hell and Wichmann 1994) microscopy does exactly this; it makes use of the highly nonlinear effect of atomic transitions to operate a sort of contrast enhancement of the illumination spot to reduce the size of the excited region. An excitation pulse is followed by a doughnut-shaped one at a wavelength which is located inside the emission band of the chromophore. The task of this second pulse is to deplete, by stimulated emission, the excited atomic level in

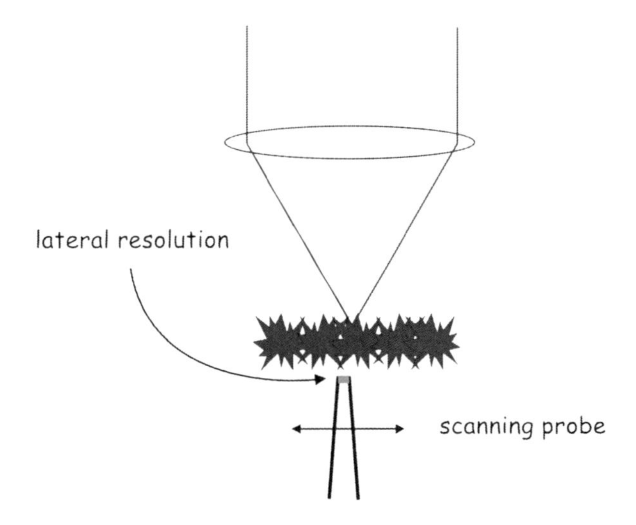

Fig. 1.32 Scanning near field optical microscopy (SNOM)

a ring-shaped region around the center of the illumination beam, leaving only a tiny internal region of the sample excited and thus able to fluoresce. A $\lambda/50$-spot size has been reported (Westphal and Hell 2005).

STED and ground state depletion (GSD) microscopies are the first implementations of a more general concept known as reversible saturable optical fluorescence transitions (RESOLFT) (Hell 2005, 2007). The Abbe formula [see (1.25)] can be rewritten as follows:

$$d = \lambda/[2\mathrm{NA}(1 + I/I_{\mathrm{sat}})^{-1/2}]. \tag{1.30}$$

I_{sat} is the transition saturation intensity and I is the applied intensity. The factor added at the denominator removes any limitation to resolution, and d can be continuously decreased.

The approach of taking advantage of the highly nonlinear saturation effect of atomic transitions has been adopted by others. An analogous solution for wide-field microscopy is saturated patterned excitation microscopy (alternatively named: saturated structured illumination microscopy) (Heintzmann and Jovin 2002; Gustafsson 2005).

1.4 Time

1.4.1 Temporal Resolution

Optical imaging is quite an instantaneous phenomenon, being a parallel process that occurs at the velocity of light. For this reason, in the 1970s and 1980s, there was an idea to use optics for analog data processing purposes rather than for the considerably slow digital computers of the time.

The bottleneck of imaging is the detection and recording procedure. Wide-field microscopy can profit from a wide choice of high-frame rate cameras if enough signal is available. Cameras featuring up to thousands of frames per second (fps, frames/s, s^{-1}) are available on the market. Mechanical laser beam scanning of confocal and nonlinear microscopy does not allow frame rate higher than 5 fps. Using resonant galvanometric scanners, velocity can be raised up to 25 fps for a 512×512 image (Pawley 2006).

Alternative actuators such as polygonal mirrors or acousto-optic deflectors can also be used (Saggau and Bansal 2008). The substantial weakness of laser scanning techniques is single point acquisition procedure followed by a sequential raster scanning. The adoption of a parallel approach, instead of the standard serial one, will dramatically improve temporal resolution. Line scanning is an example (Sheppard and Mao 1988). A line-shaped illumination followed by a slit diaphragm spatial filter is used instead of point illumination and pin-hole filtering. Once the out-of-focus light is filtered, a linear array detector records the whole intensity distribution along the image line. Scanning is needed only along one lateral direction, and up to about 120 fps for a 512×512 image can be reached. However, it is not a true confocal system because along the line direction no spatial filtering of the out-of-focus light can be operated.

The spinning disk (Nipkow disk) confocal is another example of a parallel setup (Egger and Petrán 1967; Kino and Corle 1996). A disk with lots of pin-holes located along the spirals projects a 2D array of illuminating points onto the specimen. The disk rotates at high velocity providing a full covering of the object field. The light that comes back from the sample is then spatially filtered by the same set of pin-holes or a mirror copy set.

The technique is quite popular in the field of material science where it finds application in the semiconductor industry for waver surface inspection.

The spinning disk illumination efficiency is very low because only that minimum percentage of the incoming light that pass through the pin-holes is used, while the most part is instead rejected.

To improve the filling factor, for fluorescence microscopy applications, a second disk with microlenses is inserted. Each lens focalize the illuminating light inside the corresponding pin-hole and the illumination efficiency is greatly improved (Tanaami et al. 2002).

Spinning disk confocal microscopy is a direct view system, that is, the image can be seen directly by the eyes. Alternatively, a fast CCD camera can be used and about 1,000 confocal frames/s can be acquired.

For nonlinear microscopies, such as multiphoton microscopy and higher harmonic generation microscopy, the pin-hole filtering is no longer necessary and the de-scanning process can be avoided.

The instrument setup can be highly simplified, and a wider choice of parallel scanning configurations is available (Bewersdorf et al. 1998; Nielsen et al. 2001).

In parallel scanning, too many points cannot be illuminated at a time because, if they are too close to each other, a cross-talk effect results and, in the end,

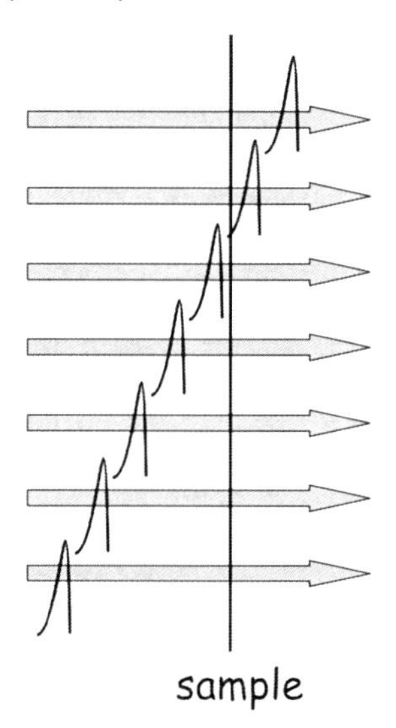

Fig. 1.33 Principle of temporally decorrelated multiphoton microscopy

confocality is lost and standard wide-field image is obtained. The finite temporal width of the laser pulse in nonlinear microscopy can be profitably exploited to solve the problem: a time shift is introduced between an illuminating point and its neighbor, and this temporal decorrelation eliminates the cross-talk effect (Fig. 1.33) (Fittinghoff et al. 2000; Oron et al. 2005; Tal et al. 2005).

1.4.2 Duration

The observation of a living sample can last until the fluorophores are completely bleached or the specimen is dead. Multiphoton microscopy is again the more convenient approach to minimize the causes of both these events. Dye excitation takes place only at the observation point, and none of the fluorescence emission signal is wasted. Moreover, the ultrafast laser long wavelength emission (around 800 nm) is less damaging for living samples. In conclusion, photobleaching and phototoxicity effects are minimized (Diaspro 2001).

1.5 Conclusions

The measurement of every physical quantity is always affected by noise, which, in the end, sets the resolution limit.

The signal-to-noise ratio quantifies the degree of accuracy of the measurement and its information content.

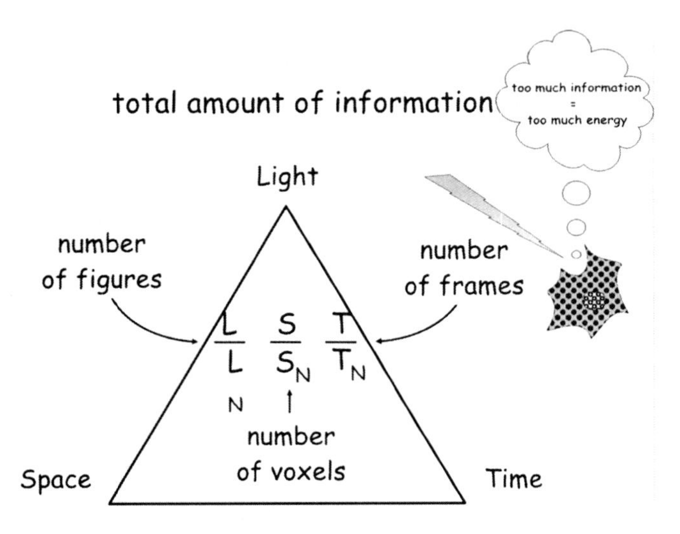

Fig. 1.34 The triangle of imaging performance

For light intensity data, it is proportional to the numbers of significant figures.

In a similar way, the ratio between the object space field and the voxel volume, whose linear dimensions are the inverse of the respective spatial resolution, gives the total number of voxels in a 3D image.

Finally, the total number of temporal frames is given by the ratio of the total observation time interval and the time spent for a single frame acquisition (which is the inverse of time resolution).

The product of these three ratios gives the total information content of the measurement. Information costs energy, in terms of light irradiance delivered onto the sample. Too much information means too much energy, which in the end will destroy the specimen. For each experiment, the total amount of information is limited and, if it is spent to enhance one aspect, this will inevitably lead to worsen the other two (Fig. 1.34) (Shotton 1995; Sheppard and Shotton 1997).

References

Axelrod D, Thompson NL, Burghardt TP (1983) Total internal inflection fluorescent microscopy. J Microsc 129:19–28

Barker WB (1930) Lens work of the ancients II: the Nineveh lens. Br J Physiol Opt 4:4–6

Bauch H, Schaffer J (2006) Optical sections by means of "structured illumination": background and application in fluorescence microscopy. Photonik Int 5:86–88

Becker W (2005) Advanced time-correlated single photon counting techniques. Springer, Berlin

Betzig E, Patterson GH, Sougrat R, Lindwasser OW, Olenych S, Bonifacino JS, Davidson MW, Lippincott-Schwartz J, Hess HF (2006) Imaging intracellular fluorescent proteins at nanometer resolution. Science 313:1642–1645

Bewersdorf J, Pick R, Hell SW (1998) Multifocal multiphoton microscopy. Opt Lett 23:655–657

Born M, Wolf E (1999) Principles of optics. Cambridge University Press, Cambridge
Bossard M (2007) 3D upgrade for optical microscopes. Imaging Microsc 9:66–69
Brewster D (1855) On an account of a rock-crystal lens and decomposed glass found in Niniveh. In: Die Fortschritte der Physik im Jahre 1852. Deutsche Physikalische Gesellschaft, vol VIII. Druck und Verlag von Georg Reimer, Berlin pp 355–356
Cuche E, Bevilacqua F, Depeursinge C (1999a) Digital holography for quantitative phase-contrast imaging. Opt Lett 24:291–293
Cuche E, Marquet P, Depeursinge C (1999b) Simultaneous amplitude-contrast and quantitative phase-contrast microscopy by numerical reconstruction of Fresnel off-axis holograms. Appl Opt 38:6994–7001
Curl CL, Bellair CJ, Harris PJ, Allman BE, Roberts A, Nugent KA, Delbridge LMD (2004) Quantitative phase microscopy – a new tool for investigating the structure and function of unstained live cells. Proc Aust Physiol Pharmacol Soc 34:121–127
Danz R, Vogelgsang A, Käthner R (2004) PlasDIC – a useful modification of the differential interference contrast according to Smith/Nomarski in transmitted light arrangement. Photonik 1:42–45
Diaspro A (2001) Confocal and two-photon microscopy: foundations. Applications and advances. Wiley-Liss, New York
Dudley JM, Genty G, Coen S (2006) Supercontinuum generation in photonic crystal fiber. Rev Mod Phys 78:1135–1184
Egger MD, Petrán M (1967) New reflected-light microscope for viewing unstained brain and ganglion cells. Science 157:305–307
Enoch JM (1998) Ancient lenses in art and sculpture and the objects viewed through them dating back 4500 years. Proc SPIE 3299:424–430
Enoch JM (2000) Duplication of unique optical effects of ancient Egyptian lenses from the IV/V Dynasties: lenses fabricated ca 2620–2400 BC or roughly 4600 years ago. Ophthalmic Physiol Opt 20:126–130
Ferraro P, Alferi D, De Nicola S, De Petrocellis L, Finizio A, Pierattini G (2006) Quantitative phase-contrast microscopy by a lateral shear approach to digital holographic image reconstruction. Opt Lett 31:1405–1407
Fittinghoff DN, Wiseman PW, Squier JA (2000) Widefield multiphoton and temporally decorrelated multifocal multiphoton microscopy. Opt Express 7:273–279
Frieden BR (1967) Optical transfer of the three-dimensional object. J Opt Soc Am 57:56–65
Goodman JW (1968) Introduction to Fourier optics. McGraw-Hill, New York
Graydon O (1998) Medieval lenses exhibit modern performances. Opto Laser Eur 56:7
Gustafsson MGL (2005) Nonlinear structured-illumination microscopy: wide-field fluorescence imaging with theoretically unlimited resolution. Proc Natl Acad Sci U S A 102:13081–13086
Gustafsson MGL, Agard DA, Sedat JW (1999) I5M: 3D widefield light microscopy with better than 100 nm axial resolution. J Microsc 195:10–16
Heintzmann R, Jovin TM (2002) Saturated patterned excitation microscopy – a concept for optical resolution improvement. J Opt Soc Am A 19:1599–1609
Hell SW (2005) Fluorescence nanoscopy: breaking the diffraction barrier by the RESOLFT concept. Nanobiotechnology 1:296–297
Hell SW (2007) Far-field optical nanoscopy. Science 316:1153–1158
Hell SW, Stelzer EHK (1992) Properties of a 4Pi-confocal fluorescence microscope. J Opt Soc Am A 9:2159–2166
Hell SW, Wichmann J (1994) Breaking the diffraction resolution limit by stimulated emission: stimulated-emission-depletion fluorescence microscopy. Opt Lett 19:780–782
Hess ST, Girirajan TP, Mason MD (2006) Ultra-high resolution imaging by fluorescence photoactivation localization microscopy. Biophys J 91:4258–4272
Hoffman R, Gross L (1975) The modulation contrast microscope. Nature 254:586–588
Inoué S (2008) Microtubule dynamics in cell division: exploring living cells with polarized light microscopy. Annu Rev Cell Dev Biol 24:1–27

Kaminksy W, Jin LW, Powell S, Maezawa I, Claborn K, Branham C, Kahr B (2006) Polarimetric imaging of amyloid. Micron 37:324–338
Kino GS, Corle TR (1996) Confocal scanning optical microscopy and related imaging systems. Academic, San Diego
Lakowicz JR (2006) Principles of fluorescence spectroscopy. Springer, New York
Lyot B (1933) Optical apparatus with wide field using interference of polarized light. C R Hebd Seances Acad Sci 197:1593–1595
Massoumian F, Juskaitis R, Neil MAA, Wilson T (2003) Quantitative polarized light microscopy. J Microsc 209:13–22
Neil MAA, Juskaitis R, Wilson T (1997) Method of obtaining optical sectioning by using structured light in a conventional microscope. Opt Lett 22:1905–1907
Nielsen T, Fricke M, Hellweg D, Andresen P (2001) High efficiency beam splitter for multifocal multiphoton microscopy. J Microsc 201:368–376
Nomarski G (1955) Microinterféromètre différentiel à ondes polarisées. J Phys Radium 16:9–13
Oron D, Tal E, Silberberg Y (2005) Scanningless depth-resolved microscopy. Opt Express 13:1468–1476
Pawley JB (2006) Handbook of biological confocal microscopy. Springer, New York
Plantzos D (1997) Crystals and lenses in the Graeco-Roman world. Am J Archaeol 101:451–464
Pohl D (1991) Scanning near-field optical microscopy (SNOM). Adv Opt Electron Microsc 12:243–312
Rieppo J, Hallikainen J, Jurvelin JS, Kiviranta I, Helminen HJ, Hyttinen MM (2008) Practical considerations in the use of polarized light microscopy in the analysis of the collagen network in articular cartilage. Microsc Res Tech 71:279–287
Rust MJ, Bates M, Zhuang X (2006) Sub-diffraction-limit imaging by stochastic optical reconstruction microscopy (STORM). Nat Methods 3:793–796
Saggau P, Bansal V (2006) Confocal microscope records living cells with high spatio-temporal resolution. SPIE Newsroom doi:10.1117/2.1200608.0363
Schmidt O, Wilms KH, Lingelbach B (1999) The Visby lenses. Optom Vis Sci 76:624–630
IMSS Istituto e Museo di Storia della Scienza (2007) http://brunelleschi.imss.fi.it/esplora/microscopio/dswmedia/storia/estoria1.html
Sheppard CJR (1986a) The spatial frequency cut-off in three-dimensional imaging. Optik 72:131–133
Sheppard CJR (1986b) The spatial frequency cut-off in three dimensional imaging II. Optik 74: 128–129
Sheppard CJR, Mao XQ (1988) Confocal microscopes with slit apertures. J Mod Opt 35:1169–1185
Sheppard CJR, Shotton DM (1997) Confocal laser scanning microscopy. BIOS Scientific, Oxford
Shotton DM (1995) Electronic light microscopy: present capabilities and future prospects. Histochem Cell Biol 104:97–137
Tal E, Oron D, Silberberg Y (2005) Improved depth resolution in video-rate line-scanning multiphoton microscopy using temporal focusing. Opt Lett 30:1686–1688
Tanaami T, Otsuki S, Tomosada N, Kosugi Y, Shimizu M, Ishida H (2002) High-speed 1-frame/ms scanning confocal microscope with a microlens and Nipkow disks. Appl Opt 41:4704–4708
Toomre D, Manstein DJ (2001) Lighting up the cell surface with evanescent wave microscopy. Trends Cell Biol 11:298–303
Wadsworth WJ, Ortigosa-Blanch A, Knight JC, Birks TA, Man TPM, Russell PSJ (2002) Supercontinuum generation in photonic crystal fibers and optical fiber tapers: a novel light source. J Opt Soc Am B 19:2148–2155
Wehner E (2003) PlasDIC, an innovative relief contrast for routine observation in cell biology. Imaging Microsc 4:23
Westphal V, Hell SW (2005) Nanoscale resolution in the focal plane of an optical microscope. Phys Rev Lett 94:143903
Wilson T, Sheppard CJR (1984) Theory and practice of scanning optical microscopy. Academic, London
Zernike F (1955) How I discovered phase contrast. Science 121:345–349

Chapter 2
The White Confocal: Continuous Spectral Tuning in Excitation and Emission

Rolf Borlinghaus

2.1 Fluorescence

In the middle of the nineteenth century, the scientific community became aware of a strange phenomenon. Under certain circumstances, one could detect color effects in otherwise homogenous or colorless solutions. First objects were solutions of chlorophyll and quinine. Solutions of chlorophyll are easily obtained from triturations of green plants. If focusing with a lens into such a solution, David Brewster (1781–1868) could detect a blood-red color at the lateral surface of the illumination cone. John William Herschel (1792–1872) realized a bluish shimmer at the glass–solution border in glass containers containing quinine solution when observed in the bright sun (Herschel 1845). Quinine is the working compound of extracts from the cinchona bark, a famous medicine for the prevention and healing of malaria. Quinine is also an ingredient in tonic water, the reason for tonic water to be promoted to the prime example specimen in demonstrations of fluorescence phenomenon to interested laymen. The famous researcher George Gabriel Stokes (1819–1903) repeated these experiments, especially with quinine solutions. He found out that the color of the illuminating light was changed when it interacted with the solution (Stokes 1852). If illuminated with short wavelength (blue), a longer wavelength (green-yellow) is returned. The change of that color, measured as displacement between excitation maximum and emission maximum, is therefore called "Stokes shift." Minerals were also known to yield such phenomena, which Stokes coined "Fluorescence," after the mineral Fluo-Spar and the word luminescence, which describes glowing effects in general.

As a matter of fact, fluorescence is not restricted to the visible range, but covers a wide range of the electromagnetic spectrum from Röntgen rays to far infrared. Nevertheless, microscopy basically uses the visible part with some minor excurses to the ultraviolet and near infrared.

R. Borlinghaus
Leica Microsystems CMS, Am Friedensplatz 3, 68165 Mannheim, Germany
e-mail: Rolf.Borlinghaus@leica-microsystems.com

A. Diaspro (ed.), *Optical Fluorescence Microscopy*,
DOI 10.1007/978-3-642-15175-0_2, © Springer-Verlag Berlin Heidelberg 2011

The most established description of the fluorescence process – that is the absorption of an energy-rich photon and the subsequent emission of a photon of lower energy – uses the term scheme that was introduced by Alexander Jablonski (1898–1980) (Jablonski 1935).

2.1.1 Fluorescent Specimen

As already mentioned, initially mainly organic compounds were objects that draw interest on research of fluorescence. Attempts to find out what could cause that behavior resulted in the discovery that certain chemical groups should be considered – in particular benzene, if linked to further groups. Such compounds were tagged "Luminophores" (Kauffmann 1900). For research with the microscope, August Köhler (1866–1948) already mentioned the "colorless dyes" (Köhler 1904). Indeed, more or less all living samples contain material that is fluorescent by nature, a phenomenon called "autofluorescence." Plant material is a very rich source for fluorochromes. The most famous and important is chlorophyll and a further wide variety of fluorescently active compounds that absorb light for the purpose of photosynthesis.

Besides substances that fluoresce without further modification, there are compounds in biological material that can be qualified for fluorescence by simple alterations. Some vitamins and hormones belong to that group.

Occasionally, it was found that some histological stains also show fluorescence, and that specific structures become visible. But it was clear that, for systematic research, completely different dyes needed to be applied. Such dyes were named "fluorochromes" (Haitinger 1934) and the process of staining is consequently called "fluorochromation". In histology, fluorochromation soon advanced to an extended science, because many fluorochromes show different emissions in different cell compartments. This effect is very well suited for "differentiation" of structures. To this group of dyes belong many compounds that are used until today – although occasionally in other applications – for example, fuchsin, rhodamine, fluoresceine, hematoxilin, eosin, acridine orange, and so on.

Differentiation by a single dye works, because emission of nearly any dye depends strongly on the molecular environment, that is, for example, the polarity, pH value, or molecules that bind to the dye. For modern fluorescence microscopy, this is both a blessing (molecular probes) and a curse (spectral shifts and quenching).

A revolution in fluorescence microscopy was triggered by the work of Coons et al. (1941) who introduced the immune-fluorescence staining (fluorescence immuno histochemistry resp. cytochemistry). The working principle of these stainings is based on the specific recognition of cellular structures, especially sugar residues and protein epitopes by antibodies. By appropriate methods, these structures can be decorated with antibodies inside cells or tissue sections. When the antibody was decorated with chemically bound fluorochromes, then exactly and only those structures will light up under the microscope. By application of different dyes for various antibodies, a whole series of structures can be visualized simultaneously. By this approach, it was possible to specifically render the structure of many cell compartments, protein distributions, or cytoskeleton elements. By staining various cell states, e.g., during the cell cycle or during differentiation, this method allowed elucidation of the development and dynamic changes in these objects. Medical diagnosis also benefits from this technique.

One more revolution was the introduction of a similar procedure by Gall and Pardue (1969) for specific sequences of DNA or RNA, the fluorescence in situ hybridization (FISH). Here, a short polynucleotide chain is synthesized and chemically bound to fluorochromes. Subsequently, this marker is hybridized with the DNA that is abundant in the cell. Of course, this is possible only with locations on the DNA that are complementary to the artificially synthesized piece. Genetic research and medical care have introduced this method in large scale, meanwhile with various modifications and most diverse application possibilities.

Not only for structural information fluorescence methods were developed, but also dyes that allowed to detect various metabolites and inorganic ions in living cells. The best known case is detection of calcium via the Ca^{2+}-chelator FURA-2. The fluorescence parameters of this compound change significantly (both excitation and emission) upon binding of Ca^{2+}-ions. When applying a calibration curve to the data, the concentration of free Ca^{2+} can be determined. These sorts of indicators are now available for a whole range of other ions and metabolites. The cumbersome and invasive method of injection of those dyes is meanwhile overcome by very ingeniously designed substances, so the living object is left unaltered in its native conditions.

Besides organic dyes, small fluorescent particles of semiconductor material have been developed (quantum dots) recently. These markers have a wide excitation spectrum and emit according to size and composition, in many different colors. Not yet sufficiently solved are biocompatibility and connection to antibodies, necessary preconditions for specific staining.

In the recent past all those concepts were radically excelled by a completely new method. A protein, which natively is abundant in jelly fish and other marine animals, expresses a natural fluorescence in the visible range: the green fluorescent protein (GFP). Initially, it was prepared from Aequatoria. Its triumphant success started when it became possible to clone the respective DNA into the genome of living cells (Chalfie et al. 1994). By appropriate genetic engineering, it enables the visualization not only of gene expression, but also of structural proteins in living and developing cells. Now living cells can be dyes without any external interference – and of course whole animals as well. The last sensational report was on a GFP-pig, expressing a greenish shimmering skin and green luminescent eyes. The application of this technique merely knows any boundaries. Meanwhile, a long list of proteins in all colors is available – with many secondary techniques to measure not only structural but also dynamic changes and processes in cells.

2.2 Fluorescence Microscopy

As mentioned above, A. Köhler was already investigating fluorescence for microscopy. The first commercially available fluorescence microscopy was produced and sold by Carl Reichert in Vienna. The systems at that time were basically ordinary microscopes with little variations that allowed connecting fluorescence devices. The microscope itself appears like a small appendage to the huge illumination apparatus, for example, coal or metal arc lamps with 10–30 A current drain.

To visualize fluorescence phenomena, it is first necessary to separate, using appropriate filters, an excitation band from the white light sources and guide it onto the sample for illumination. For this purpose, short-pass or band-pass filters are employed. Short-pass filters transmit light only below a specified wavelength; band-pass filters transmit only in a specified segment of the spectrum.

In the beginning, these instruments were dimensioned for illumination with near UV light or deep blue light, as fluorescence was more or less a synonym for UV illumination. Small and efficient sources for UV light were only available with the construction of compact high-pressure mercury arc lamps. These lamps also offer a high power density in the visible range and are evolved to a standard for the excitation of immunohistochemical stains.

To excite with UV light, filters made of black glass (Woods filter) were used. These filters have a good transmission between 300 and 420 nm. However, caution is advised as transmission increases at around 700 nm in these filters. This light is not visible, but can cause spurious effects in photographic exposures. By time, many color glasses were developed that were available in various thicknesses for many different purposes.

A revolution in filter technology was the invention of dielectric coatings that allowed depositing multiple layers on glass substrates that create all sorts of spectral bands. Such filters are still standard in today's fluorescence applications.

In general, microscopes come in two different types: transmitted light and incident light microscopes. Both types are suited for fluorescence, but the incident light version offers a significantly better separation of excitation and fluorescence light. The separation power is important, as the fluorescence intensity is usually at least 1,000 times less as compared to illumination intensity.

In case of incident light microscopes, the illumination light has to be guided into the objective lens by a reflecting mirror, while the emission is collected from the same side of the lens and has to pass that mirror for subsequent recording. The simplest case of the reflecting mirror is a "gray splitter", which reflects partially all colors equally and transmits complementarily. A 30/70 gray splitter, for example, has 30% reflectivity and 70% transmission. Consequently, the excitation energy and the fluorescence is partially lost, which is most severely a problem on the emission side and the reason not to use 50/50 splitters. Illumination is less precious as emission, so 70% are sacrificed in order to harvest 70% of the fluorescence. Even 05/95 splitters are sometimes used, if fluorescence is very weak or bleaches fast. With the introduction of dielectric coatings, mirrors became available that reflect and transmit respectively different areas of the white spectrum. These "dichroitic splitting mirrors" were developed in large diversity and are used for a whole variety of dyes and dye combinations.

To remove residual excitation light completely from emission, a further filter is inserted between beam splitter and detector: the emission filter (barrier filter). Here, appropriate band-pass or long-pass filters are used.

In current instruments, excitation filter, beam splitter, and emission filter are combined in a "filter cube" that is easily exchanged. So, all the different regimes for illumination and detection can be switched quickly and simply. The extensive motorizations in modern research microscopes leave this concept dispensable and allow a wide flexibility by independent combination of the different plan-optical elements – without loss of time.

2.3 Confocal Fluorescence

A first application of a microscopic setup that allows to measure intensities in thicker samples without disturbance from other focal planes was published in 1951 (Naora 1951). Today, confocal microscopy is one of the most common tools in biological research (Diaspro 2002) for a graphical explanation of confocal microscopy (Fig. 2.1). Optical sectioning, as the fruit of a confocal setup, performs effectively only in incident light. By the currently rapid development of fluorescence techniques for biological research, also confocal microscopy mutated within 20 years from a dubious considered rare application to one of the most important daily routine methods. In confocal imaging, the illumination light has to be focused diffraction limited into the sample. For that reason, the light source has to be selected very carefully. Technical reasons leave lasers as the only suitable light source for confocal microscopes, as only they provide complete collimation at sufficient luminance.

Still, the well-established gas lasers are the most popular ones – especially argon ion lasers and helium–neon lasers. The use of krypton and cadmium lasers has more or less vanished. The most important advantage of gas lasers is the emission of more than one line, and therefore, fewer units needed.

Modern developments are solid-state and diode lasers that offer a much higher degree of freedom in selection of the emission wavelength. Diode lasers are also used inside solid-state lasers as pump lasers. These lasers are quite compact and small, but have the disadvantage of emitting only a single line.

In the chapter on fluorescence specimens, it was shown that the huge varieties of dyes cover the whole spectrum of visible light. Thus, to achieve a differentiated representation of all fluorochromes, it is highly desirable to have the individual dye excited by arbitrarily selectable wavelengths. This is not possible with traditional lasers. Here, significant excitation gaps remain in the spectrum. To increase the density of lines, a large conglomeration of lasers is necessary, which needs complex and laborious coupling. These setups are of course prone to misalignments, are

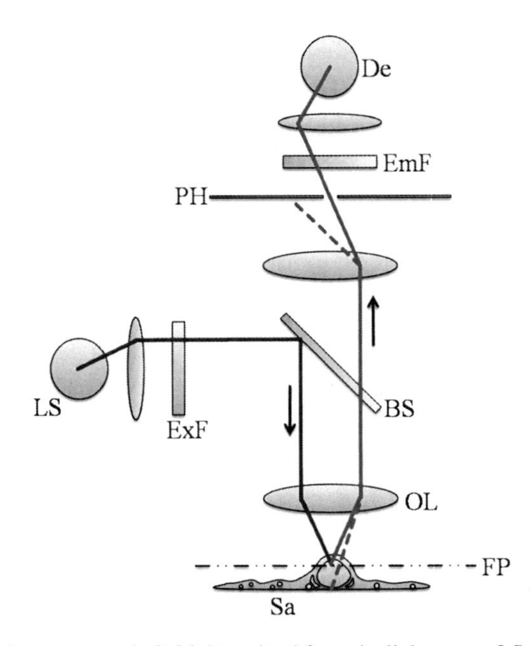

Fig. 2.1 Confocal fluorescence. *Left*: Light emitted from the light source LS passes the excitation filter ExF and is diffraction limited focused by the objective lens OL into the focal plane FP. *Right*: Emission from the focal plane (*solid line*) passes through objective lens and beam splitter BS and is focused to a spot in the intermediate image plane, where the pinhole PH is located. Focal emission can pass the pinhole and is filtered by an emission filter EmF before reaching the detector De. Emission from other planes than the focal plane (*broken line*) cannot pass the pinhole and thus is blocked from detection. The pinhole acts as spatial filter for the z-dimension and thus creates optical sections. Conventional lasers emit only one or a few narrow lines. Excitation filter, beam splitter, and emission filter are classically fixed-parameter devices, usually colored mirrors or glass filters. The white confocal concept uses spectrally tunable devices in all these places

quite expensive, and have high energy consumption. The efficiency is very low and therefore lost heat has to be removed by cooling devices, often accompanied by displeasing noise.

The most comfortable way to achieve optimal excitation would of course be a system that allows dialing the color by a simple knob or slider. So, for each dye and for any combination of dyes, the optimal line or set of lines would be available – a dream?

2.4 A Tunable Laser

Such a dream has become true. A classical laser emits only tiny narrow lines with a bandwidth of usually less than one nanometer. The basis for a tunable laser is a white laser. A laser that emits white light has a high energy density over a wide range of the visible spectrum. In combination with an acousto-optical element permitting to select narrow bands of a few nanometer widths out of the white emission, a tunable laser was realized (Birk and Storz 2001).

Novel fiber technologies were the basic principle that led to the invention of white lasers (Knight et al. 1996). The main component of such an instrument is a fiber whose core is an ensemble of many symmetrically arrayed cavities. This fiber type is called "photonic crystal fiber." Initially, they were developed for telecommunication purposes, consequently for infrared light applications. If a short pulse of high energy is coupled into a photonic fiber, then at the glass–air interface many nonlinear photonic processes cause various recombinations of photons. The result is the conversion of the line spectrum into a wide band, therefore called a "supercontinuum." The width of that band depends on the crystal structure and on the length of the fiber. By selecting appropriate structures, it was possible to extend the broadening into the visible range.

Initially, a seed laser is used to generate short pulses in the range of picoseconds at about 100 MHz repetition frequency in the infrared. These seed lasers are also fiber lasers. A series of diode lasers is fiber-coupled with the seed laser to amplify the intensity of the pulsed light to some 10 W of light (Fig. 2.2). These high energy pulses are then fed into the photonic crystal fiber that consequently emits about 1–2 W of visible light in the range of 450–700 nm. The power density arrives at 1 mW/nm visible, which is a good value to record high signal-to-noise images in confocal microscopy.

Although there are some rare exceptions that use the full white spectrum for illumination, the most typical application is fluorescence that requires selection of an appropriate spectral band for excitation, if a white light source is used. This is equivalent to classical widefield fluorescence microscopy using, for example,

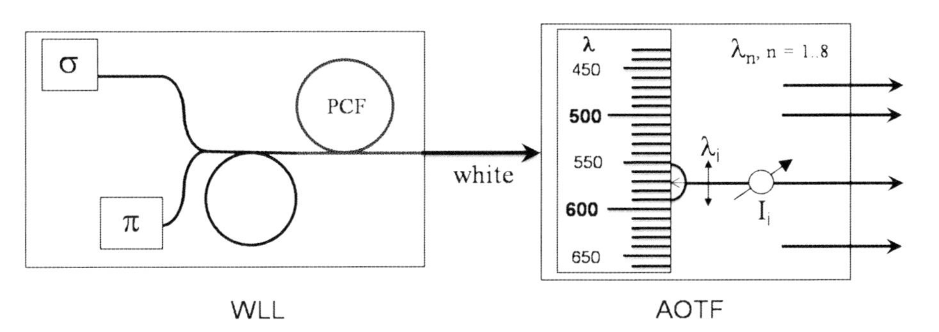

Fig. 2.2 White light laser and spectral tuning. *WLL*: The white light laser is a compound fiber-based collimated light source that consists of a seed laser σ, a high-power pump laser π, and the photonic crystal fiber PCF. The emission is a continuous spectrum with close to white characteristics in the visible range. *AOTF*: The acousto-optical tunable filter allows to simultaneously pick (currently up to eight) independent narrow bands (bandlets) from the white spectrum. For each bandlet the center wavelength λ_i is steplessly tunable, and also is the intensity (I_i). This device allows creating any illumination pattern in the wavelength–intensity space that might be required for fluorescence and is especially suited for confocal microscopy

a mercury high-pressure lamp and appropriate filters. Also, for selection of bands from the white laser, filters could be a solution. A much more efficient, flexible, and elegant solution is the employment of an acousto-optical device for the purpose. An acousto-optical crystal is capable of deflecting out of a white beam one or a series of narrow bands of only a few nanometers bandwidth. The crystal is excited mechanically and the wave grid effectively acts as a diffractive element that causes specific wavelength to be deflected. The deflected light exits the crystal in an angle (first order) to the principal beam (zeroth order). The deflected wavelength is controlled by the mechanical excitation frequency, and the intensity by the amplitude of the mechanical excitation. Casually speaking, the crystal works as a "photonic track switch". These devices are therefore called "acousto-optical tunable filters" (AOTFs).

By applying appropriate measures when designing the crystal and the electronic control, the light at the crystal's exit is collinear to the microscope's optical axis and thus suitable for illumination in the microscope. AOTFs are used in confocal microscopy already since 1992 to select laser lines and control their intensity. In comparison to plan-optical solutions (with filter glasses), an AOTF offers even more advantages. Many excitations of the crystal can be superimposed interference-free, so that a number of laser lines are simultaneously selectable. Current instruments use typically eight channels. A further advantage is the control of the amplitude that influences the intensity of the deflected light. In consequence, the device is a "dimmer" for many colors at the same time – each color's intensity can be controlled independently. And the whole regime of deflected lines can be reprogrammed within a matter of microseconds. The fast switch of illumination regimes is a necessary precondition for some applications to work at all, such as the illumination of hand-selected regions of interest or sequential illumination by different colors of single lines, which build up the image. Such methods help to reduce crosstalk or are used to excite ratio dyes in a timely correlated manner.

The combination of a white emitting laser and an acousto-optical tunable filter is consequently the ideal light source for confocal fluorescence microscopy (Fig. 2.2). Up to eight lines may be selected continuously in color and in intensity, where the switch time of illumination regimes is in the range of the duration of a single picture element.

2.5 Tunable Beam Splitting

The next step in incident light fluorescence microscopy is coupling the excitation light into the incident light beam path to illuminate the sample. Even though the classical solution by gray splitter or dichroic mirrors will work in connection with a tunable laser, these concepts invalidate the benefits of the new invention.

The simplest solution is of course a gray splitter that would also allow tuning the excitation wavelength continuously without the need of changing the splitting mirror. The major disadvantage, though, is the significant loss of both the applied laser energy and the emitted fluorescence light, which is highly undesirable. For that reason, a gray splitter solution would be a little attractive makeshift.

The implementation of dichroitic splitting mirrors is also not ideal, as already mentioned above. They work more efficiently, but are very inflexible as many splitters would be necessary to interchange. For a reasonably dense covering over 250 nm, at least ten different splitting mirrors would be necessary, mounted on a wheel or slider. This is mechanically fragile, slow, and expensive. Not to mention that for the typical case, that is multiple staining, multiple splitters for a huge number of combinations need to be available as well. In order to serve for the mentioned eight lines, allowing excitation only at distances of 36 nm (compared to a continuous, i.e., stepless tuning!), 2^8 different splitters, i.e., 256 different elements, would be necessary to be mounted and operated – not really a serious suggestion.

Here again, by employment of acousto-optical elements, a potent and very elegant solution was found: the acousto-optical beam splitter (AOBS) (Birk et al. 2002). Basically, this is an AOTF in reverse operation, which, however, requires the resolution of a couple of sophisticated technical problems.

If the acousto-optical crystal is tuned for a selected wavelength, then upon illumination with white light, this wavelength will be deflected into the first order. All other colors will pass the crystal straight and leave at the zeroth order. As the light pass is symmetrical, it is as well possible to guide light of a very narrow band into the first order, which will exit coaxially at the original entrance of the crystal. If adjusted correctly, this light could be used to irradiate the sample. The emitted fluorescence will never have the same wavelength as the excitation, rather always be shifted to the red. On its way back through the crystal, the emission is therefore not deflected into first order and can be collected completely by the detector. As the first-order bands in acousto-optical devices are always very narrow (typically around 1–2 nm), the excitation is very efficient and the losses for the precious fluorescence emission are very low. This is especially beneficial when compared to multiple-band dichroitic mirror splitter systems.

The most striking benefits of this method in combination with a tunable laser are of course the fact that the reflection peaks (the very narrow bands that allow light to pass from first order to the entrance of the crystal) can be controlled directly and without any detour, together with the selection of the excitation peak (Fig. 2.3). As soon as a certain frequency is applied to the AOTF in order to provide a selected excitation, the AOBS crystal that is coupled electronically to the AOTF crystal will be reprogrammed synchronously without any interactive control needed by the operator. This enables the continuous tuning of the excitation, prevents losses due to improper beam splitters, avoids losses in time due to mechanical switch of splitting mirrors, and at the same time represents a highly transmissive optical component that needs to be capable of simultaneous injection of various tunable excitations and transmission of the consequent fluorescence emissions.

And still there is one more merit: the operator does not need to follow all this reasoning. "By accident we used the wrong beam splitter" does not happen anymore.

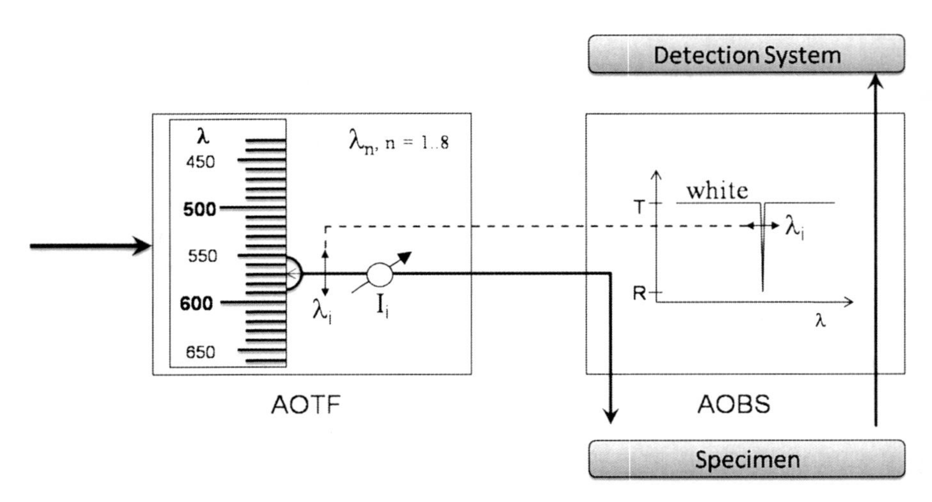

Fig. 2.3 White beam splitting by acousto-optical beam splitter (AOBS). The color- and intensity-shaped excitation light (multiple bandlets) is guided by the AOBS into the specimen. From there, emission light is transmitted into the detection system. The transmission is white, with the exception of the narrow bandlets that are used for excitation. Both devices, AOTF and AOBS, are controlled synchronously in parallel (*dotted line*) when the wavelength of the illumination pattern is modified. No further interaction ("selection of dichroic") is necessary. The white transmission is better than 95% and allows immediate recording of emission spectra that are not chromatically modified by beam splitters

2.6 Tunable Spectral Detectors

To record fluorescence flexibly as offered by a tunable laser and a tunable beam splitter, it is also desirable to have emission filters tunable. Basically, tunable band filters already exist for a long time: spectral analysis of light started when Joseph von Fraunhofer (1787–1826) invented the spectrometer in 1814. By spreading the light spectrally by means of a dispersive element (here: a prism) and collecting only the desired band by blocking the unwanted parts by barriers, the task can be done easily. This is the concept of all modern spectrophotometers. The challenge was to allow the recording of multiple bands simultaneously and at the same time reduce losses as much as possible. This task was solved by an ingenious combination of comparably simple devices. A multiband detector allows the simultaneous recording of multiple parts of the spectrum, and the bands are adjustable individually and steplessly (Engelhardt 1997).

For signal reasons, a prism is chosen over a grid, since a grid has many disadvantages in terms of photon inefficiency as they create many orders and show severe polarization effects. The slit is made from two independently movable barriers, very similar to a conventional spectrophotometer. In a commercial spectrometer, a lot of effort is made to avoid any reflections from the barriers, as this may cause stray light and consequently distort the measurements. In a multiband detector, the edges of the barriers are designed as highly reflective mirrors.

The initially unwanted parts of the spectrum are reflected and directed to successive detectors that are designed with similar barriers. Repeated combination of this setup allows to record a series of bands simultaneously (Fig. 2.4). The losses at the mirrors are very low ($<1\%$) and the slit itself has no absorption at all, of course.

This configuration constitutes a detection system, where many bands can be recorded simultaneously and the cut-off edges of the bands are individually and steplessly tunable. This is the ideal detector for multiple fluorescence samples. And it is the natural extension to a laser with multiple, tunable emission peaks.

As a side effect, the multiband detector also allows to record emission spectra, as it anyway consists of a series of spectrophotometers. The original intension was to record multiple bands simultaneously at negligible losses

2.7 Optimal Excitation

The abundance of many different dyes applied in biological research for fluorescence microscopy does make a flexible solution for exciting these dyes indispensable. Thus, the most important benefit of a tunable laser is obvious: setting the excitation wavelength to the most optimal position – in the simplest case to the excitation maximum of the dye.

Commercially available dyes are often named by excitation wavelength. Nevertheless, this name does not necessarily indicate the excitation maximum, but probably the best classical laser line that should be used to excite that dyes. The routinely used dye "Alexa488," for example, was baptized according to the most often used line emitted by the argon ion plasma, that is 488 nm. The excitation maximum of Alexa488 is around 500 nm; at 488 nm, the absorption dropped to only 80%.

A second commonly used dye, Alexa546, actually does absorb maximally at 561 nm. There is a good laser line from solid-state lasers at 561 nm. However, this does not imply that this would also be the best color for excitation. As a matter of fact, a tunable laser easily enables the dialing of the excitation according to the excitation maximum, but one has to keep in mind that between excitation and emission a certain distance has to be maintained in order not to distort the measurement. For this reason, in many cases it is recommendable to excite at shorter wavelengths than maximum, and especially if the Stoke's Shift is comparably short and the emission spectrum consequently has severe overlap with the excitation spectrum. In those cases one can still collect the larger part of the emission by shifting the excitation to the blue and moving the edge of the emission band accordingly. This procedure is usually not a disadvantage, as the laser is rarely used at full power (Fig. 2.5).

Thus, to compensate for the lesser absorption, one may just increase the laser intensity to reach a better signal-to-noise ratio in the recorded data. A combination of a tunable laser, an electronically coupled tunable beam splitter, and a tunable band detector – this optimal setting of excitation is found very conveniently by

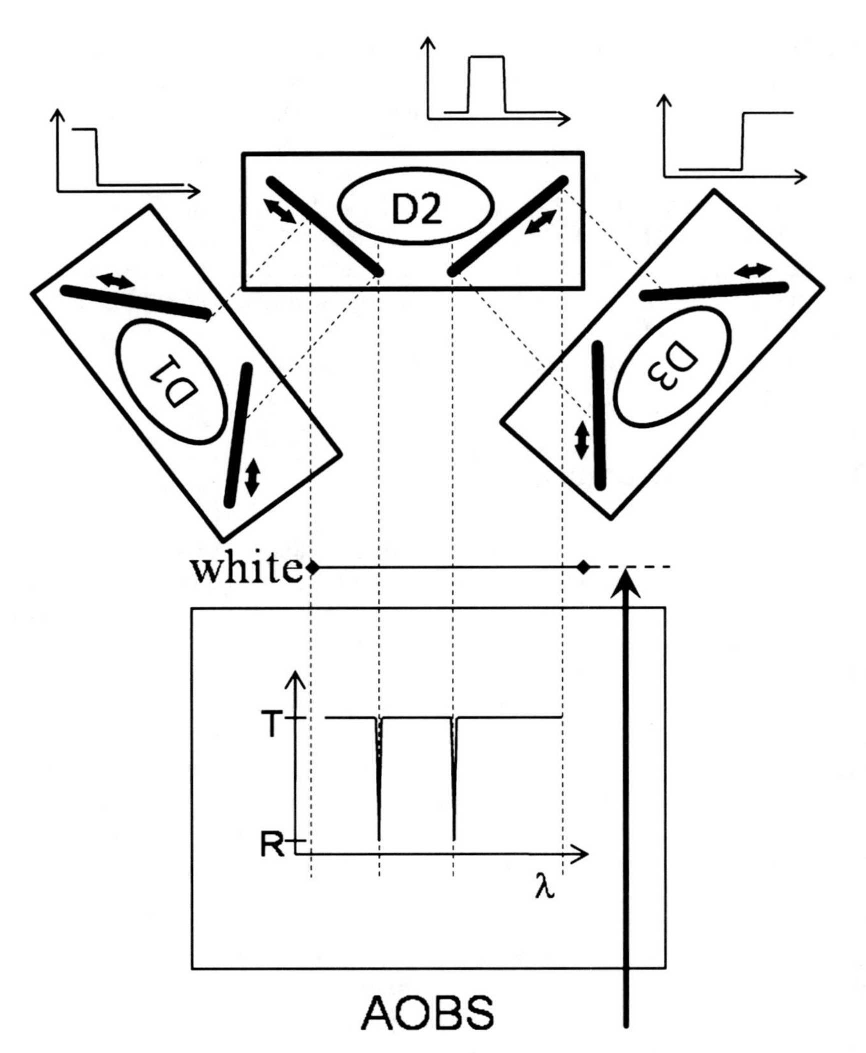

Fig. 2.4 Spectral multiband detection. The emission from the specimen is passed by the AOBS to a detection system that is made of a sequence of classical spectrophotometer devices. The emission light is dispersed by a prism (not shown) and directed to a detector D2. In front of this detector, a spectral band is selected by two movable barriers that only pass the desired band. Parts of the spectrum with shorter wavelength are reflected to the next detector D1, as the barriers are made of high-reflecting mirrors. The detector D1 is also equipped with an equivalent band selection device. The same is true for that fraction of the spectrum that has longer wavelength than required for D2. This way of cascading any required number of band selector devices allows distributing the spectrum in any possible fractions to different detectors. This is an immediate tool for simultaneous recording of multiparameter fluorescence at lowest possible losses and maximum flexibility. The photometer slits have no chromatic properties, as compared to filters

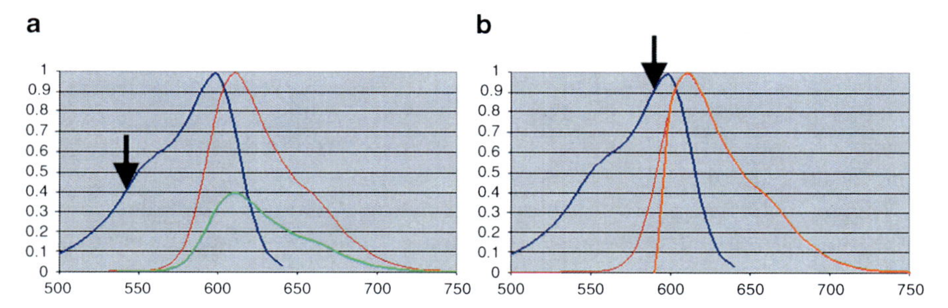

Fig. 2.5 Optimal fluorescence excitation by tunable devices. (**a**) A theoretical fluorescence dye with excitation spectrum (*blue*) and emission spectrum (*red*) is excited with a fixed wavelength laser (*arrow*). The intensity of the emission equals the area under the green curve, if a white detection system is used (collection of all emission from about the excitation wavelength to the *red border* – here: 750 nm). (**b**) Excitation tune to an optimal wavelength, where the dye is well excited and the band for recording emission is not cutting away any significant part of the emission spectrum. Recorded emission intensity is ca 2.5 times as compared to example (**a**)

recording a few test images. You need just to vary the parameters during scanning until you find the best result, interactively within a matter of seconds. Of course, the system also is capable of incrementing the parameters automatically and record a series of images. And you may wish to select the desired parameters from the image stack afterwards.

2.8 Optimal Excitation for Multiple Stainings

A tool for stepless tuning of excitation wavelength is of even more importance for samples that are multiply stained. Here, the "cross-talking" between channels is a severe problem, when two or more fluorescence channels are recorded simultaneously. On the excitation side, a laser line meant to excite a shorter wavelength dye often will also excite the long-wavelength dye, although at a lower cross-section. This causes a better signal collected in the emission band for the long-wavelength dye, but at the same time the emission from that same dye may be increased in the short-wavelength channel, if the emission spectrum is wide enough. This of course will corrupt the separation of the two signals. This effect is called "bleed though" or "cross-talking." To tackle that problem, the short-wavelength excitation may be tuned further into the blue range – by a tunable laser. The excitation for the blue dye will be reduced, too, but usually not to the same degree as the excitation for the red dye. How those ratios change is, of course, strongly dependent on the dyes used.

As the laser available is usually more than sufficient to excite the dye at maximum absorption, the reduced blue signal may be corrected for by increasing the short-wavelength excitation intensity. A laser that is steplessly tunable both in wavelength and intensity is the ideal source for these optimizations.

Similarly, it may occur that the long-wavelength excitation also excites the blue dye, causing a wrong blue signal in the red channel (without the benefit of a better signal in the blue channel, as emission is always of longer wavelength than excitation). For correction, the excitation wavelength must be tuned further red. This case is more critical, as it implies as well the need to narrow the collection band for the red signal.

With a tunable laser, it is very easy to balance the signals for multiple dyes and, at the same time, to reduce the unwanted signal in channels of complementary dyes. And in combination with a spectral band detector, it is possible to adapt the collected emission bands in order to increase signal to noise and further reduce crosstalk. A method that employs tunable laser lines would be rather laborious without an automatic synchronous tracing of the beam split parameters and the collection edges of the emission bands. With respect to the selection of appropriate filters and beam splitters, in many cases it would be just impossible.

When excitation wavelengths are optimized, the precision of separation may be enhanced further by the known procedures – if any are necessary. Here, fit in the already mentioned adaptation of the emission bands, the method of sequential scanning of the single channels, and – if nothing else will help – mathematical treatment by unmixing algorithms. The best method is of course: use simple to separate dyes – if any possible.

2.9 Förster Resonance Energy Transfer Problems

These days, fluorescence microscopy knows many new methods that well exceed the mere recording of images for rendering morphologic structures. An example that benefits particularly from tunable excitation wavelengths is the Förster resonance energy transfer (FRET) method. In FRET pairs of fluorochromes, the shorter wavelength dye is called the donor (D) and the longer wavelength dye is called the acceptor (A). In order for FRET to occur, the emission of D needs to spectrally overlap with the excitation of A. Then, the energy from D may be transferred to A without involving a photon (radiationless transfer). As radiationless transfer occurs only if the molecules are spatially very close (a few nanometers), the occurrence of FRET gives some evidence for the spatial colocalization of the fluorophores. If the fluorophores are bound to biological structure elements, e.g., proteins, then this conclusion also holds for the biological molecules that are investigated.

Detection of FRET is possible by two different methods: FRET-AB (for "FRET acceptor bleaching") and FRET-SE (for "FRET-sensitized emission"). FRET-SE tries to measure directly the acceptor's emission that was caused by donor excitation. Under normal (non-FRET) conditions, only the donor would emit photons (EmD) when excited with appropriate wavelength (ExD). If radiationless transition occurs, then a second color is emitted, the acceptor emission (EmA) upon the single excitation by ExD. Separation of both emission channels is not trivial, as one has to quantitatively measure the comparably weak FRET signal against the strong donor emission. A separation that is just pleasing the eye is not sufficient. As in living objects the fluorochromized molecules move comparably fast, it is essential to record both emission channels simultaneously. As excitation is done by one single wavelength, there is a good chance to find emission of donor not only in the short-wavelength band for EmD, but also in the emission channel for EmA. The latter should only render the FRET signal. Due to the strong donor emission, this spill into the FRET channel causes significant disturbance in measurement accuracy of the faint FRET signal. The errors due to overlap of emissions of both channels may be corrected for by adaptation of the emission-band characteristics and by mathematical correction. Normally, for these kinds of experiments, a series of control measurements have to be recorded in order to capture all the parameters that cause crosstalk. These parameters are then applied to the data to get a significant measure for the radiationless transfer. Under good conditions, even the distance of the "fretting" molecules might be estimated.

A further source of errors that can have an impact on the precision of the FRET measurement is direct excitation of the acceptor molecule by the donor excitation wavelength. This of course causes directly a "bogus FRET signal," as there is no way to discern whether the emission from the acceptor was caused by direct excitation or as consequence of a FRET event. Here, the tunable laser offers the solution: the excitation wavelength may be selected from the bluer range of the spectrum until the effect is as small as possible or vanishes entirely. This way, by fine-tuning excitation, distinctly improved qualitative data may be obtained.

In FRET-AB, which is preferentially appropriate for fixed samples, a quantifiable donor image is recorded. Then, the acceptor molecules are destroyed by application of a high dose of light with a wavelength at high absorption of the acceptor ("bleaching"). If prior to bleaching FRET has occurred, then radiationless transfer is now interrupted and the donor emission should increase. Therefore, a second donor image is recorded and the intensity is compared to the first recording prior to bleaching. As was shown by Valentina Caorsi (Bianchini et al. 2010), a serious problem is caused by the fact that during bleaching of the acceptor, also donor molecules may be destroyed if donor absorption is too strong at the bleaching wavelength. Then, the donor fluorescence is reduced again after bleaching , and the increase caused by FRET is compensated for or at least reduced. The absorption properties in a real sample differ usually significantly from data published in literature, where in most cases spectra of the dye in solvent are published. The tunable laser offers here a means to measure the real absorption properties (spectrum) of the donor and to compare that with data from the literature. And in addition, it offers as well a solution to the problem: the wavelength for bleaching may be moved as far as necessary into the red part of the spectrum, until the donor will not absorb and the donor molecules will not be bleached and sensible FRET data can be obtained.

2.10 Excitation Spectra In Situ

The tunability of the white laser consequentially offers a possibility to automatically increment the excitation wavelength and record data at each color. The result is a stack of data in the excitation wavelength dimension. By selection of regions of interests in the image plane, the excitation spectra of various structures can be displayed graphically and the spectral data are available for further evaluation as well. As already mentioned, the optical properties of fluorochromes strongly depend on the local environment that the molecule is sensing. This fact was the precondition to use fluorescent dyes as pH indicators, polarity sensors, or potential-sensitive probes. The modern bioindicators are as well based on this interaction of the fluorochromes and the local environment – where the environment could as well be a second fluorochrome. Here, in many cases the sensing mechanism involves a molecule that binds to the probed metabolite. Binding then causes a conformational change or cleavage, rendering the attached fluorochromes to a different environment, or changing the distance to a second fluorochrome (FRET-based indicators).

As there are a tremendous number of modern fluorescence-based indicators, especially the continuously newly developed mutants of fluorescent proteins, the measurement of the spectral properties inside the sample became indispensable.

Excitation spectra also help to separate dyes by unmixing methods based on excitation. That is of particular interest with dyes that have strongly overlapping emissions or emissions that are more or less identical. As a consequence, a new method is available: spectral excitation unmixing.

2.11 λ^2-Maps

With the freedom to record spectra by varying both emission and excitation, a new and additional analytical method is available: two-dimensional maps that correlate excitation wavelength and emission wavelength (λ^2-maps). For this purpose, emission spectra are recorded in a series of excitation wavelengths. With systems that provide automatic incrementation of excitation color and the corresponding emission bands, this becomes an easy task. The obtained data are displayed as intensity distributions in the spectral landscape.

λ^2-mapping is significantly beneficial when samples contain multiple stains (i.e., more than approximately four) or are composed of a complex mixture of fluorescent molecules. In complex samples, such as autofluorescing biofilms or eightfold-stained samples, this type of analysis is applied with great success. (Borlinghaus et al. 2006; Bianchini et al. 2010).

2.12 Fluorescence Lifetime Imaging

As mentioned above, the white laser is pumped by an IR laser that is pulsed at around 100 MHz. Consequently, also the white light is pulsed at the same frequency. The pulse width is in the range of some 20 ps. These parameters make this white light source an ideal tool for measurement of fluorescence lifetime. Classical fluorescence lifetime imaging (FLIM) is performed by detecting the delay time between a light pulse that excites the fluorochromes and the arrival of the first photon at the sensor. As this delay is distributed statistically, measurements have to be repeated many times (at the same position) and the data are plotted in a time histogram. The typical decay time τ is then extracted from these data by standard curve-fitting algorithms. A scanning microscope allows doing this calculation at each picture element. The resulting image consequently does not show intensities as values, but fluorescence decay times and is hence called a τ-map. With the tunable laser that is described above, it is not only possible to select the wavelength for optimal excitation, but also to record and evaluate sequences of τ-maps against excitation wavelength. This yields an entirely new quality of information: the correlation of fluorescence lifetime and excitation color, which was not accessible by now.

By means of a spectral detector it is also possible to correlate fluorescence lifetime and emission color (a method called "spectral FLIM"). In a combination of all these methods, the excitation tuning provides a third dimension when measuring fluorescence lifetimes. And it also opens up the possibility of correlating excitation and emission wavelength by fluorescence lifetime ($\lambda^2\tau$-maps).

2.13 Unlimited Spectral Performance

As shown in this article, the use of a tunable visible laser that allows multiple lines simultaneously and fast switching (μs) is a crucial improvement for confocal fluorescence microscopy. This is particularly true, when the corresponding parameters for beam splitting and emission band selection are continuously adaptable to the selected excitation (Borlinghaus 2007). Then, such an instrument is a completely new tool allowing new insights by new methods and applications that have not been available so far. This is true for structural imaging, for measurements of dynamic processes in living material, and for the new fluorescence applications in biomedical research, like FLIM.

References

Bianchini P, Caorsi V, Ronzitti E, Brolinghaus R, Diaspro A (2010) Advantages of a super continuum white light source applied to confocal laser scanning microscopy (CLSM) (in preparation)
Birk H, Storz R (2001) US Patent 6.611.643
Birk H, Engelhardt J, Storz R, Hartmann N, Bradl J, Ulrich H (2002) Programmable beam splitter for confocal laser scanning microscopy. Progr Biomed Opt Imag SPIE 3(13):16–27

Borlinghaus R (2007) Colours count: how the challenge of fluorescence was solved in confocal microscopy. In: Mendez-Vilas A, Diaz J (eds) Modern research and educational topics in microscopy, vol 3. Formatex, Badajoz, pp 890–899

Borlinghaus R, Gugel H, Albertano P, Seyfried P (2006) Closing the spectral gap – the transition from fixed-parameter fluorescence to tunable devices in confocal microscopy. Proc SPIE 6090:159–164

Chalfie M, Tu Y, Euskirchen G, Ward WW, Prasher DC (1994) Green fluorescent protein as a marker for gene expression. Science 263:802–805

Coons AH, Creech HJ, Jones RN (1941) Immunological properties of an antibody containing a fluorescent group. Proc Soc Exp Biol Med 47:200–202

Diaspro A (2002) Confocal and two-photon microscopy: foundations, applications, and advances. Wiley-Liss, New York

Engelhardt J (1997) US Patent 5.886.784

Gall JG, Pardue ML (1969) Formation and detection of RNA-DNA hybrid molecules in cytological preparations. Proc Natl Acad Sci U S A 63:378–383

Haitinger M (1934) Die methoden der fluoreszenzmikroskopie. In: Abderhalden E (ed) Abderhaldens Handbuch der Biol Arbeitsmehtoden. Abt II Teil 2, Urban and Schwarzenberg, Berlin and Wien

Herschel J (1845) On a case of superficial colour presented by a homogeneous liquid internally colourless. Philos Trans R Soc Lond 135:143–145

Jablonski A (1935) Über den mechanismus der photolumineszenz von farbstoffphosphoren. Z Phys 94:38–46

Kauffmann H (1900) Untersuchungen über das ringsystem des benzols. Ber Dtsch Chem Ges 33:1735

Knight JC, Birks TA, Russell TS, Atkin DM (1996) All-silica single-mode optical fiber with photonic crystal cladding. Opt Lett 21:1547–1549

Köhler A (1904) Mikrophotographische untersuchungen mit ultraviolettem licht. Z Wiss Mikrosk 21:55

Naora H (1951) Microspectrophotometry and cytochemical analysis of nucleic acids. Science 114 (2959):279

Stokes GG (1852) On the change of refrangibility of light. Philos Trans R Soc Lond 142:463–562

Chapter 3
Second/Third Harmonic Generation Microscopy

Shakil Rehman, Naveen K. Balla, Elijah Y. Y. Seng, and Colin J. R. Sheppard

3.1 Introduction

When the energy density at the focal spot of a microscope is sufficiently large, nonlinear optical effects such as, harmonic generation, sum-frequency generation, coherent Raman scattering, parametric oscillations, and multi-photon fluorescence can be observed. These optical phenomena can be used in a nonlinear optical microscope to study the biological material. Nonlinear optical microscopy may be divided into incoherent and coherent modes. Incoherent nonlinear microscopy is characterized by the optical signal (like fluorescence) with a random phase, whose power is proportional to the concentration of radiating molecules. The principle of nonlinear fluorescence microscopes is based on the simultaneous absorption of two

S. Rehman (✉)
Division of Bioengineering, National University of Singapore, 9 Engineering Drive, Singapore 117576, Singapore
and
Singapore Eye Research Institute, 11, Third Hospital Avenue, #05-00, Singapore 168751, Singapore
e-mail: shakil@nus.edu.sg

N.K. Balla
Division of Bioengineering, National University of Singapore, 9 Engineering Drive, Singapore 117576, Singapore
and
Computation and Systems Biology, Singapore-MIT Alliance, National University of Singapore, Singapore 117576, Singapore

E.Y.Y. Seng
Division of Bioengineering, National University of Singapore, 9 Engineering Drive, Singapore 117576, Singapore

C.J.R. Sheppard
Division of Bioengineering, National University of Singapore, 9 Engineering Drive, Singapore 117576, Singapore
and
Department of Diagnostic Radiology, National University of Singapore, 5 Lower Kent Ridge Road, Singapore 119074, Singapore

A. Diaspro (ed.), *Optical Fluorescence Microscopy*,
DOI 10.1007/978-3-642-15175-0_3,

or more photons, as in, two-photon excited fluorescence microscopy (Denk et al. 1990) and three-photon excitation fluorescence microscopy (Maiti et al. 1997; Schrader et al. 1997). The coherent mode of nonlinear microscopy is characterized by the optical signal whose phase is determined by factors like phase of excitation field and the geometrical distribution of radiating molecules. Some of the coherent nonlinear microscopic techniques are second harmonic generation (SHG) microscopy (Campagnola et al. 1999; Gannaway and Sheppard 1978; Gauderon et al. 1998; Sheppard and Kompfner 1978), coherent anti-Stokes Raman scattering (CARS) microscopy (Duncan et al. 1982), and third harmonic generation (THG) microscopy (Barad et al. 1997; Müller et al. 1998; Squier et al. 1998; Yelin and Silberberg 1999). All of these nonlinear microscope modalities have benefited from the fact that it is easier than ever before to generate ultrashort laser pulses at specific wavelengths.

SHG is a nonlinear optical process which can take place in a microscope using ultrashort laser pulses in the near-infrared range. The amplitude of SHG is proportional to the square of the incident light intensity. In 1978, Sheppard first proposed the idea that two-photon excited fluorescence (2PEF) and SHG can be used for nonlinear microscopy. Later, in 1990, Denk et al. used ultrashort pulse lasers to demonstrate the 2PEF microscopy that involves near-simultaneous absorption of two photons to excite a fluorophore, followed by an incoherent emission of fluorescence. The nonlinear process of SHG takes place through an interaction between the electric field and its spatial derivative. Emission by this process is highly anisotropic, coherent, and phase-coupled to the excitation and is therefore subject to phase-matching effects between the electric fields associated with the process. In SHG, two photons are converted into a single photon at twice the excitation energy emitted coherently. SHG takes place in systems lacking a center of symmetry. In biological materials, cellular membranes possess such asymmetrically distributed molecular structure. Supramolecular structures within cells and tissues that can produce SHG signals are collagen and actin filaments. Being a nonlinear process, SHG signal is maximum at the focus of the microscope, resulting in the intrinsic three-dimensional sectioning without the use of a confocal aperture that greatly reduces out-of-focus plane photobleaching and phototoxicity. Near-infrared wavelength excitation allows excellent depth penetration in biological material; therefore, this method is well-suited for studying intact tissue samples. Information about the organization of chromophores, including dyes and structural proteins, at the molecular level can be extracted from SHG imaging data in several ways. SHG signals have well-defined polarizations, and hence SHG polarization anisotropy can also be used to determine the absolute orientation and degree of organization of proteins in tissues.

THG is a nonlinear optical process which has been intensely studied since the early days of nonlinear optics (Bloembergen 1965; Shen 1984). This process has been applied in imaging of lipids in cells and tissues (Débarre et al. 2006), subcellular structure in neurons (Yelin and Silberberg 1999), cell nuclei (Yu et al. 2008), and simultaneous SHG, THG, and 2PEF imaging of live biomaterials (Chu et al. 2001). All materials have some third-order nonlinearity depending on the material property, symmetry, and the incident light. At the focus of microscope objective lens, when a discontinuity or inhomogeneity is encountered, like an interface between two media, the symmetry along the optical axis is broken and a third harmonic signal can be obtained. High optical sectioning can be achieved because of the nonlinear process of THG taking place only at the focal plane. Being the property of all materials, third harmonic signal can be used for noninvasive microscopy, particularly in biological materials without the need of fluorescence labeling.

3.2 Nonlinear Optics Background

Nonlinear optical phenomena occur when the response of a material to the applied optical field depends nonlinearly on the strength of the field. The applied field may cause changes in the distribution or motion of the internal electric charges such as electrons, ions, or nuclei within a molecular system, resulting in a field-induced electric dipole moment which, in turn, acts as a new source to emit a secondary wave. This is the fundamental process of optically field-induced polarization in a molecular system and re-emission of a secondary electromagnetic wave.

The induced polarization in a medium depends on the strength of the applied field and can be written as

$$P(t) = \varepsilon_0 \chi^{(1)} E(t)\,, \tag{3.1}$$

where is ε_0 is the permittivity of free space, $\chi(1)$ the linear susceptibility of the medium, E is the incident field, and P is the induced polarization. The electric susceptibility χ indicates the ability of the electric dipoles in a dielectric to align themselves with the electric field. A general expression for the induced polarization can be written in a power series expansion in the magnitude of the electric field as

$$P(t) = \varepsilon_0 \chi^{(1)} E(t) + \varepsilon_0 \chi^{(2)} E^2(t) + \varepsilon_0 \chi^{(3)} E^3(t) + \cdots. \tag{3.2}$$

$\chi^{(2)}$ and $\chi^{(3)}$ are the second-order and third-order nonlinear susceptibilities, respectively. For materials with inversion symmetry, all the even-order nonlinear susceptibility coefficients are equal to zero. In general, P and E are vector quantities, therefore, nth-order susceptibility $\chi^{(n)}$ becomes a $(n+1)$ rank tensor. First term in (3.2) represents the linear polarization and it can be shown that the linear susceptibility $\chi^{(1)} = n^2 - 1$, where n is the refractive index. Second term in (3.2) is referred to as second-order nonlinear polarization $P^{(2)}(t) = \varepsilon_0 \chi^{(2)} E^2(t)$ and $P^{(3)}(t) = \varepsilon_0 \chi^{(3)} E^3(t)$ as third-order nonlinear polarization and so on. Each of these polarization terms gives rise to a different physical phenomenon. For example, second-order polarization is used to produce SHG and sum/difference frequency generation, and third-order polarization is responsible for THG, Raman scattering,

Brillouin scattering, self-focusing, and optical phase conjugation. Furthermore, second-order nonlinear optical interactions can only occur in noncentrosymmetric crystals, that is, in crystals having no inversion symmetry, while third-order nonlinear phenomena can occur in any medium regardless of whether it possesses inversion symmetry or not.

Consider that an electric field applied to a medium is of the form

$$E(t) = E_0 + E_\omega \cos(\omega t)\,, \tag{3.3}$$

where E_0 is the dc field and E_ω is the amplitude of the field at frequency ω. Substituting this value in (3.2) and solving for various polarizations, we get cross-terms that represent the nonlinear phenomena. For example, the second-order polarization can be written as

$$P^{(2)}(t) = \varepsilon_0\chi^{(2)}E^2(t) = \varepsilon_0\chi^{(2)}\left[E_0^2 + \frac{E_\omega^2}{2} + \frac{E_\omega^2\cos(2\omega t)}{2} + 2E_0E_\omega\cos(\omega t)\right]. \tag{3.4}$$

The four terms of (3.4) represent $\chi^{(2)}$ based nonlinear optical processes and are explained as follows.

$\varepsilon_0\chi^{(2)}E_0^2$: DC hyper-polarizability.

$\varepsilon_0\chi^{(2)}E_\omega^2/2$: DC optical rectification. Applied field produces a static voltage in the medium that is proportional to the applied field.

$\varepsilon_0\chi^{(2)}E_\omega^2\cos(\omega t)/2$: Polarization generated at frequency 2ω that can be used to obtain a signal at twice the frequency of the input field and is responsible for SHG.

$\varepsilon_0\chi^{(2)}2E_0E_\omega\cos(\omega t)$: Linear electro-optic effect (Pockels effect) represents the modification of the refractive index because of an applied DC electric field and can be used for optical switching and for phase modulation of light.

By substituting (3.3) in third-order polarization term of (3.2), we get

$$\begin{aligned} P^{(3)}(t) &= \varepsilon_0\chi^{(3)}E^3(t) \\ &= \varepsilon_0\chi^{(3)}[E_0 + E_\omega\cos(\omega t)]^3. \end{aligned} \tag{3.5}$$

Cross-terms obtained from this equation provide $\chi^{(3)}$ based optical processes, some of which are explained as follows:

$\varepsilon_0\chi^{(3)}E_0^2E_\omega\cos(\omega t)$: Quadratic electro-optic effect also known as the Kerr effect. This occurs in isotropic media such as gases and liquids as well as solids and is not as widely used as the Pockels effect because of quadratic dependence with the applied field.

$\varepsilon_0\chi^{(3)}E_0^2E_\omega^2\cos(2\omega t)$: DC-induced SHG. Because of symmetry reasons SHG is not possible in isotropic materials. In this case, applied electric field breaks the symmetry, making SHG possible.

$\varepsilon_0\chi^{(3)}E_\omega^3\cos(3\omega t)$: THG at three times the frequency of the applied field. THG is allowed in all materials including isotropic ones.

$\varepsilon_0\chi^{(3)}E_{\omega}^2\cos(\omega t)$: Optical (AC) Kerr effect. Third-order susceptibility permits a term at frequency ω caused by the applied field at the same frequency. This term gives rise to self-focusing and self-phase modulation.

Here, we would like to give the readers a feel for how these nonlinearities arise, through the example of an anharmonic oscillator suggested by Bloembergen (1965). The approximate motion of a bound electron in a dielectric medium can be expressed with the equation of motion of an anharmonic oscillator as

$$\frac{\mathrm{d}^2x}{\mathrm{d}t^2}+\gamma\frac{\mathrm{d}x}{\mathrm{d}t}+\omega_0^2x+ax^2=F\,, \tag{3.6}$$

where F is the external force acting on the electron, x is the displacement from its mean position, ω_o is the natural frequency of the electron, γ is the damping parameter, and ax^2 is the nonlinear term with a being a nonlinear coefficient. Let us assume an incident electromagnetic wave consisting of two oscillating electric fields of amplitude E_1 and E_2 and frequencies ω_1 and ω_2, of the form $E_1\mathrm{e}^{-i\omega_1 t}+E_2\mathrm{e}^{-i\omega_2 t}+\mathrm{c.c}$, where c.c refers to the complex conjugate terms. The external force F would then be $-eE/m$ where e and m are the charge and mass of an electron. According to the perturbation theory, we can write the solution of (3.3) as

$$x=\lambda x^{(1)}+\lambda^2x^{(2)}+\lambda^3x^{(3)}+\cdots. \tag{3.7}$$

For the above equation to hold for any value of λ, the differential equation should be satisfied for each power of λ. So we get a series of equations as follows.

First order in λ: $\mathrm{d}^2x^{(1)}/\mathrm{d}t^2+\gamma\mathrm{d}x^{(1)}/\mathrm{d}t+\omega_0^2x^{(1)}=-eE/m$

Second order in λ: $\mathrm{d}^2x^{(2)}/\mathrm{d}t^2+\gamma\mathrm{d}x^{(2)}/\mathrm{d}t+\omega_0^2x^{(2)}=-a[x^{(2)}]^2$

Third order in λ: $\mathrm{d}^2x^{(3)}/\mathrm{d}t^2+\gamma\mathrm{d}x^{(3)}/\mathrm{d}t+\omega_0^2x^{(3)}=-ax^{(1)}x^{(2)}$ and so on.

Solving for $x^{(1)}$ we get, $x^{(1)}=\left[(eE_1/m)/(\omega_1^2-\omega_0^2)\right]\mathrm{e}^{i\omega_1 t}+\left[(eE_2/m)/(\omega_1^2-\omega_0^2)\right]\mathrm{e}^{i\omega_2 t}+\mathrm{c.c.}$

To solve for the second-order term, let us substitute the value of $x^{(1)}$ in the differential equation for $x^{(2)}$ and get

$$\begin{aligned}x^{(2)}=&\frac{\frac{e^2}{m^2}E_1^2}{(\omega_1^2-\omega_0^2)^2\left[(2\omega_1)^2-\omega_0^2\right]}\mathrm{e}^{i2\omega_1 t}+\frac{\frac{e^2}{m^2}E_2^2}{(\omega_2^2-\omega_0^2)^2\left[(2\omega_2)^2-\omega_0^2\right]}\mathrm{e}^{i2\omega_2 t}\\&+\frac{\frac{e^2}{m^2}E_1^2E_2^2}{(\omega_1^2-\omega_0^2)^2(\omega_2^2-\omega_0^2)^2\left[(\omega_1+\omega_2)^2-\omega_0^2\right]}\mathrm{e}^{i(\omega_1+\omega_2)t}\\&+\frac{\frac{e^2}{m^2}E_1^2E_2^2}{(\omega_1^2-\omega_0^2)^2(\omega_2^2-\omega_0^2)^2\left[(\omega_1-\omega_2)^2-\omega_0^2\right]}\mathrm{e}^{i(\omega_1-\omega_2)t}+\mathrm{c.c.}\end{aligned} \tag{3.8}$$

Here, we have omitted the DC electric field as it is not relevant in this example. The expression for $x^{(2)}$ contains two harmonic terms, a sum frequency term

$(\omega_1 + \omega_2)$ and a difference frequency term $(\omega_1 - \omega_2)$ which explain the second-order optical phenomena as defined by (3.2). In a similar fashion, we can derive the third-order harmonics by solving for $x^{(3)}$. The second-order polarization, due to a single electron of charge e, is given by $ex^{(2)}$. If the number density of electrons in the given dielectric is N_e, the bulk second-order polarization will be $P^{(2)} = N_e ex^{(2)}$. As we already know the form of second-order polarization from (3.2), $P^{(2)}(t) = \varepsilon_0 \chi^{(2)} E^2(t)$. Therefore, we can get an expression for second-order nonlinear susceptibility as

$$\chi^{(2)} = \frac{N_e ex^{(2)}}{\varepsilon_n E^2}. \tag{3.9}$$

This derivation is given in scalar form for simplicity. In a similar fashion, one can solve for third-order nonlinear susceptibility for this problem.

3.3 Second Harmonic Generation

SHG is a three-wave mixing process based on second-order polarization as shown in Fig. 3.1a. In this picture, two photons of frequency ω are converted into one photon of twice the frequency at 2ω. The energy levels represented by solid lines are atomic or molecular levels and dotted lines represent the virtual states. These virtual energy states are combined states of the molecular system and the photons of the incident radiation. This property of nonlinear harmonic generation via virtual states leaves no residual energy in the medium and the emitted photon has exactly same amount of energy as that of the absorbed photons. The conservation of energy in such a nonlinear process provides the noninvasive property of imaging in the microscopic applications of biological material.

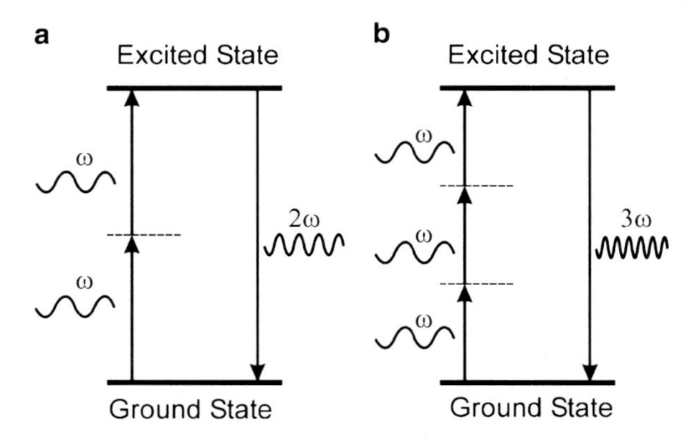

Fig. 3.1 Energy level description of SHG (**a**) and THG (**b**) processes

Because of symmetry reasons, elements of the second-order susceptibility cancel out in an isotropic homogeneous medium; hence, second-order nonlinear optical effect can occur in a noncentrosymmetric crystalline structure only. The first experiment carried out to demonstrate optical SHG was performed in 1961 (Franken et al. 1961). Some other physical nonlinear processes based on $\chi^{(2)}$ are linear electro-optic effect (Pockel's effect), optical rectification, and DC polarizability. The second order-induced polarization can be written as (He and Liu 1999)

$$P_i^{(2)}(2\omega) = \varepsilon_0 \sum_{jk} \chi_{ijk}^{(2)}(\omega,\ \omega) E_j(\omega) E_k(\omega)\,, \tag{3.10}$$

where $E(\omega)$ is the incident wave at frequency ω and the induced polarization is at 2ω that produces a new radiation at double the frequency. Here, $\chi_{ijk}^{(2)}$ is a third-rank tensor responsible for the SHG; for centrosymmetric materials, $\chi_{ijk}^{(2)}$ vanishes because of inversion symmetry. The indices *ijk* are summed over linear polarization directions of x, y, and z. The induced polarization $P_x^{(2)}(2\omega)$ produces an electromagnetic field $E_x(2\omega)$ at twice the frequency of the input field. If we assume the incident electric field of the form $E_i \sin(\omega t)$, then the resulting polarization becomes

$$P_i^{(2)}(2\omega) = \varepsilon_0 \sum_{jk} \chi_{ijk}^{(2)}(\omega,\ \omega) E_j(\omega) E_k(\omega) \sin^2(\omega t)\,, \tag{3.11}$$

$$= \frac{1}{2} \varepsilon_0 \sum_{jk} \chi_{ijk}^{(2)}(\omega,\ \omega) E_j(\omega) E_k(\omega) [1 - \cos(2\omega t)]\,. \tag{3.12}$$

The first term in (3.9) gives rise to a DC polarization within the material and the second term with $\cos(2\omega t)$ corresponds to a polarization wave that oscillates at twice the fundamental frequency ω and that acts as a source for the second harmonic output field. Because of symmetry selection rules, the elements of the nonlinear susceptibility tensor $\chi_{jk}^{(2)}$ vanish for materials with inversion symmetry. This symmetry rule also applies equally well for all even powers of nonlinear susceptibilities given by (3.2). Therefore, second harmonic effect is not observed for an isotropic and centrosymmetric material. For effective nonlinear harmonic generation effect, the specimen must be relatively transparent to the wavelength of fundamental illumination light and the generated higher harmonics.

Conversion efficiency in SHG relates to spatiotemporal overlap of waves at fundamental and second harmonic frequencies, as well as to their group and phase velocity matching. For effective energy transfer from the fundamental wave to the second harmonic waves, both the energy and momentum conservation must be satisfied. Energy conservation states that $\hbar\omega_0 + \hbar\omega_0 = \hbar 2\omega_0$, where $\hbar = h/2\pi$, with h being Planck's constant. This may as well be written as $\omega_0 + \omega_0 = 2\omega_0$. Similarly, the momentum conservation should be satisfied through the momentum relation given by $\hbar k_0 + \hbar k_0 = \hbar 2k_0$, or $k_0 + k_0 = 2k_0$ for collinear upconversion. These two conservation requirements lead to phase-matching requirement for the refractive index of the nonlinear medium at the two frequencies as $n(2\omega_0) = n(\omega_0)$.

For isotropic or cubic materials, this condition is not satisfied, as $n(2\omega_0) > n(\omega_0)$ in a normal dispersion. This condition is satisfied only for an anomalous dispersion or in a birefringent crystal. This can be done by selecting a special direction with in a crystal so that the dispersion effect of the refractive index can be compensated by the birefringence effect. Two methods are used for collinear phase matching by selecting polarizations of the two waves. In type I phase matching, $n^{o}(\omega_0) + n^{o}(\omega_0) = 2n^{e}(2\omega_0)$, where the incident fundamental wave consists of ordinary polarization while the second harmonic wave contains an extraordinary polarization component. The superscripts of refractive indices, n^{o} and n^{e} represent the ordinary and extraordinary wave components. In type II phase matching, $n^{o}(\omega_0) + n^{e}(\omega_0) = 2n^{e}(2\omega_0)$, where the fundamental wave consists of two polarizations (one ordinary and the other extraordinary polarization) while the second harmonic wave contains extraordinary polarization component.

The above statements are true for a negative uniaxial crystal and roles of ordinary and extraordinary waves are reversed for a positive uniaxial crystal.

3.3.1 SHG Microscopy

The first application of SHG in optical microscopy was imaging the structure of nonlinearities in a crystal of ZnSe, in which, a large area of the specimen was illuminated by a laser and a second harmonic signal was imaged in an optical microscope (Hellwarth and Christensen 1974). The scanning mode of a SHG microscope was implemented by (Gannaway and Sheppard 1978).The first biological application of SHG microscopy was imaging a rat-tail tendon (Freund and Deutsch 1986). The development of femtosecond mode-locked lasers provided stable sources of high-intensity excitation required for the efficient generation of nonlinear responses. Now SHG microscopy is used in a variety of biological applications. Optical microscopy with SHG is widely used for contrast generation in collagen (Campagnola et al. 2002; Chu et al. 2007; Freund et al. 1986; Lin et al. 2005), myocytes (Barzda et al. 2005; Campagnola et al. 2002), plants, and chloroplasts (Cox et al. 2005; Mizutani et al. 2000; Prent et al. 2005).

SHG microscopy is based on the noncentrosymmetric organization of microstructures in a sample, and the SHG signal can be generated as a result of a broken symmetry at an interface or because of a noncentrosymmetric arrangement within bulk structures. At an interface, biological membranes can produce detectable SHG signal as in the case of a lipid bilayer (Moreaux et al. 2000). Molecular structures having a symmetric distribution of chirality in the membrane do not give rise to SHG signal; only an ordered asymmetric distribution of chiral molecules in the membranes is responsible for SHG (Campagnola et al. 1999). SHG can also be produced in biogenic crystal structures, for example, in calcite or starch granules (Mizutani et al. 2000) and biophotonic crystalline and semi-crystalline structures in living cells (Chu et al. 2002).

The phase-matching conditions in nonlinear crystals are usually obtained by angle or temperature tuning (Boyd 2003). The SHG signal from biogenic crystals is generated in a similar way as in nonlinear crystals. The difference between macroscopic and microscopic measurements of SHG is that in microscopic experiments the fundamental light is tightly focused with a high Numerical Aperture (NA) objective lens providing a cone of light with a wide range for incident angles satisfying the phase-matching conditions for some of the rays.

3.3.2 Applications

SHG has been applied to diverse microscopic modalities such as, recording holograms using second harmonic signals scattered from nanocrystals and an independently generated second harmonic reference beam (Pu et al. 2008). By far, the most studied biological structure by SHG microscopy is collagen. Figure 3.2 shows the collagen structure in the cross-section of sclera of a 42 postnatal-days-old rabbit's eye. The images show the forward and backward scattered SHG signal.

Collagen and myosin fibrils of muscles are known to give a good second harmonic signal. Here, we show the effectiveness of SHG microscopy in a 2-month-old zebra fish. The whole fish was mounted on a microscope well slide using 0.5% low-melting agarose. The image was formed by an overlapping second harmonic image over the transmission image as shown in Fig. 3.3. The second harmonic signal from the skin is due to the collagen layer within the skin. The second harmonic signal from within the fish body is due to muscles. A higher zoom into one of these regions (inset) Fig. 3.3b, exposes the fibrillose structure of myosin in the muscles. The image was obtained with an objective ×40, 0.6 NA and using Ti–sapphire laser at a wavelength of 822 nm.

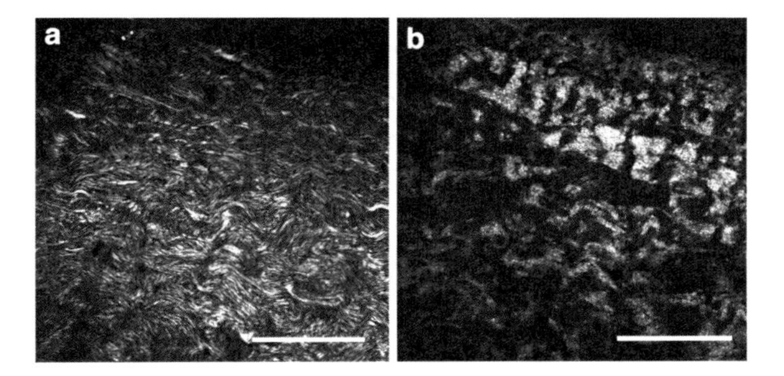

Fig. 3.2 Backward (**a**) and forward (**b**) scattered SHG signal from the surface cuts of sclera, with NA 0.75 objective, scale bar is 50 μm

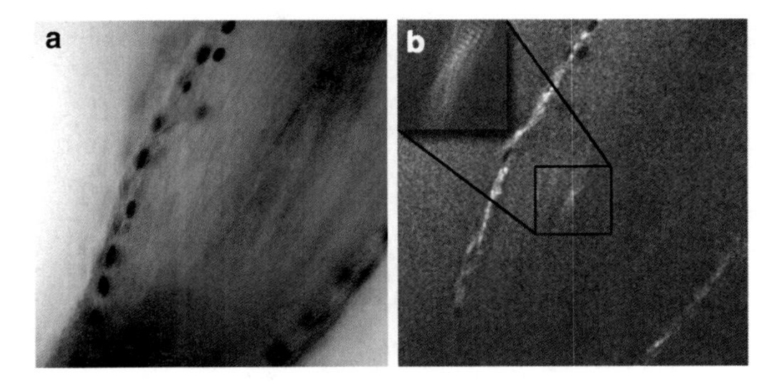

Fig. 3.3 SHG image of a zebra fish (**a**) transmission image (**b**) SHG image, inset shows the enlarged view of an area identifying actin filaments

3.3.3 Polarization Dependence of SHG

In SHG microscopy, it is often advantageous to place an analyzer before the detector so as to obtain images that reflect the polarization state of the SHG signal. For focusing with low Numerical Aperture objectives (<NA 0.5 for a dry objective), we may use a paraxial approximation to the electric field at the focus. A linearly polarized beam focused with a low NA objective will thus give an electric field that is linearly polarized at the focus.

Taking collagen as an example, we can express the induced SHG polarization as being related to the input polarization as

$$\begin{bmatrix} P_x^{2\omega} \\ P_y^{2\omega} \\ P_z^{2\omega} \end{bmatrix} = \begin{bmatrix} 0 & 0 & 0 & 0 & d_{15} & 0 \\ 0 & 0 & 0 & d_{24} & 0 & 0 \\ d_{31} & d_{31} & d_{33} & 0 & 0 & 0 \end{bmatrix} \begin{bmatrix} E_x E_x \\ E_y E_y \\ E_z E_z \\ 2E_y E_z \\ 2E_x E_z \\ 2E_y E_y \end{bmatrix}, \tag{3.13}$$

where the 3×6 matrix represents the $\chi^{(2)}$ tensor of collagen (Fukada and Yasuda 1964; Roth and Freund 1979), d_{ij} represents the tensor coefficients, and z is parallel to the long axis of the collagen fibril. Typical values of d_{33}/d_{31} range from 0.8 to 2.4, with the lower values being found primarily in young rats and the higher values in adult rats (Roth and Freund 1979; Williams et al. 2005).

Figure 3.4 shows the schematic representation of a possible setup for imaging collagen (e.g., rat-tail tendon). It may be seen that for a linearly polarized beam at an angle α to the long axis (z - axis), the induced SHG polarization is

$$P_z^{2\omega} = d_{33}E^2\cos^2\alpha + d_{31}E^2\sin^2\alpha\,, \tag{3.14}$$

$$P_x^{2\omega} = d_{15}E^2 \sin 2\alpha. \tag{3.15}$$

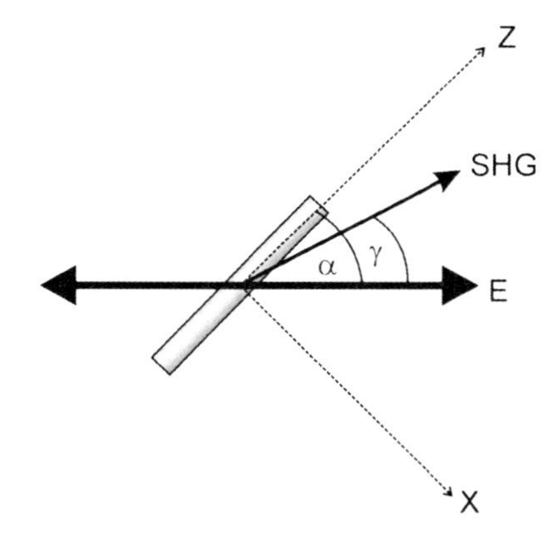

Fig. 3.4 Illustration of an incident electric field on a collagen fibril and SHG

It is thus possible to induce SHG from collagen with an input beam polarized either parallel or perpendicular to the z - axis as in (3.11). Interestingly, (3.11) shows that if the collagen is excited by an input beam either parallel or perpendicular to the z - axis, the emitted second harmonic signal generated is always polarized parallel to the long axis of the tendon as shown in Fig. 3.4. The dependence of SHG on the input polarization has thus been used to study more about the individual tensor elements, especially in biological samples where they may prove to have clinical significance (Chu et al. 2004; Roth and Freund 1979; Stoller et al. 2002; Williams et al. 2005; Yew and Sheppard 2007).

High NA objectives are often used in imaging as they provide better resolution. On the other hand, the paraxial approximation used for low NA objectives is no longer valid and one will need to use a vectorial theory of focusing with the higher NA objectives (Richards and Wolf 1959). A good example is that of the radially polarized beam (Quabis et al. 2000) which, when focused with a high NA objective, produces a strong axial component of the electric field at the focus. The azimuthally polarized beam (Youngworth and Brown 2000), on the other hand, produces an electric field at the focus that is solely transverse when focused with a high NA objective. The importance of such a vectorial theory relative to that of SHG can be seen if one were to section rat-tail tendon in a transverse manner such that the d_{33} tensor element can only be excited with a strong axial component of the electric field (Yew and Sheppard 2006, 2007). This can be seen in Fig. 3.5 where a radially and azimuthally polarized beam is focused with a high NA objective. For simplicity the former can be taken to result in a strong, axially directed electric field component at the focus while the latter has only transversal components of the electric field. It is seen in Fig. 3.5b that there is still SHG detected despite the electric field being only in the transverse plane at the focus. Equation 3.10 indicates that the SHG is being generated from $d_{32}E_xE_x$ and $d_{31}E_yE_y$, whereas the SHG in Fig. 3.5a comes from $d_{31}E_xE_x$, $d_{32}E_yE_y$ and $d_{33}E_zE_z$. At the same time, one would note that the SHG is polarized in the z-direction and is thus also radially polarized after being collimated

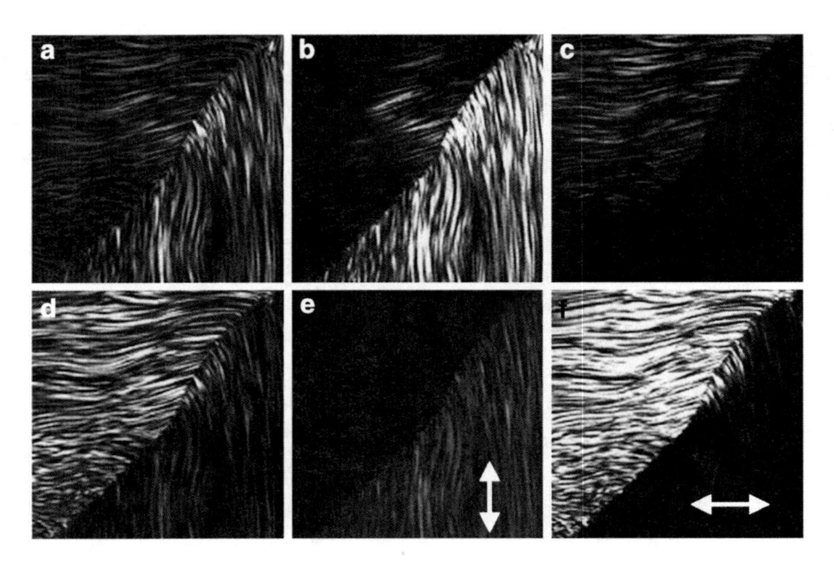

Fig. 3.5 SHG images of rat-tail tendon for a fundamental beam polarized vertically (**a**)–(**c**). SHG images without analyzer (**d**), and with an analyzer: (**e**) and (**f**) indicated by double-headed arrows, oriented vertically for image (**b**), and horizontally for image (**c**). Collagen fibrils excited by an excitation beam polarized either parallel or perpendicular to the long axis emit SHG polarized parallel to the long axis of the fibrils

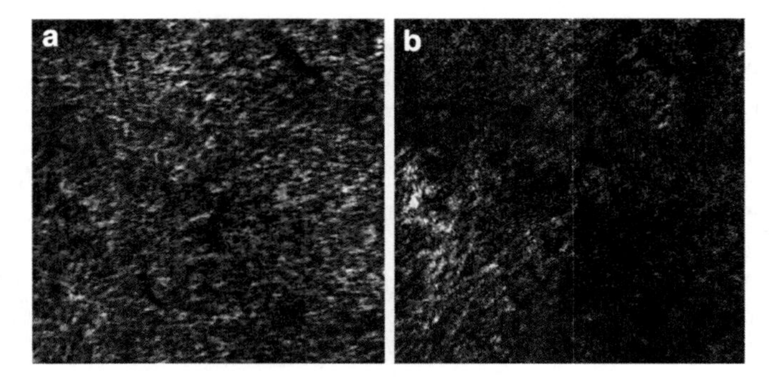

Fig. 3.6 SHG images of transversely sectioned rat-tail tendon. The long axis of the tendon is into the paper. The effect of focusing a radially polarized beam with a high NA objective is seen in (**a**) where the d33 tensor element is preferentially excited through the strong axial component at the focus. In (**b**) the converse is true when focusing an azimuthally polarized beam with a high NA objective, as this results in only transversal electric field components at the focus

(Yew and Sheppard 2006, 2007). This further emphasizes the need for combining a full vectorial theory of focusing with the tensor properties of SHG. The effect of focussing a radially polarized beam at high NA are shown in Fig. 3.6.

3.4 Third Harmonic Generation

THG is a four-wave mixing process based on third-order nonlinear polarization as shown in Fig. 3.1b. Unlike the second-order processes, the third-order process is possible in all media, with or without inversion symmetry. The amplitude of nonlinear susceptibility $\left|\chi^{(3)}\right| \ll \left|\chi^{(2)}\right|$, resulting in the magnitude of THG being much smaller than that of SHG. THG was first reported in 1962 in a centrosymmetric calcite crystal (Terhune et al. 1962). Other physical optical processes based on third-order susceptibility $\chi^{(3)}$ are the Kerr effect, DC-induced SHG, and AC Kerr effect.

The third-order induced polarization can be written as

$$P_i^{(3)}(3\omega) = \varepsilon_0 \sum_{jkl} \chi_{ijkl}^{(3)}(\omega,\ \omega,\ \omega) E_j(\omega) E_k(\omega) E_l(\omega)\,. \tag{3.16}$$

where $E(\omega)$ is the incident wave at frequency ω and $\chi^{(3)}$ is a fourth-rank tensor and indices $ijkl$ are summed over linear polarization directions of x, y, and z. The induced polarization is at 3ω that produces a new radiation at three times the frequency of the incident wave. Cubic nonlinearities are responsible optical processes like self-focusing and self-defocusing of optical beams, and also give rise to interesting effects, such as, optical bistability, phase conjugation, and optical spatial and temporal solitons. The presence of a cubic nonlinearity along with the quadratic nonlinearity can produce additional interactions that can be written as

$$\begin{aligned} \omega_0 + \omega_0 - \omega_0 \rightarrow \omega_0;\ \omega_0 + 2\omega_0 - 2\omega_0 \rightarrow \omega_0 \\ 2\omega_0 + \omega_0 - \omega_0 \rightarrow 2\omega_0;\ 2\omega_0 + 2\omega_0 - 2\omega_0 \rightarrow 2\omega_0\,. \end{aligned} \tag{3.17}$$

These relations hold for a fundamental optical frequency ω_0 under plane wave approximation. Although the fundamental and second harmonic profiles remain approximately Gaussian at low conversion efficiencies, their profiles may be different from Gaussian because of the depletion of the fundamental signal.

The phase-matching condition for THG can be written as $3k_1 - k_3 = 0$. Here, $\mathbf{k}_3$ and $\mathbf{k}_1$ are the wave vectors of the third harmonic field and the fundamental field, respectively. This phase-matching condition can also be written in terms of refractive index n as $[n(\omega) - n(3\omega)] = 0$. In general, the refractive index of a medium is always a function of frequency (the dispersion effect), making $n(\omega) \neq n(3\omega)$ and only in some special cases $n(\omega) = n(3\omega)$. With a focused excitation beam, THG from a homogenous bulk medium is canceled when the phase-matching condition is satisfied. This interesting phenomenon has been explained on the basis of Gouy phase shift, across the focus of the excitation beam (Cheng and Xie 2002). Along its propagation direction, a Gaussian beam acquires a phase shift which differs from that for a plane wave with the same optical frequency. This difference is called the

Gouy phase shift and is given by $\varphi(z) = -\arctan(z/z_R)$, where z_R is the Rayleigh range. This gives a phase shift of π within the Rayleigh range for a Gaussian beam.

As THG is a coherent process, the emitted optical field from all the molecules is added in contrast with the addition of intensities as in the case of an incoherent process such as fluorescence. When the phases of the interacting optical fields are properly matched, a condition termed as phase-matching, the total signal intensity is proportional to the square of the number of scattering molecules (Boyd 2003). When phase-matching is not right, the generated signal is significantly low in magnitude. As the condition of phase-matching is dependent on the relative geometry of the illuminating beam, the signal, and the medium, signals generated by a coherent process are typically small (Cheng and Xie 2002).

Third harmonic signals can be effectively generated from an interface or from an object with a size comparable to the FWHM (Full Width at Half Maximum) of the axial excitation intensity profile. The signal from a bulk medium is canceled by a wave-vector mismatch associated with the Gouy phase shift of the focused excitation field. This permits THG imaging of small features with a high signal-to-background ratio. The THG radiation from a small object or an interface perpendicular to the optical axis exhibits a sharp radiation pattern along the optical axis in the forward direction. For an interface parallel to the optical axis, the role of the Gouy phase shift is to deflect the phase-matching direction, that is, the THG radiation maximum direction, off the optical axis.

The method of THG microscopy in most samples is on the basis of a peculiarity of the phase-matching conditions. The third harmonic signal vanishes completely from most bulk samples, and is generated only when the illuminating beam is focused at a small inclusion or an interface between two materials. A signal is generated even when the two materials are index matched, as the THG process relies on the nonlinear susceptibility rather than on the index of refraction.

As third harmonic signal is produced by a tightly focused laser beam at an interface, it can be used to image biological and nonbiological specimens. The production of the THG signal is restricted to the condition of breaking the axial

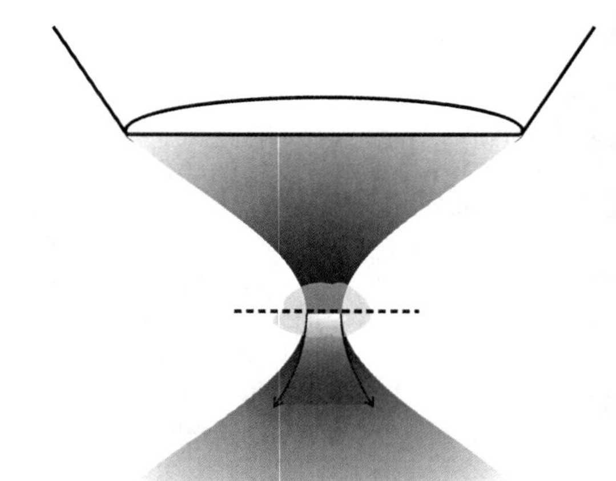

Fig. 3.7 THG signal shown by *arrows*, from a plane (*dotted line*) at the focus of the excitation laser beam. The focal plane is at an interface inside the object

focal symmetry by a change in material properties. The localized production of third harmonic light at such boundaries provides inherent optical sectioning. Therefore, it is possible to obtain three-dimensional images of microscopic objects by detecting THG signal from different planes perpendicular to the axis of laser beam propagation. This principle of THG is schematically shown in Fig. 3.7. The axial symmetry is broken by some interface or change in refractive index owing to the Gouy phase shift.

3.4.1 THG Microscopy

Third harmonic light is produced by a laser beam tightly focused at an interface. It is possible to image both biological and non-biological specimens with inherent optical sectioning of THG microscopy. Third harmonic light in a material is produced given that the axial focal symmetry can be broken by a change in the material properties like interfaces and boundaries because of refractive index or nonlinear susceptibility changes. This localized production of third harmonic light at material interfaces provides the inherent optical sectioning desired in a three-dimensional imaging at higher axial resolution. It is possible to produce three-dimensional images by THG from different planes perpendicular to the axis of beam propagation. Compared to other modes of single or multi-photon laser fluorescence microscopy, no exogenous fluorophores are needed in THG microscopy.

It was demonstrated by Tsang that under tight focusing conditions THG can be generated through an interface within the focal volume of the excitation beam (Tsang 1995). Later, it was shown that whenever there is either a change in refractive index or third-order nonlinear susceptibility, third harmonic light is produced, and as a result of this interface effect, third harmonic imaging is possible and can be applied to study transparent samples having low contrast (Barad et al. 1997). Volumetric imaging has also been done in both biological and non-biological

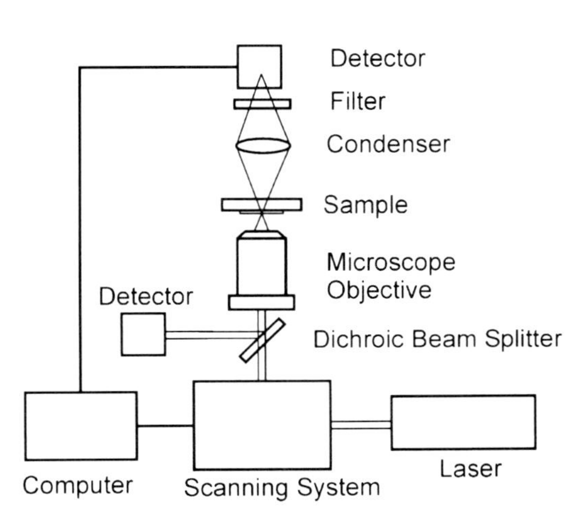

Fig. 3.8 Schematic layout of a typical nonlinear optical laser scanning microscope

specimens demonstrating the dynamical imaging properties of THG in live samples (Müller et al. 1998; Squier et al. 1998).

A typical nonlinear microscope system is shown in Fig. 3.8, where a standard optical microscope can be modified into a THG microscope. The laser source is usually a high power laser with femtosecond pulses. The laser beam is coupled through one of the microscope ports and is focused into the sample by the microscope objective. The focal point is scanned in the *xy* plane using two optical scanners, and along the *z*-axis using the motorized stage of the microscope. The third harmonic light is collected by the condenser and detected by a photomultiplier tube (PMT) after filtering out the fundamental wavelength using a band-pass interference filter. The signal from the PMT is amplified, digitized and coupled into a computer, which synchronizes the scanning process and the data collection.

3.4.2 Applications

THG near the focal point of a tightly focused laser beam can be used to probe small structures of transparent samples at the interfaces and inhomogeneities. Because of the coherent nature of third harmonic process, the axial resolution of THG microscopy is, however, equal to the confocal parameter of the fundamental beam (Barad et al. 1997). The signal level in THG microscopy can be optimized by the influence of sample structure and beam focusing (Débarre et al. 2005), where they controlled the signal level by modifying the Rayleigh range of the excitation beam and applied this method for the contrast modulation in THG images of Drosophila embryos. Besides structural and beam shape dependence, the THG is also sensitive to the local differences in the refractive index, third-order susceptibility, and dispersion. This aspect of THG was used to image lipids that are present in many biological cells and tissues (Debarre et al. 2006). In this study, a multimodal technique for microscopy was used to image lipid bodies, fluorescent compounds in tissues and extra-cellular matrix by combining two-photon fluorescence, SHG, and THG microscopy. Another multimodal nonlinear microscopic technique based on a femtosecond Cr:Forsterite laser was used to simultaneously generate SHG, THG, two-photon, and three-photon fluorescence images. Multi-photon excited fluorescence (MPEF) provides functional information of molecules, while SHG and THG can be used to image organized biological subcellular structures and interfaces. THG and SHG involve only the virtual state without energy deposition; therefore, they cause no photodamage or bleaching in the biological specimens (Chu et al. 2001).

THG microscopy is shown to be particularly suitable for imaging biogenic crystals and polarization sensitive imaging of crystalline structure in biological samples (Oron et al. 2003, 2004). THG has been used for the characterization of saline solutions and structural changes in collagen (Shcheslavskiy et al. 2004). It is shown that THG epidetection is generally possible when the sample structure is embedded in a scattering, non-absorbing tissue with thickness greater than the reduced scattering mean free path (Débarre et al. 2007).

Recently, a multimodal microscopic technique was implemented by a combination of THG, SHG, and MPEF image contrast methods on the same microscope (Gualda et al. 2008). In this study, specific cellular and anatomical features of the nematode *Caenorhabditis elegans* were imaged with laser pulses at 1,028 nm from a Ti:sapphire laser to excite the biological samples and the emitted THG signal was detected in the near-UV at 343 nm.

3.5 Laser Sources for SHG and THG Microscopy

First laser used in nonlinear microscopy of a biological tissue, was a Q-switched ruby laser at 694 nm, to study the collagen structure but resulted in strong absorption of the SHG signal at 347 nm (Fine and Hansen 1971). Later, a Q-switched Nd–YAG laser at 1,064 nm and nanosecond pulses was used to observe connective tissues (Freund et al. 1986). More recently, femtosecond pulse laser beams are being used for SHG and THG microscopy because of their high peak intensities.

The most common and current industrial standard excitation source for nonlinear microscopy is the femtosecond pulse Ti:sapphire laser. With average power of several watts, around 80 MHz repetition rate, its wavelength range is only limited by the bandwidth of the Ti:sapphire gain medium to between approximately 690 and 1,070 nm. This wavelength range is sometimes called therapeutic window, as it can penetrate deep into the biological tissue without causing significant photodamage, and is not absorbed by water which may otherwise result in heating.

For THG microscopy, a Ti:saphire laser seems to be ideal but strong absorption of the signal generated in the UV region limits its application to thin biological specimens. To avoid UV absorption of the generated THG signal in thick biological samples, THG microscopy is done with wavelengths longer than 1,200 nm. Cr:Forsterite laser at excitation wavelength of 1,230 nm and repetition rate of 110 MHz is widely used for THG microscopy (Chu et al. 2001, MC Chan et al. 2008). Other lasers used for THG microscopy include optical parametric oscillator (OPO) working at a wavelength of 1,500 nm and repetition rate of ~80 MHz, synchronously pumped by a femtosecond Ti:sapphire laser (Canioni et al. 2001; Yelin and Silberberg 1999); an optical parametric amplifier (OPA) at 1,200 nm and 250 kHz repetition rate pumped by a Ti:sapphire laser (Müller et al. 1998; Squier et al. 1998); a fiber laser at 1,560 nm with a repetition rate of 50 MHz (Millard et al. 1999); and some new laser sources being used for nonlinear microscopy are Yb: glass, Nd:glass, Cr:LiSAF, and fiber lasers.

3.6 Conclusion

The lateral resolution in SHG/THG microscopy remains close to the diffraction limit of the imaging optics. In general, the interaction volume in the specimen decreases with an increasing order of the nonlinearity of the process. For multi-photon absorption processes, this effect is counterbalanced by a relative increase in the excitation wavelength. Hence, the real advantage of such a technique lies not in the improved resolution but in the noninvasive and in vivo imaging without any sample staining. SHG/THG microscopy because of its noninvasive nature is suitable for in vivo imaging of live specimens without any preparation. The intensity of the SHG signal depends on square of the incident light intensity, while the intensity of THG signal depends on the third power of the incident light intensity. This nonlinear dependence on the intensity leads to localized excitation and is ideal for intrinsic optical sectioning in scanning laser microscopy. SHG microscopy is suitable for imaging stacked membranes and arranged proteins with organized structures. The THG microscopy is applicable to imaging cellular or subcellular interfaces. The main advantage of THG/SHG microscopy is due to the virtual nature of higher harmonic generation, in which, no saturation or bleaching results in the generated signal. Continuous viewing without compromising the sample viability can be achieved because of this property of nonlinear process, in which no energy is released. SHG/THG microscopy can be applied to live specimen without compromising its viability, while high resolution morphological, structural, functional, and cellular information of biomedical specimens can be obtained. With the flexibility of combining with fluorescence based microscopes, SHG/THG microscopy is likely to become a major imaging modality in biomedical fields. Also, given the status of laser development, it is to be expected that nonlinear optical microscopic techniques such as SHG/THG will be utilized for high resolution microscopy in a variety of applications.

References

Barad Y, Eisenberg H, Horowitz M et al (1997) Nonlinear scanning laser microscopy by third-harmonic generation. Appl Phys Lett 70(8):922–924

Barzda V, Greenhalgh C, Aus der Au J et al (2005) Visualization of mitochondria in cardiomyocytes by simultaneous harmonic generation and fluorescence microscopy. Opt Express 13:8263–8276

Bloembergen N (1965) Nonlinear optics, 4th edn. Benjamen, New York

Boyd RW (2003) Nonlinear optics. Academic, San Diego

Campagnola PJ, Wei M-D, Lewis A et al (1999) High-resolution nonlinear optical imaging of live cells by second harmonic generation. Biophys J 77(6):3341–3349

Campagnola PJ, Millard AC, Terasaki M et al (2002) Three-dimensional high-resolution second-harmonic generation imaging of endogenous structural proteins in biological tissues. Biophys J 82(1):493–508

Canioni L, Rivet S, Sarger L et al (2001) Imaging of Ca^{2+} intracellular dynamics with a third-harmonic generation microscope. Opt Lett 26(8):515–517

Chan MC, Chu SW, Tseng CH et al (2008) Cr:Forsterite-laser-based fiber-optic nonlinear endoscope with higher efficiencies. Microsc Res Tech 71:559–563

Cheng J-X, Xie XS (2002) Green's function formulation for third-harmonic generation microscopy. J Opt Soc Am B 19(7):1604–1610

Chu S-W, Chen I-H, Liu T-M et al (2001) Multimodal nonlinear spectral microscopy based on a femtosecond Cr:forsterite laser. Opt Lett 26:1909–1911

Chu S-W, Chen I-H, Liu T-M et al (2002) Nonlinear bio-photonic crystal effects revealed with multimodal nonlinear microscopy. J Microsc 208(3):190–200

Chu S-W, Chen S-Y, Chern G-W et al (2004) Studies of $\chi^{(2)}/\chi^{(3)}$ tensors in submicron-scaled bio-tissues by polarization harmonics optical microscopy. Biophys J 86(6):3914–3922

Chu S-W, Tai S-P, Chan M-C et al (2007) Thickness dependence of optical second harmonic generation in collagen fibrils. Opt Express 15(19):12005–12010

Cox G, Moreno N, Feijó J (2005) Second-harmonic imaging of plant polysaccharides. J Biomed Opt 10:0240131–0240136

Debarre D, Supatto W, Pena A-M et al (2006) Imaging lipid bodies in cells and tissues using third-harmonic generation microscopy. Nat Meth 3(1):47–53

Débarre D, Supatto W, Beaurepaire E (2005) Structure sensitivity in third-harmonic generation microscopy. Opt Lett 30:2134–2136

Débarre D, Suppato W, Pena A-M et al (2006) Imaging in lipid bodies in cells and tissues using third-harmonic generation microscopy. Nat Meth 3(1):47–53

Débarre D, Olivier N, Beaurepaire E (2007) Signal epidetection in third-harmonic generation microscopy of turbid media. Opt Express 15(15):8913–8924

Denk W, Strickler JH, Webb WW (1990) Two-photon laser scanning fluorescence microscopy. Science 248:73–76

Duncan MD, Reintjes J, Manuccia TJ (1982) Scanning coherent anti-Stokes Raman microscope. Opt Lett 7(8):350

Fine S, Hansen WP (1971) Optical second harmonic generation in biological systems. Appl Opt 10:2350–2353

Franken PA, Hill AE, Peters CW et al (1961) Generation of optical harmonics. Phys Rev Lett 7:118

Freund I, Deutsch M (1986) Second-harmonic microscopy of biological tissue. Opt Lett 11:94–96

Freund I, Deutsch M, Sprecher A (1986) Connective tissue polarity: optical second-harmonic microscopy, crossed-beam summation, and small-angle scattering in rat-tail tendon. Biophys J 50:693–712

Fukada E, Yasuda I (1964) Piezoelectric effects in collagen. Jpn J Appl Phys 3:117–121

Gannaway JN, Sheppard CJR (1978) Second-harmonic imaging in the scanning optical microscope. Opt Quant Electron 10:435–439

Gauderon R, Lukins PB, Sheppard CJR (1998) Three-dimensional second-harmonic generation imaging with femtosecond laser pulses. Opt Lett 23(15):1209–1211

Gualda EJ, Filippidis G, Voglis G et al (2008) In vivo imaging of cellular structures in *Caenorhabditis elegans* by combined TPEF, SHG and THG microscopy. J Microsc 229(1):141–150

He G-S, Liu SH (1999) Physics of nonlinear optics. World Scientific, Singapore

Hellwarth R, Christensen P (1974) Nonlinear optical microscopic examination of structure in polycrystalline ZnSe. Opt Commun 12(3):318–322

Lin SJ, Hsiao CY, Sun Y et al (2005) Monitoring the thermally induced structural transitions of collagen by use of second-harmonic generation microscopy. Opt Lett 30:622–624

Maiti S, Shear JB, Williams RM et al (1997) Measuring serotonin distribution in live cells with three-photon excitation. Science 275:530–532

Millard AC, Wiseman PW, Fittinghoff DN et al (1999) Third-harmonic generation microscopy by use of a compact femtosecond fiber laser source. Appl Opt 38:7393–7397

Mizutani G, Sonoda Y, Sano H et al (2000) Detection of starch granules in a living plant by optical second harmonic microscopy. J Lumin 87–89:824–826

Moreaux L, Sandre O, Blanchard-Desce M et al (2000) Membrane imaging by simultaneous second-harmonic generation and two-photon microscopy. Opt Lett 25:320–322

Müller M, Squier J, Wilson KR et al (1998) 3D-microscopy of transparent objects using third-harmonic generation. J Microsc 191:266–274

Oron D, Tal E, Silberberg Y (2003) Depth-resolved multiphoton polarization microscopy by third-harmonic generation. Opt Lett 28(23):2315–2317

Oron D, Yelin D, Tal E et al (2004) Depth-resolved structural imaging by third-harmonic generation microscopy. J Struct Biol 147(1):3–11
Prent N, Cisek R, Greenhalgh C et al (2005) Application of nonlinear microscopy for studying the structure and dynamics in biological systems. Proc SPIE 5971:5971061–5971068
Pu Y, Centurion M, Psaltis D (2008) Harmonic holography: a new holographic principle. Appl Opt 47:A103–A110
Quabis S, Dorn R, Eberler M et al (2000) Focusing light to a tighter spot. Opt Commun 179:1–7
Richards B, Wolf E (1959) Electromagnetic diffraction in optical systems. II Structure of the image field in an aplanatic system. Proceedings of the Royal Society of London Series A 253(1274):358–379
Roth S, Freund I (1979) Second harmonic generation in collagen. J Chem Phys 70:1637–1643
Schrader M, Bahlmann K, Hell SW (1997) Three-photon-excitation microscopy: theory, experiment and applications. Optik 104:116–124
Shcheslavskiy V, Petrov GI, Saltiel S et al (2004) Quantitative characterization of aqueous solutions probed by the third-harmonic generation microscopy. J Struct Biol 147(1):42–49
Shen YR (1984) The principles of nonlinear optics. Wiley, New York
Sheppard CJR, Kompfner R (1978) Resonant scanning optical microscope. Appl Opt 17(18): 2879–2882
Squier JA, Muller M, Brakenhoff GJ et al (1998) Third harmonic generation microscopy. Opt Express 3:315–324
Stoller P, Kim BM, Rubenchik AM et al (2002) Polarization-dependent optical second-harmonic imaging of a rat-tail tendon. J Biomed Opt 7:205–214
Terhune RW, Maker PD, Savage CM (1962) Optical harmonic generation in calcite. Phys Rev Lett 8(10):404–406
Tsang TYF (1995) Optical third-harmonic generation at interfaces. Physical Review A 52: 4116–4125
Williams RM, Zipfel WR, Webb WW (2005) Interpreting second harmonic generation images of collagen in fibrils. Biophys J 88:1377–1386
Yelin D, Silberberg Y (1999) Laser scanning third-harmonic-generation microscopy in biology. Opt Express 5(8):169–175
Yew EYS, Sheppard CJR (2006) Effects of axial field components on second harmonic generation microscopy. Opt Express 14:1167–1173
Yew EYS, Sheppard CJR (2007) Second-harmonic generation polarization microscopy with tightly focused linearly and radially and azimuthally polarized beams. Opt Commun 275(2): 453–457
Youngworth K, Brown T (2000) Focusing of high numerical aperture cylindrical-vector beams. Opt Express 7:77–87
Yu C-H, Tai S-P, Kung C-T et al (2008) Molecular third-harmonic-generation microscopy through resonance enhancement with absorbing dye. Opt Lett 33:387–389

Chapter 4
Role of Scattering and Nonlinear Effects in the Illumination and the Photobleaching Distribution Profiles

Zeno Lavagnino, Francesca Cella Zanacchi, and Alberto Diaspro

4.1 Introduction

One of the most promising research activity in life sciences concerns the study of molecular transport properties and mechanisms, such as signaling and trafficking (Palamidessi et al. 2008), in thick biological samples. Two-photon excitation (2PE) (Mayer 1931; Denk et al. 1990; Diaspro et al. 2006) microscopy gives us a powerful tool for deep imaging in biological samples at submicron resolution. Unfortunately, using such a technique, light diffusion in the excitation and in the emission optical pathway has to be taken into account since it could be responsible for undesired side effects. Biological tissues are highly scattering media and this leads to a progressive attenuation of the excitation intensity profile at the geometrical focus position as the light travels within the sample and the localization of the maximum 2PE intensity can shift closer to the surface far from the focal region (Ying et al. 1999) and the 2PE imaging depth limit appears strongly limited by near-surface fluorescence (Ying et al. 2000). In some perturbation techniques, a high-intensity level delivered on the sample is required and this can lead to this significant undesired excitation of the surface layers of the sample. Photobleaching effects also can potentially occur near the top of the sample.

Z. Lavagnino
Department of Physics, LAMBS, University of Genova, Genova, Italy
and
Italian Institute of Technology, Genova, Italy

F. Cella Zanacchi and A. Diaspro (✉)
Italian Institute of Technology, Genova, Italy
e-mail: alberto.diaspro@iit.it

A. Diaspro (ed.), *Optical Fluorescence Microscopy*,
DOI 10.1007/978-3-642-15175-0_4,

4.2 Intensity Distribution of a Gaussian Beam

It is possible to obtain the real three-dimensional (3D) intensity distribution of a beam focused by a lens (Born and Wolf 1959). We can write the amplitude of the point spread function for a light source of wavelength λ:

$$h(u,v) = -i\frac{2\pi n S \sin^2\alpha}{\lambda} e^{\frac{iu}{\sin^2\alpha}} \int_0^1 J_0(v\rho) e^{\frac{-iu\rho^2}{2}} \rho \mathrm{d}\rho \, , \tag{4.1}$$

where n is the refractive index of the medium, α the angular aperture of the lens, J_0 is the zero-order Bessel function, and u and v are the normalized optical coordinates.

$$v = \frac{2\pi \mathrm{NA} r}{\lambda}, \quad u = \frac{2\pi (\mathrm{NA})^2 z}{n\lambda} \, . \tag{4.2}$$

The three-dimensional intensity distribution is the product of the amplitude PSF and its complex conjugate and can be written as (Sheppard 1984):

$$\left|h^2(u,v)\right| = \left|2\int_0^1 J_0(v\rho) e^{\frac{-iu\rho^2}{2}} \rho \mathrm{d}\rho\right|^2 . \tag{4.3}$$

Such a distribution can be approximatively defined taking into account the intensity distribution produced by a Gaussian focused beam:

$$I_{\mathrm{ill}}(\rho,z) = I_0 \left[\frac{\omega_0}{\omega(z)}\right]^2 e^{\frac{-2\rho^2}{\omega^2(z)}} \, , \tag{4.4}$$

where ω_0 is the beam waist, z_0 is the Rayleigh range (Fig. 4.1), ρ is the radial coordinate, z is the axial coordinate, I_0 is the Gaussian beam intensity in the focal region, z_s is the axial coordinate of the focal plane, and $\omega(z)$ is described by:

$$W(z) = \omega_0 \left\{1 + \left[\frac{(z - z_s)}{z_0}\right]^2\right\}^{\frac{1}{2}} . \tag{4.5}$$

In the case of single-photon excitation, the 3D fluorescence intensity distribution is directly proportional to the 3D illumination intensity distribution.

Unlike of what happens in the single-photon excitation case, the two-photon 3D fluorescence intensity distribution depends quadratically on the illumination distribution:

$$I_{\mathrm{2P}}(\rho,z) \propto I_{\mathrm{ill}}^2(\rho,z) \, . \tag{4.6}$$

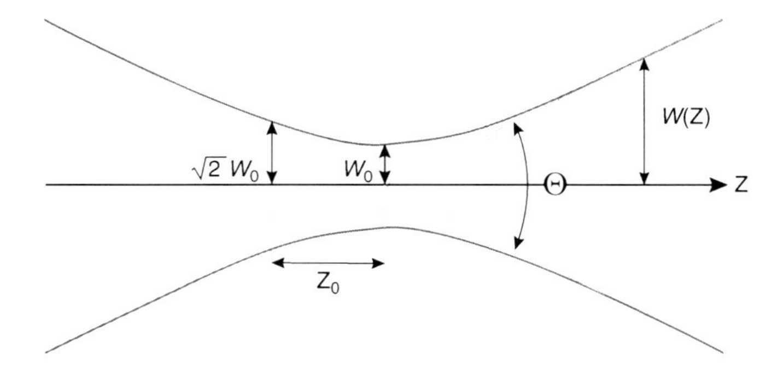

Fig. 4.1 Beam propagation scheme (z_0 is the Rayleigh range and ω_0 represents the beam waist)

4.3 Intensity Distribution Modified by Scattering

Scattering phenomenon occurs where there is an inhomogeneous distribution of the refractive index in a medium. In the case of biological samples, localized particles or different compartments can be the cause of this inhomogeneity.

In fact, in many biological samples, the main diffusive centers consist of particles with size ranging from 8 nm to 10 μm. If the scattering particle dimension is smaller than the incident wavelength, the intensity distribution of the scattered light depends on the inverse of the fourth power of the incident light wavelength and Rayleigh scattering occurs.

However, because most of the diffusive centers in biological samples have dimensions comparable with the light wavelength, this can lead to an anisotropic scattered intensity distribution (Mie scattering regime) (Mie 1908).

Even though Mie's theory has been defined for scattering by a single spherical particle, it can be extended to a larger number of particles under two conditions: they must be randomly distributed and they have to be separated with higher distances than the incident light wavelength.

The attenuation of the ballistic (nondiffused) component of the light in a non-absorbing medium is characterized by the extinction coefficient μ. The inversion of the extinction coefficient leads us to introduce the scattering mean free path, that is, the average distance covered by a photon between a scattering event and the next one.

The total power carried on a sample section of thickness dz will be attenuated constantly:

$$\frac{\mathrm{d}P}{P} = -\mu_{\mathrm{ext}}\mathrm{d}z\,, \tag{4.7}$$

where P is the total power carried by the beam and μ_{ext} is the total extinction coefficient, resulting from the sum of two contributions, scattering (s) and absorption (a):

$$\mu_{\mathrm{ext}} = \mu_{\mathrm{s}} + \mu_{\mathrm{a}}\,. \tag{4.8}$$

In single-photon excitation regime, both the excitation and emission intensity distributions are affected by scattering effects and we should consider an average extinction coefficient:

$$\mu_{\text{ext}}^{1\text{P}} \approx \mu_{\text{em}} + \mu_{\text{exc}} \,. \tag{4.9}$$

Considering only the ballistic component of the incident light (neglecting the diffused contribution), the effect due to scattering phenomenon on the light propagation can be described as an exponential intensity loss.

For a uniform scattering medium, the intensity distribution in the focus of a lens results:

$$I_{\text{ex}}(\rho, z) = I_0 \left[\frac{\omega_0}{\omega(z)}\right]^2 \exp\left[-\frac{2\rho^2}{\omega^2(z)}\right] \exp(-\mu_{\text{ext}}^{1\text{P}} z) \,. \tag{4.10}$$

The higher wavelength used in 2PE makes this technique a suitable tool for performing imaging in depth in scattering samples. In fact, for a given scattering sample, the mean free path length under two-photon excitation is about six times the mean free path length of single-photon excitation.

Nevertheless, scattered light brings to a limited penetration depth (Theer and Denk 2006) inducing a strong reduction in signal-to-noise ratio (SNR), and this aspect represents a crucial point for nonlinear imaging in vivo.

The 2PE intensity distribution modified by scattering is written as follows:

$$I_{\text{ex}}(\rho, z) \propto I_0^2 \left[\frac{\omega_0}{\omega(z)}\right]^4 \exp\left[-\frac{4\rho^2}{\omega^2(z)}\right] \exp(-\mu_{\text{ext}}^{2\text{P}} z) \,. \tag{4.11}$$

Because of its quadratic dependence on the illumination distribution, the 2P fluorescence intensity distribution's extinction coefficient is defined as follows:

$$\mu_{\text{ext}}^{2\text{P}} = 2\mu_{\text{em}} + \mu_{\text{exc}} \,, \tag{4.12}$$

where μ_{em} is the contribution due to emission and μ_{exc} is the contribution due to excitation.

Considering transparent biological samples, the absorption contribution can be neglected (De Grauw et al. 2002) (i.e., $\mu_{\text{s}} \gg \mu_{\text{a}}$). Under this assumption, we can consider the total extinction coefficient in 2PE regime as follows:

$$\mu_{\text{s}} \approx \mu_{\text{ext}}^{2\text{P}} = 2\mu_{\text{em}} + \mu_{\text{exc}} \,. \tag{4.13}$$

In the end, we can now write the intensity distribution along the optical axis as follows:

$$I_{\text{ex}}(0, z) \propto I_0^2 \left[\frac{\omega_0}{\omega(z)}\right]^4 \exp(-\mu_{\text{s}} z) \,. \tag{4.14}$$

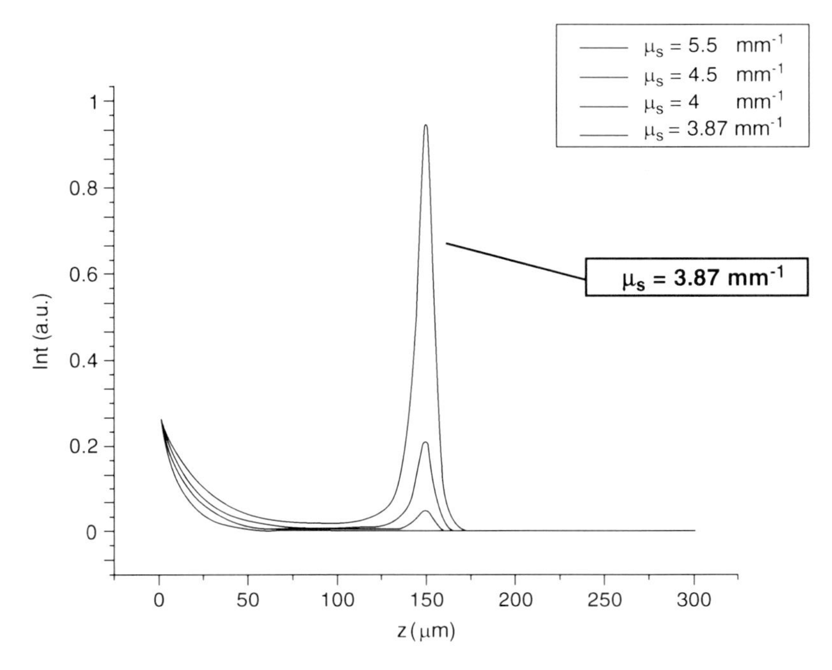

Fig. 4.2 Axial profile computation of the 2PE intensity distribution with different scattering coefficients ($\mu_s = 3.87\ \mathrm{mm}^{-1}$ is a value comparable with the scattering coefficient of many biological samples)

Recent works (Theer and Denk 2006) showed that the 2PE intensity distribution modified by scattering [as defined in (4.6)] can induce significant effects close to the sample surface (Ying et al. 1999) resulting in a decrease in the imaging penetration depth.

Computations (Matlab 7.0. Mathworks, Inc.) of the intensity axial profile [cf. (4.14)] have been performed for different scattering coefficients (Fig. 4.2) to verify whether scattering-induced excitation of near-surface sample layers occurs. The objective numerical aperture used for the computation is 0.8. Such a NA value constitutes a good compromise because it leads to a good resolution in correspondence with a significant imaging depth. The focal plane was set 150 μm deep in the sample, and the scattering coefficient ranges from 3.87 to 5.5 mm^{-1}.

Increasing the scattering properties of the sample, a significant loss in the focal plane occurs, resulting in a growth of the excitation intensity of the surface layers, shifting the excitation peak near the top of the sample.

4.4 Photobleaching Effects Induced by Scattering

The main aim of this work is to verify if the excitation of the superficial layers of the sample, induced by the scattering properties of the sample itself, is enough to induce a photobleaching process of fluorescent molecules.

When photobleaching phenomenon occurs, the fluorophore shifts to the first excited triplet state where interaction with molecules (such as oxygen) causes structural modification inducing the permanent loss of its fluorescent properties. When high-intensity radiation is delivered on the sample, fluorescent molecules lose permanently their capability to emit any fluorescent signal, moving to an irreversible "dark" state. In first approximation, the photobleaching process under two-photon excitation can be described as a first-order reaction (Patterson and Piston 2000):

$$\frac{\mathrm{d}F(x,y,z,t)}{\mathrm{d}t} = -k_{2\mathrm{PE}}\langle I_{\mathrm{bl}}^2(x,y,z)\rangle F(x,y,z,t)\,, \tag{4.15}$$

where $k_{2\mathrm{PE}}$ is the bleach parameter and depends on the fluorophore characteristics and on the system properties used to perform photobleaching, and I_{bl} is the laser intensity distribution delivered on the sample.

Considering photobleaching as a first-order reaction [cf. (4.15)] and considering a Gaussian excitation beam modified by scattering [cf. (4.11)], the distribution of photobleached molecules (Mazza et al. 2007) can be described as:

$$F_0(\rho,z) \approx F_{\mathrm{in}}\mathrm{e}^{-k_{2\mathrm{PE}}I_0^2\left[\frac{\omega_0}{\omega(z)}\right]^4 \mathrm{e}^{-\frac{4\rho^2}{\omega^2(z)}} e^{-\mu_s z}}\,. \tag{4.16}$$

In the same way along the optical axis, (4.16) is reduced as follows:

$$F_0(0,z) \approx F_{\mathrm{in}}\mathrm{e}^{-k_{2\mathrm{PE}}I_0^2\left[\frac{\omega_0}{\omega(z)}\right]^4 \mathrm{e}^{-\mu_s z}}\,. \tag{4.17}$$

To verify the presence of photobleaching effects close to the surface of the sample, we performed some tests on immobile fluorescent samples which were obtained by ionic cross-linking of poly-allylamine hydrochloride (PAH, Sigma-Aldrich), a positively charged polyelectrolyte, covalently labeled with fluorophores Alexa 555 and Cy5 (Invitrogen). The gelification process is obtained by mixing a volume of PAH solution (60 kDa, 90 mg/ml in water) with an equal volume of cross-linker (sodium phosphate – NaH_2PO_4) solution (170 mg/ml in water). 3D uniformity of the fluorescent sample was tested (Mazza et al. 2007) according to the method developed by Zwier et al. (2004). Changing the phosphate concentration, the gel's size can be controlled and its thickness can be reduced by decreasing phosphate concentration. Furthermore, the process can be repeated, overlapping PAH and phosphate layers, until the desired gel thickness has been reached. Two different kinds of fluorescent molecules have been covalently labeled to the gel: Alexa Fluor 555 was chosen for its photostability properties and CY5 has been chosen to have a higher quantum efficiency fluorescent labeling.

We performed experimental tests on 250-μm-thick gels to verify the presence or the absence of superficial effects due to photobleaching induced by scattering.

The focal plane was set 150 μm deep in the sample, to be sure we can overlook some contribution due to lack of fluorescence in the last layers of the sample. The measure of the 2P excitation spectrum of the ALEXA555 fluorophore was realized to determine the best wavelength to excite the molecule with this technique.

This is also the wavelength which permits the best photobleaching efficiency ($L = 750$ nm).

The acquisition of the images was performed with a ×40 objective (water immersion) with NA = 0.8. To vary the scattering properties of the sample, we introduced a 0.3% volume concentrated nonfluorescent silica beads ($d = 0.453$ μm, comparable with the excitation wavelength) which brought the scattering coefficient to $\mu_s = 3.87\ \text{mm}^{-1}$ and the anisotropy factor $g = 0.75$. The following images (cf. Fig. 4.3) show the distribution of bleached molecules in two gels, one with the concentrated beads considered previously and the other without beads.

As anyone can see no photobleaching effects due to scattering properties of the sample are remarkable in the superficial layers of the gel. We repeated the measurements with a different fluorophore, with higher quantum yield, CY5. We performed imaging with the same parameters and the same concentration of beads as before and what we obtained is shown in Fig. 4.4.

Experimental tests on a gel labeled with a higher quantum yield fluorophore (CY5) cannot allow us to see any photobleaching phenomena in the superficial layers of the sample due to scattering effects. Since experimental tests did not show that 2PE deep imaging in a scattering sample ($\mu_s = 3.87\text{mm}^{-1}$) can induce photobleaching of the molecules localized in the region close to the surface, according to the conditions presented above, we performed simulations of the photobleaching intensity axial profile [cf. (4.17)] by varying the scattering coefficient of the sample. Calculation of the photobleaching intensity axial profile has been performed to determine the range of the scattering coefficient for which a considerable photobleaching effect can be observed.

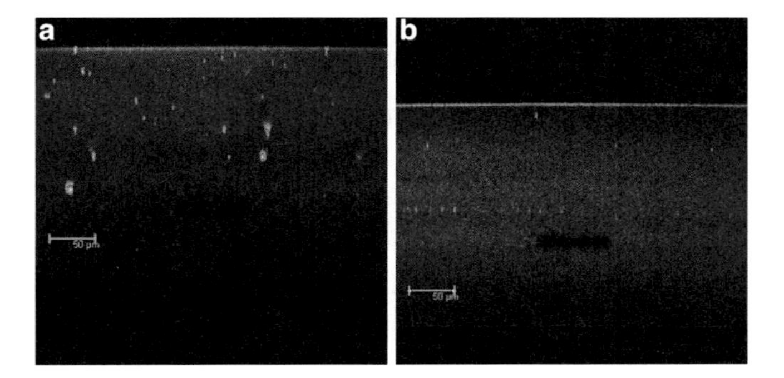

Fig. 4.3 Images of the xz section of the photobleached molecule distribution on the polyelectrolyte gel labeled with Alexa 555 (**a**) containing beads (0.3% volume concentration) (**b**) without diffusing centers. The bleached region is 70 × 70 μm^2 large and set 150 μm deep in the sample. No photobleaching effects occur near the top of the sample (scale bar is 50 μm)

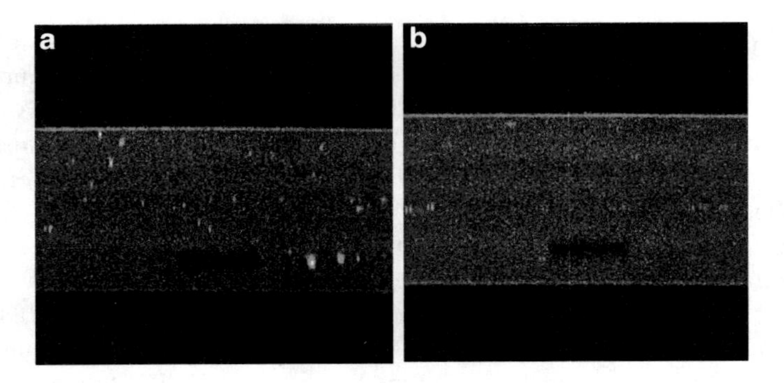

Fig. 4.4 Images of the *xz* section of the photobleached molecule distribution on the polyelectrolyte gel labeled with CY5 (**a**) containing beads (0.3% volume concentration) (**b**) without diffusing centers. The bleached region is $70 \times 70\ \mu m^2$ large and set 150 μm deep in the sample. Even with higher quantum yield, no photobleaching effects occur near the top of the sample (scale bar is 50 μm)

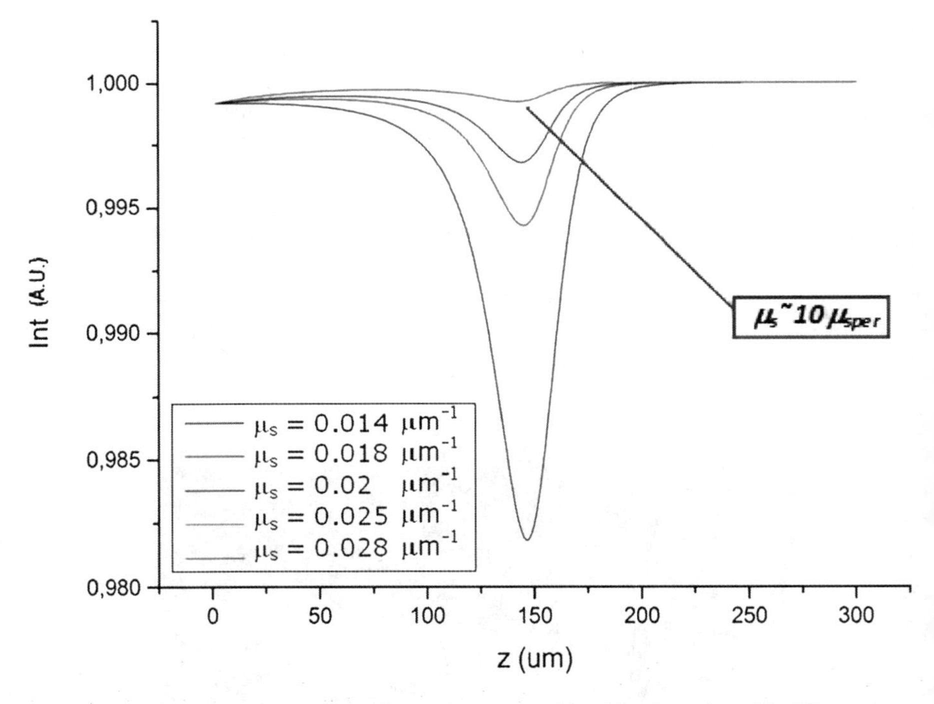

Fig. 4.5 Axial profile computations of two-photon photobleaching intensity with different scattering coefficients. To see a significant photobleaching effect near the surface, one order of magnitude higher values have to be considered

Results (cf. Fig. 4.5) show that photobleaching effects on the surface layers occur only if one order of magnitude higher scattering coefficients are considered.

4.5 Conclusions

The longer wavelengths used in two-photon excitation microscopy provide one of the main advantages of this technique leading to a reduced scattering cross-section. Nevertheless, increasing the scattering coefficients leads both to a shift in the intensity closer to the surface and to a significant reduction in the intensity peak at the focal region. These effects represent, even in the 2PE regime, a strong limitation to the imaging penetration depth in diffusive samples.

Results reported here confirm that scattering effects can be neglected for thin and moderately thick biological samples. Calculation of the two-photon excitation intensity distribution showed that, for scattering coefficients close to biological sample values, the intensity generated close to the surface starts to play a significant role. Experimental measurements of immobile fluorescent samples (polyelectrolyte gel) do not reveal, for the scattering coefficient considered ($\mu_s = 3.87\ \text{mm}^{-1}$), any kind of photobleaching effect close to the sample surface. A calculation of the photobleaching intensity axial profile [cf. (4.14)] has been performed to test the range of the scattering coefficients which provide a considerable photobleaching effect. Computations confirm that to induce any photobleaching process close to the surface, scattering coefficients of one order of magnitude higher than the one used for the experimental test should be considered (Fig. 4.5).

References

Born M, Wolf E (1959) Principles of optics. Cambridge University Press, Cambridge

De Grauw C, Frederix P, Gerritsen H (2002) Aberrations and penetration in in-depth confocal and two-photon-excitation microscopy. In: Diaspro A (ed) Confocal and two-photon microscopy: foundations, applications and advances. Wiley-Liss, Wilmington, pp 160–163

Denk W, Strickler J, Webb W (1990) Two photon laser scanning fluorescence microscopy. Science 248(4951):73–76

Diaspro A, Chirico G, Collini M (2006) Two-photon fluorescence exitation and related thechniques in biological microscopy. Q Rev Biophys 15:1–70

Mayer MG (1931) Uber elementarakte mit zwei quantensprungen. Ann Phys 9:273–295

Mazza D, Cella F, Vicidomini G, Krol S, Diaspro A (2007) Role of three-dimensional bleach distribution in confocal and two-photon fluorescence recovery after photobleaching experiments. Appl Opt 46:7401–7411

Mie G (1908) Beitrge zur optik trber medien, speziell kolloidaler metallsungen. Ann Phys 4:377–445

Palamidessi A, Frittoli E, Garr M, Faretta M, Mione M, Testa I, Diaspro A, Lanzetti L, Scita G, Fiore PPD (2008) Endocytic tracking of rac is required for the spatial restriction of signaling in cell migration. Cell 134:135–147

Patterson GH, Piston DW (2000) Photobleaching in two photon excitation microscopy. Biophys J 78(4):2159–2162

Sheppard WTC (1984) Theory and practice of scanning optical microscopy. Academic, New York

Theer P, Denk W (2006) On the fundamental imaging-depth limit in two-photon microscopy. J Opt Soc Am A Opt Image Sci Vis 23:3139–3149

Ying J, Liu F, Alfano RR (1999) Spatial distribution of two-photon-excited fluorescence in scattering media. Appl Opt 38:224–229

Ying J, Liu F, Alfano RR (2000) Effect of scattering on nonlinear optical scanning microscopy imaging of highly scattering media. Appl Opt 39:509–514

Zwier JM, Rooij GJV, Hofstraat JW, Brakenhoff GJ (2004) Image calibration in fluorescence microscopy. J Microsc 216:15–24

Chapter 5
New Analytical Tools for Evaluation of Spherical Aberration in Optical Microscopy

Isabel Escobar, Emilio Sánchez-Ortiga, Genaro Saavedra, and Manuel Martínez-Corral

5.1 Introduction

The use of thick specimens is one of the goals in the analysis and the processing of three-dimensional (3D) samples by optical scanning techniques, as scanning microscopes, optical data storage systems, or laser trapping devices. The required tightly focussed spot volumes are usually achieved by use of high-NA immersion objective lenses. In many practical situations, the refractive index of the samples does not match that of the immersion medium. Consequently, when imaging deep inside the specimen, an important amount of spherical aberration (SA) is introduced, producing a spreading in the focusing response (Török et al. 1997; Sheppard 1998). To solve this well-known problem, the good-quality high-NA microscope objectives incorporate some kind of correction collar (Schwertner et al. 2005). However, when imaging 3D samples, the use of this collar permits the correction of the SA only for a given section of the sample. Thus, although sequential compensation of this aberration can be performed during the depth scanning, it is not possible to achieve simultaneously a global correction for the whole scanned sample. This kind of distortions has been the focus of important research efforts in biology, where poor-quality images are obtained (Wilson and Carlini 1988; Hell et al. 1993; Török et al. 1995a, b; Booth and Wilson 2000), in optical data storage systems, where limitations in the voxel size are experienced (Braat 1997; Day and Gu 1998; Stallinga 2005a, b), or in laser trapping technology, where the efficiency of the confinement is compromised (Ke and Gu 1998; Reihani et al. 2006).

Consequently, the design of pupil elements that increase the tolerance of the system to the SA is of great interest. We present in this contribution a novel formalism for the assessment of the effect of the SA in high-NA optical systems. This model allows us to optimize the design of pupil filters to decrease the sensitivity of the focusing/imaging systems to this aberration, providing a response which is quasi-invariant to these sample induced distortions.

I. Escobar, E. Sánchez-Ortiga, G. Saavedra, and M. Martínez-Corral (✉)
Optics Department, Universitat de València, 46100 Burjassot, Spain
e-mail: manuel.martinez@uv.es

A. Diaspro (ed.), *Optical Fluorescence Microscopy*,
DOI 10.1007/978-3-642-15175-0_5,

5.2 Basic Theory

To describe our approach, we start by considering a high-NA objective lens that is illuminated by a monochromatic, scalar plane wave. In the simple scheme presented in Fig. 5.1, we represent the objective through its principal surfaces (Martinez-Corral and Saavedra 2009). The back principal surface S_2, is, as in the paraxial case, a plane surface. The front principal surface, S_1, is a sphere of radius f centred at the focal point. As is well known, in most high-NA objectives the aperture stop is inserted at the back-focal plane. Then, an ideal microscope objective transforms a monochromatic plane wave into a truncated spherical wavefront. The amplitude transmittance of the aperture stop is mapped onto S_1 and its effect can be analysed in different ways (Sheppard and Gu 1993). According to the scalar, non-paraxial Debye's formulation, the amplitude distribution in the neighbourhood of the focal point can be expressed as (Gu 2000)

$$U(r_N, z_N) = \int_0^{\alpha} \sqrt{\cos\theta}\, p(\theta) \exp[i2\pi W(\theta)]\, J_0\left(2\pi r_N \frac{\sin\theta}{\sin\alpha}\right) \times \exp\left[-i2\pi z_N \frac{\sin^2(\theta/2)}{\sin^2(\alpha/2)}\right] sin\theta\, d\theta\,. \tag{5.1}$$

In this equation, $p(\theta)$ accounts for the amplitude transmittance at the objective exit pupil, and α is the maximum value for the aperture angle θ. For the projection of the pupil transmittance, we have assumed that the sine condition holds. Lateral and axial positions are expressed through normalized coordinates

$$r_N = \frac{n}{\lambda} r \sin\alpha \;\; \text{and} \;\; z_N = \frac{2n}{\lambda} z \sin^2\left(\frac{\alpha}{2}\right), \tag{5.2}$$

respectively, where r and z are cylindrical coordinates with origin at the focal point. We have included a phase factor, $W(\theta)$, which in the forthcoming analysis will account for phase distortion – *aberration* – occurred during the focusing. Let us consider now that the field exiting from the high-NA objective is focused, at a depth t, through a planar interface between two dielectric materials with refractive

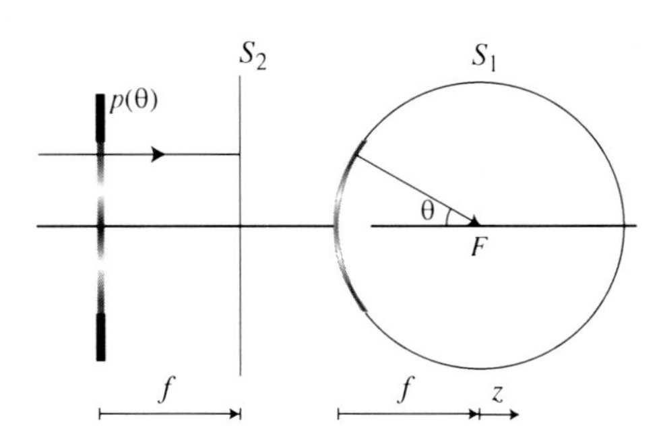

Fig. 5.1 Scheme of focusing by a high-NA objective

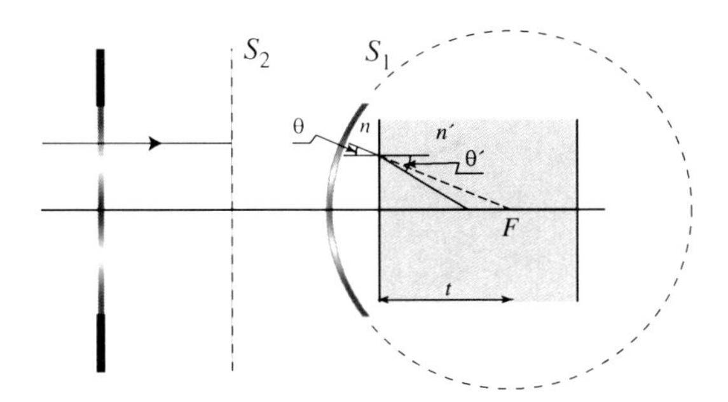

Fig. 5.2 Scheme of focusing into two media separated by a planar interface

indexes n and n', as depicted in Fig. 5.2. It have been shown elsewhere (Török et al. 1995a; Booth et al. 1998) that the phase delay suffered by the ray is given by

$$W(\theta; t) = \frac{t}{\lambda}(n' \cos\theta' - n\cos\theta), \tag{5.3}$$

where θ' represents the refraction angle associated to θ, as shown in Fig. 5.2. Since this function is dependent only on the azimuthal angle θ, it describes only SA. Expanding this expression into power series of $\sin(\theta/2)$ up to fourth-order approximation, we obtain (Sheppard and Cogswell 1991; Sheppard 1995)

$$W(\theta; t) = \frac{t}{\lambda}(n' - n)\left[1 + \frac{2n}{n'}\sin^2\left(\frac{\theta}{2}\right) + 2(n' + n)\frac{n^2}{n'^3}\sin^4\left(\frac{\theta}{2}\right)\right]. \tag{5.4}$$

Thus, the amplitude distribution in the neighbourhood of the focus of a high-NA beam that is focused deep through a stratified medium is given by

$$\begin{aligned} U(r_N, z'_N; w_{40}) = & \int_0^\alpha \sqrt{\cos\theta}\, p(\theta)\, J_0\left(2\pi r_N \frac{\sin\theta}{\sin\alpha}\right) \\ & \times \exp\left\{ i2\pi \left[w_{40}\frac{\sin^4(\theta/2)}{\sin^4(\alpha/2)} - z'_N \frac{\sin^2(\theta/2)}{\sin^2(\alpha/2)}\right]\right\} sin\theta d\theta , \end{aligned} \tag{5.5}$$

where the coefficients for the refractive defocus and the primary SA are

$$w_{20} = 2t(n' - n)\frac{n}{n'}\sin^2\left(\frac{\alpha}{2}\right) \quad \text{and} \quad w_{40} = 2t\left(n'^2 - n^2\right)\frac{n^2}{n'^3}\sin^4\left(\frac{\alpha}{2}\right), \tag{5.6}$$

respectively. In (5.5) we have omitted some irrelevant constant phase factors and, besides, we have defined the reduced axial coordinate as $z'_N = z_N - w_{20}$. Note that more accurate calculations should take into account the influence of the transmission coefficients of the interface (Török and Varga 1997; Haeberlé et al. 2003). However, the use of these more precise models would not change the results significantly.

Our aim here is the design of beam-shaping elements for reduction of SA impact in the focusing properties of the system. Although this effect should be studied throughout the 3D focused field, it is difficult to extract any intuitive information to guide the design from the complete function in (5.5). Instead, we fix our attention in the axial behaviour of the system. In fact, it is well-known that SA does not degrade the images of 2D objects significantly, provided that one selects the penetration depth adequately (Saavedra et al. 2009). On the other hand, we experience that SA strongly degrades the 3D intensity point-spread function. Thus, a design based on the optimization of the axial response of the focusing system seems adequate.

To this end, we first particularize (5.5) to points in the optical axis by taking $r_N = 0$. Second, for the sake of convenience, we perform the following nonlinear mapping

$$\zeta = \frac{\sin^2(\theta/2)}{\sin^2(\alpha/2)} - 0.5\,, \quad q(\zeta) = p(\theta)\sqrt{\cos\theta}\,. \tag{5.7}$$

With this transformation, (5.5) becomes

$$\begin{aligned} U_a(w'_{20}; w_{40}) &= U(r_N = 0, w'_{20}; w_{40}) \\ &= \int_{-0.5}^{0.5} q(\zeta)\exp\left(i2\pi w_{40}\zeta^2\right)\exp(-i2\pi w'_{20}\zeta)\mathrm{d}\zeta\,, \end{aligned} \tag{5.8}$$

where some irrelevant constant factors have been omitted. In this equation, we have defined the reduced defocus coefficient as $w'_{20} = z'_N - w_{40}$. Note that (5.8) establishes a Fourier-transform relationship between the axial amplitude response of the system and the mapped pupil function $q(\zeta)$ when no SA is induced. This simple link is perturbed by the presence of the phase factor depending on the SA coefficient w_{40}, modifying the effective pupil of the system. The effect of this "perturbation" is studied in the following section.

5.3 Evolution of the Second-Order Moment

Although (5.8) is a very compact formula, which allows the calculations of the axial spherically aberrated focused field, it is preferable to work with some global or average parameters that give partial but relevant information of the beam. This analysis provides a powerful, simple mathematical tool to tackle the design of strategies for compensation of the effect of SA on the system response. In particular, we find in this section the transformation laws for the width of the axial intensity distribution depending on the SA induced in the focusing process.

A classical parameter used to assess the width or extension of a function around its central value is its variance. In our case, the variance of the axial intensity distribution is given by

$$\sigma^2_{w'_{20}}(w_{40}) = \frac{\langle w'^2_{20}\rangle_{w_{40}}}{\langle w'^0_{20}\rangle_{w_{40}}} - \frac{\langle w'_{20}\rangle^2_{w_{40}}}{\langle w'^0_{20}\rangle^2_{w_{40}}}, \tag{5.9}$$

where

$$\langle w'^n_{20}\rangle_{w_{40}} = \int_{-\infty}^{\infty} w'_{20}{}^n |U_a(w'_{20}; w_{40})|^2 dw'_{20} \tag{5.10}$$

stands for the *n*th-moment of the axial irradiance response.

Now, we can profit from the Fourier-transform relationship stated in the previous section to express this width as a function of global parameters of the mapped pupil $q(\zeta)$. It is easy to show that

$$\langle w'^n_{20}\rangle_{w_{40}} = \frac{1}{(-i2\pi)^n}\int_{-\infty}^{\infty} \frac{\mathrm{d}^n q(\zeta; w_{40})}{\mathrm{d}\zeta^n} q^*(\zeta; w_{40}) d\zeta, \tag{5.11}$$

being $q(\zeta; w_{40}) = q(\zeta)\exp(i2\pi w_{40}\zeta^2)$. For the moments up to second order, this formula leads to

$$\begin{aligned}
\langle w'^0_{20}\rangle_{w_{40}} &= \langle w'^0_{20}\rangle_0 = \langle \zeta^0\rangle, \\
\langle w'^1_{20}\rangle_{w_{40}} &= \langle w'^1_{20}\rangle_0 + 2\langle \zeta^1\rangle w_{40}, \\
\langle w'^2_{20}\rangle_{w_{40}} &= \langle w'^2_{20}\rangle_0 - \frac{i}{\pi}\left[2\int_{-\infty}^{\infty} \zeta \frac{\mathrm{d}q(\zeta)}{\mathrm{d}\zeta} q^*(\zeta)\mathrm{d}\zeta + \langle \zeta^0\rangle\right] w_{40} + 4\langle \zeta^2\rangle w_{40}^2,
\end{aligned} \tag{5.12}$$

where

$$\langle \zeta^n\rangle = \int_{-\infty}^{\infty} \zeta^n |q(\zeta)|^2 d\zeta. \tag{5.13}$$

Substituting (5.12) in (5.9), we finally find that

$$\begin{aligned}
\sigma^2_{w'_{20}}(w_{40}) &= \sigma^2_{w'_{20}}(0) + 4\sigma^2_{\zeta} w_{40}^2 \\
&\quad + \left\{\frac{2}{\pi\langle w'^0_{20}\rangle_0}\Im m\left[\int_{-\infty}^{\infty} \zeta \frac{dq(\zeta)}{d\zeta} q^*(\zeta) d\zeta\right] - \frac{4}{\langle w'^0_{20}\rangle^2_0}\langle \zeta\rangle\langle w'_{20}\rangle_0\right\} w_{40}.
\end{aligned} \tag{5.14}$$

Note that a further simplification can be performed in the case of a real pupil function $q(\zeta)$, leading to

$$\sigma^2_{w'_{20}}(w_{40}) = \sigma^2_{w'_{20}}(0)\left[1 + 4\frac{\sigma^2_\zeta}{\sigma^2_{w'_{20}}(0)}w^2_{40}\right]. \tag{5.15}$$

In both (5.14) and (5.15) we obtain a parabolic behaviour of the width of the axial response as a function of w_{40}. The narrowness of this parabola is related with the tolerance of the axial response to SA: the faster the variation of the width of the response, the more sensible to SA the system is. From this result, it is straightforward to compare the "sensitivity" to SA that two different mapped pupils $q(\zeta)$ provide to the system, and, consequently, select properly the one that "desensitizes" most the axial response.

However, the above results have been obtained by implicitly assuming some continuity and differentiability properties for the pupil functions that are not usually fulfilled in real systems. The more general case of hard-edge functions is analysed in the next section.

5.4 Generalized Second-Order Moment

Up to now, we have developed the transformation laws for the width of the axial intensity distribution depending on the SA. However, such an analytical procedure fails when the system contains hard-edge diffracting elements. The problem arises when we try to calculate the second-order moment of the axial irradiance response $\langle w'^2_{20}\rangle$, which diverges.

To overcome this problem we make use of Martínez-Herrero and co-workers reasoning (Martínez-Herrero and Mejías 1993), and we define a generalized second-order moment for the axial irradiance response as

$$\begin{aligned}\langle w'^2_{20}\rangle_{G,w_{40}=0} &= \langle w'^2_{20}\rangle_{G,0} \\ &= \frac{1}{4\pi^2}\int_{-D}^{D}\frac{\mathrm{d}q(\zeta)}{\mathrm{d}\zeta}^2\mathrm{d}\zeta + \frac{1}{\pi^2 D}\left[|q(D)|^2 + |q(-D)|^2\right],\end{aligned} \tag{5.16}$$

where $[-D, D]$ denotes the compact support of the function $q(\zeta)$, that is, the shortest interval for which $q(\zeta) = 0\ \forall\zeta$ outside. In the above expression, we have assumed a pupil function $q(\zeta)$ with

$$\langle\zeta\rangle = \langle w'_{20}\rangle_0 = 0. \tag{5.17}$$

Real and even, mapped pupil functions fulfil this condition, but we will later generalize this equation for any pupil function.

As can be found in Martínez-Herrero and Mejías (1993), this generalized moment represents in fact a measure of the second-order moment $\langle w'^2_{20} \rangle$, but within a truncated integration interval that includes a significant energy of the spectrum of $q(\zeta)$.The main advantage of this particular definition for these generalized moments is that an analysis of its propagation laws through lineal optical systems leads to the same transformation formula as for the usual second-order moments (Martínez-Herrero et al. 1995; Lü and Luo 2000). In this way, for instance, for hard-edge real pupil elements [fulfilling (5.16)], we can obtain the following generalized transformation law for the width of the axial intensity distribution

$$\sigma^2_{w'_{20},G}(w_{40}) = \sigma^2_{w'_{20},G}(0)\left[1 + 4\frac{\sigma^2_\zeta}{\sigma^2_{w'_{20},G}(0)} w^2_{40}\right], \tag{5.18}$$

where the second-order moment of the axial irradiance response is given now by (5.16). This result is, thus, a proper generalization of the transformation law developed in the above section applied now to the generalized variance of the axial response of the system.

Finally, let us expand this procedure for general pupil functions with $\langle \zeta \rangle, \langle w'_{20} \rangle_0 \neq 0$. To obtain the analogous expression to (5.16) in this case, we define an auxiliary function $\widehat{q}(\zeta)$ in terms of $q(\zeta)$ as

$$\widehat{q}(\zeta) = q\left(\zeta + \frac{\langle \zeta \rangle}{\langle \zeta^0 \rangle}\right) \exp\left(-i2\pi\zeta \frac{\langle w'_{20} \rangle_0}{\langle w'^0_{20} \rangle_0}\right), \tag{5.19}$$

which trivially satisfies the conditions in (5.16). It is also easy to demonstrate that the parameter $\sigma^2_{w'_{20},G}$ for $q(\zeta)$ can be obtained by applying the definition in (5.16) to $\widehat{q}(\zeta)$. Thus

$$\begin{aligned}\sigma^2_{w'_{20},G}(0) = {} & \frac{1}{4\pi^2\langle w'^0_{20} \rangle_0} \int_{-D}^{D} \left|\frac{\mathrm{d}q(\zeta)}{\mathrm{d}\zeta}\right|^2 \mathrm{d}\zeta + \frac{1}{\pi^2 D \langle w'^0_{20} \rangle_0}\left(|q(D)|^2 + |q(-D)|^2\right) \\ & + \left[\frac{\langle w'_{20} \rangle_0^2}{\langle w'^0_{20} \rangle_0^2} - \frac{\langle w'_{20} \rangle_0}{\pi \langle w'^0_{20} \rangle_0^2} \Im m \int_{-D}^{D} \frac{\mathrm{d}q(\zeta)}{\mathrm{d}\zeta} q^*(\zeta)\mathrm{d}\zeta\right].\end{aligned} \tag{5.20}$$

From this expression, transformation law in (5.18) can be applied for real pupils, or the analogous to (5.14) for non-real functions.

The above formulation constitutes a very useful tool for the analysis of the influence of SA in high-NA optical instruments. But what is more important is that it provides us with a powerful technique for the design of beam-shaping elements for reduction of SA impact, as we show in the next section.

5.5 Design of Beam-Shaping Elements for Reduction of SA Impact

In previous sections, we obtained the transformation laws for the width of the axial intensity distribution depending on SA. Next, we pay attention to (5.18) for a truncated real pupil function $q(\zeta)$. Analysing this expression is trivial to realize that $\sigma^2_{w'_{20},G}(0)$ determines the minimum effective width of the axial intensity distribution, whereas, the parameter σ^2_ζ assesses the modification "speed" of the width of the axial irradiance as the coefficient w_{40} increases. From that interpretation, since our aim throughout this section is the design of pupil elements that increase the tolerance of the system to the sample induced SA, the variance of the function $q(\zeta)$, σ^2_ζ, is a proper assessment parameter. From this reasoning, we can consider the merit function

$$\Gamma_{w_{40}} = \frac{\sigma_{\zeta,c}}{\sigma_{\zeta,a}}, \tag{5.21}$$

which we called tolerance to the SA. $\sigma_{\zeta,a}$ and $\sigma_{\zeta,c}$ are the standard deviations of the mapped pupil for the apodized and non-apodized system, respectively. In this way, any pupil mask that provides $\Gamma_{w_{40}}>1$ will increase the robustness against SA with respect to the non-apodized system. For the sake of simplicity, we will consider hereafter that the original (non-apodized) shape of the pupil of the system is circular, which is otherwise the most common case in microscopy, this assumption leads to a particular value $\sigma_{\zeta,c}$.

As an example of optimization procedure following this formalism, we propose now the use of the family of binary masks known as shaded ring (SR) filters. These filters are composed of three annular zones with two different transmittances and each mask is uniquely specified by two construction parameters (μ, η), as defined in Fig. 5.3. It is straightforward to show that for these filters the tolerance to the SA is given by

$$\Gamma_{w_{40}}(\mu, \eta) = \sqrt{\frac{(1-\eta)^2(1-\mu)+\mu}{(1-\eta)^2(1-\mu^3)+\mu^3}}, \tag{5.22}$$

and they all satisfy $\Gamma_{w_{40}}(\mu, \eta)>1$. Therefore, any of these pupil masks makes the system to be less sensitive to the SA.

One "free" parameter to choose in the selection of a particular member of the above family is the width $\sigma^2_{w'_{20},G}(0)$ of the axial intensity distribution provided in absence of SA. It is interesting to use filters with similar response to the clear aperture when $w_{40} = 0$. Thus, we select filters that fulfil

$$\sigma^{2(a)}_{z_N,G}(0) = \sigma^{2(c)}_{z_N,G}(0) = 0.4\,. \tag{5.23}$$

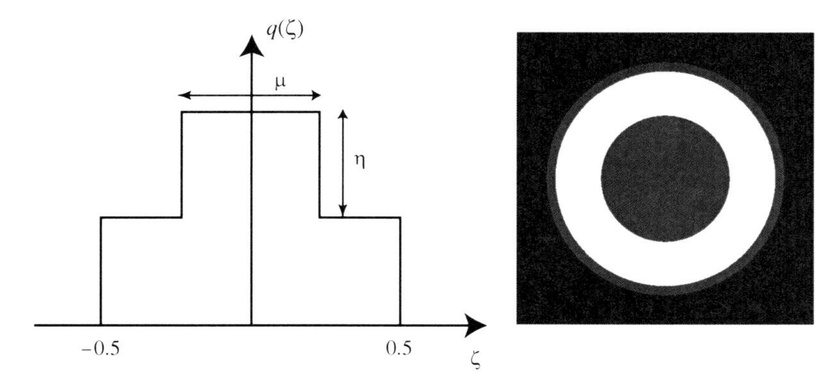

Fig. 5.3 Amplitude transmittance of the SR filter

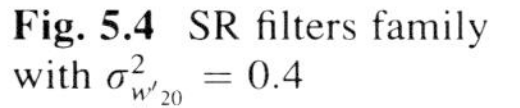
Fig. 5.4 SR filters family with $\sigma^2_{w'_{20}} = 0.4$

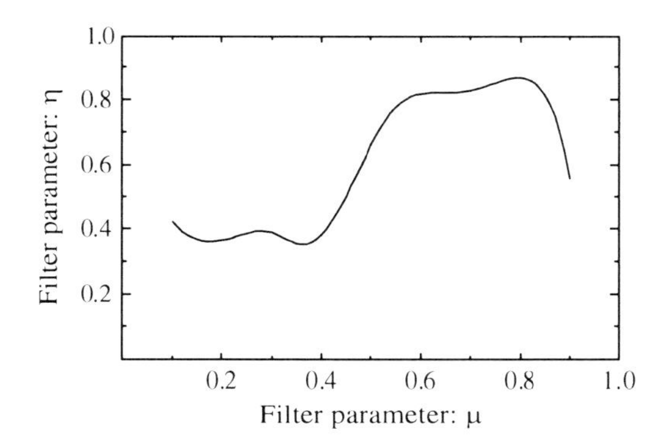

In this way, the apodized system —always with $\Gamma_{w'_{40}}>1$— will maintain a variance of the axial intensity distribution smaller than the corresponding to the non-apodized system for any amount of SA. Note that (5.23) restricts our choice to a SR filters subset characterized by an implicit relationship between μ and η, as shown in Fig. 5.4. In Fig. 5.5 we show the tolerance to SA for this filters subset. The optimal design of the system was done, in this case, by selecting the filter that corresponds to the maximum of this curve. This maximum corresponds to $\mu = 0.55$, whose partner through the representation in Fig. 5.4 is $\eta = 0.77$.

To verify our optimal design, in Fig. 5.6 we represent the evolution of the variance of the axial intensity distribution against the SA for both the clear aperture and the optimum SR filter. This filter is called spherical aberration tolerance (SAT) filter. Note that for any amount of the coefficient w_{40}, the variance of the SAT filter is always lower than that of the circular aperture. This behaviour is easy to see in Fig. 5.7, where we show the axial intensity distribution for $w_{40} = 0, -1, -2$ and -3. We can actually check that in the absence of SA, the SAT filter is almost as good as the circular pupil. However, as the SA increases, the response of the circular aperture degrades very fast, whereas the response of the SAT filter remains fairly unchanged.

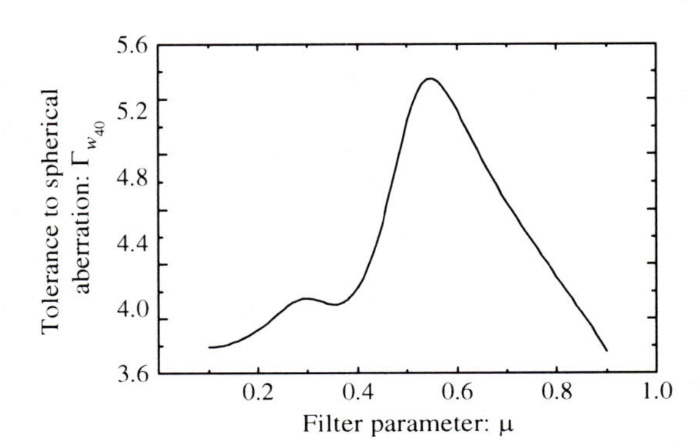

Fig. 5.5 Tolerance to spherical aberration for SR filters with $\sigma^2_{w'_{20}} = 0.4$

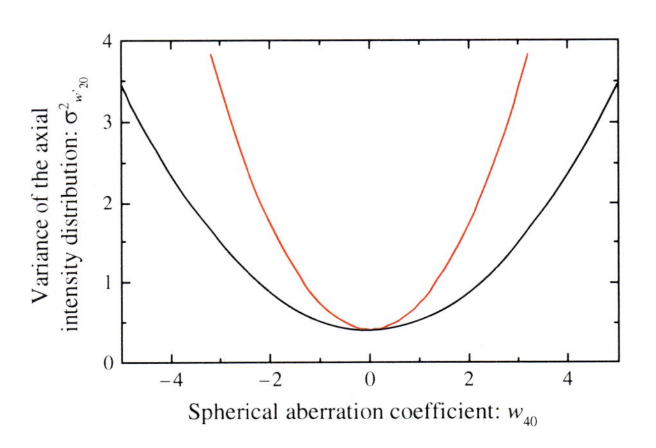

Fig. 5.6 Evolution of the second-order moment of the axial intensity distribution: *red line* clear aperture and *black line* SR filter

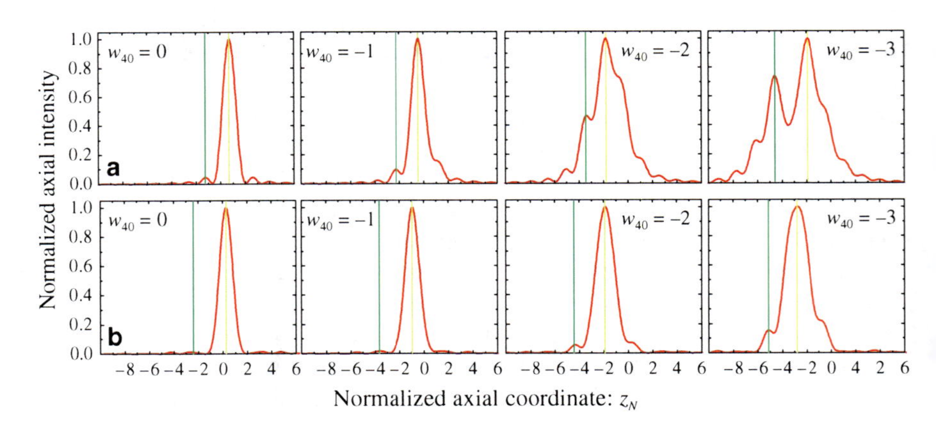

Fig. 5.7 Axial intensity distribution for several amount of spherical aberration: (**a**) clear aperture and (**b**) SAT filter

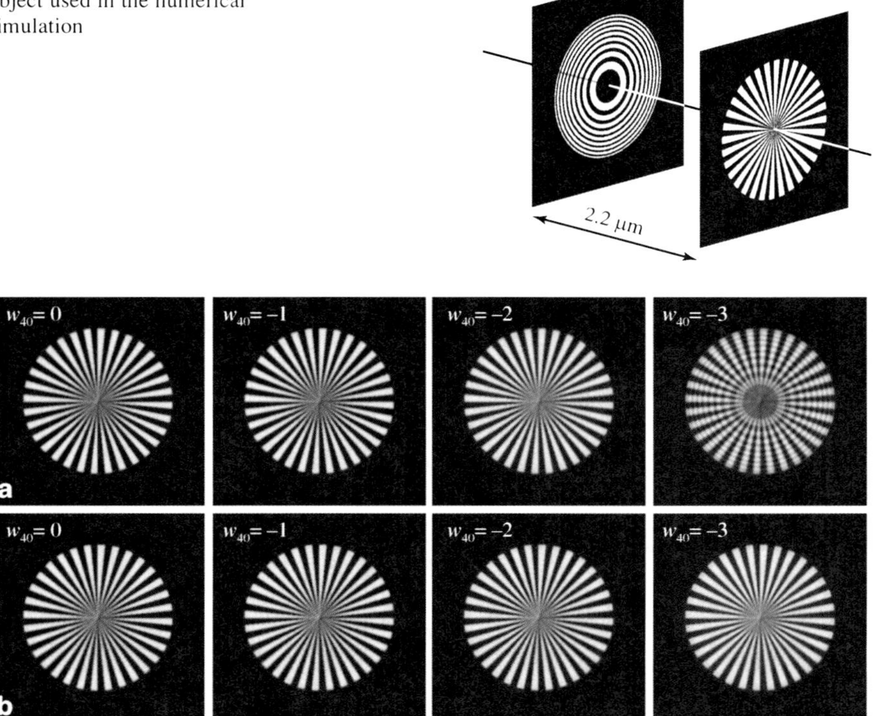

Fig. 5.8 Scheme of the 3D object used in the numerical simulation

Fig. 5.9 Three-dimensional image obtained with: (**a**) the clear circular pupil as the aperture stop; (**b**) the SAT filter as the aperture stop

Finally, we perform now a numerical simulation of a confocal imaging experiment. As the 3D object we assume a simple synthetic object composed by two plane plates, as shown in Fig. 5.8. We consider that the spoke target is placed in the best focus plane in all images (yellow line in Fig. 5.7). In Fig. 5.9 we show the image obtained in that plane with both the circular aperture and the SAT filter as the objective aperture stop. Note that, as it is well-known, scanning confocal microscopes are characterized by high optical sectioning. However, we clearly see that the bigger the induced SA, the higher the influence of the adjacent plate in the image of the spoke target obtained with the circular aperture. On the other hand, we get a clear image for any amount of SA with the SAT filter.

5.6 Experimental Results

In order to demonstrate the theoretical prediction described in previous sections, we present now the results of an experiment carried out to measure the response of the above apodized system when a perfect planar reflector mirror is used as specimen.

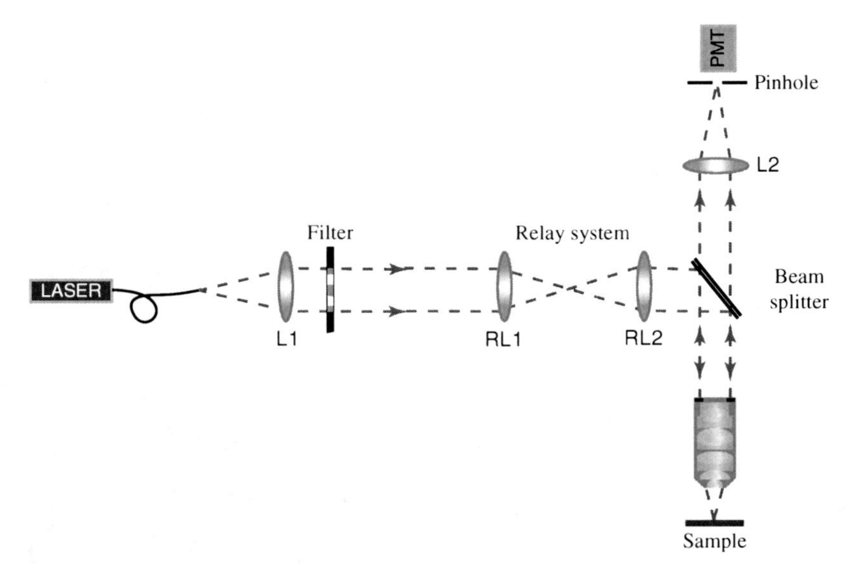

Fig. 5.10 Schematic geometry of the experimental setup

This function is often used to assess the axial resolution of an imaging system (Sheppard 1987). In Fig. 5.10 we show a scheme of the experimental setup. Light from a He–Ne laser ($\lambda = 632.8$ nm) is guided through a single-mode optical fibre to the optical axis. The output beam fibre is regarded as a point source. The light from the source is collimated by a lens L_1 and, after passing through a relay system, it is then focused via a microscope objective, with NA = 1.2 (water), onto a mirror which is controlled by a piezo driver and thus scanned along the axial direction. The signal reflected from the scanned mirror is finally focused on a pinhole located in front of the detector, which is a photo-multiplier tube (PMT). A standard coverslip is glued to the mirror surface.

Our aim is to modify the exit pupil of the system with the SAT filter to increase the tolerance to SA. Since the exit pupil is located somewhere inside the objective, we use a relay system formed by RL1 ($f_1 = 200$ mm) and RL2 ($f_2 = 175$ mm) to image the filter onto it. The filter was fabricated with high-contrast photographic film (Kodak© Technical Pan™).

In order to induce the SA in the experimental setup in a controlled way, we use the correction collar (cc) of the microscope objective. If the position of this collar is the one that compensates the SA induced by the coverslip – cc = 0.14 in our case – the system will be free of aberrations. However, for other positions of the collar, the system will have an over- or under-compensation that will produce SA.

The obtained experimental data are shown in Figs. 5.11 and 5.12. It can be seen that as the aberration is increased, the response becomes more asymmetric and the sidelobes on one side become stronger with the clear aperture. In contrast, the SAT filter designed for increasing the tolerance to the SA provides the system with an

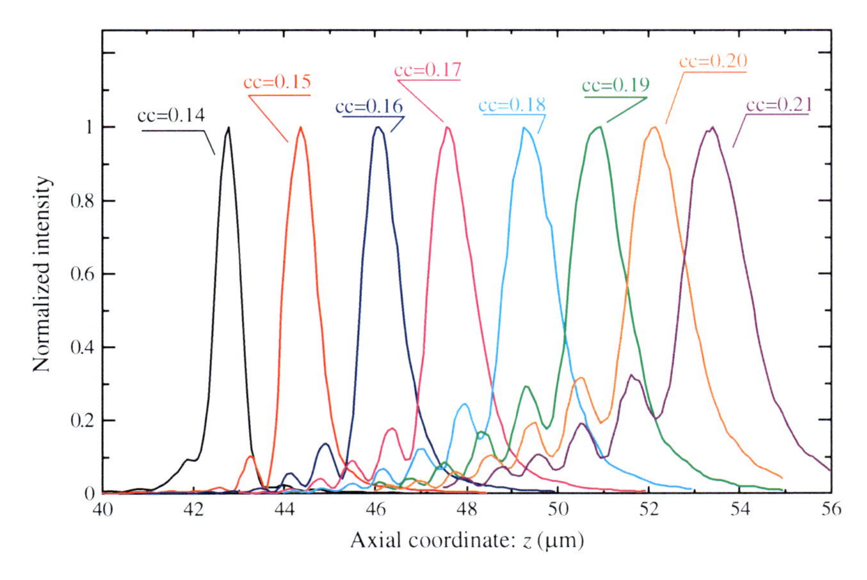

Fig. 5.11 Experimental axial response to a planar mirror with the clear aperture for several amount of spherical aberration: *black line* $w_{40} = 0$, *red line* $w_{40} = 0.7$, *dark blue line* $w_{40} = 1.4$, *pink line* $w_{40} = 2.1$, *turquoise blue line* $w_{40} = 2.8$, *dark green line* $w_{40} = 3.5$, *orange line* $w_{40} = 4.2$, *dark purple line* $w_{40} = 4.9$

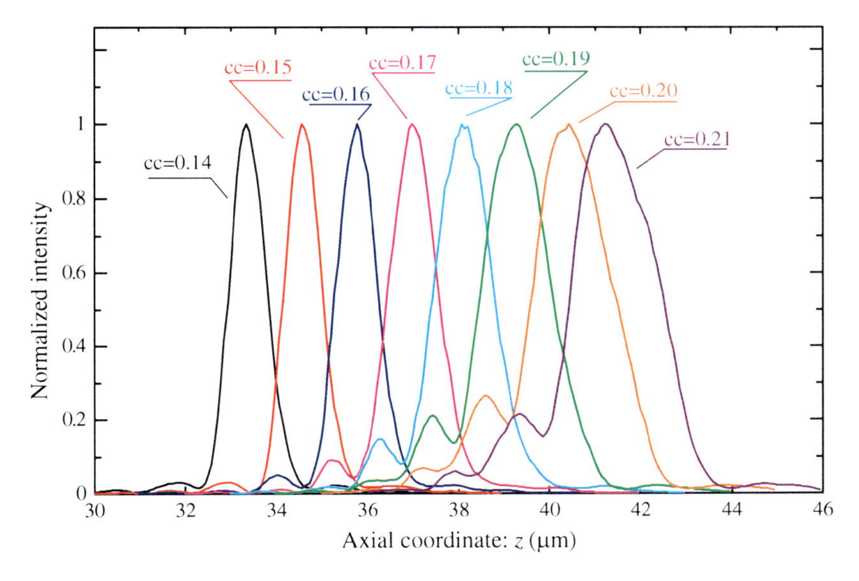

Fig. 5.12 Experimental axial response to a planar mirror with the SAT filter for several amount of spherical aberration: *black line* $w_{40} = 0$, *red line* $w_{40} = 0.7$, *dark blue line* $w_{40} = 1.4$, *pink line* $w_{40} = 2.1$, *turquoise blue line* $w_{40} = 2.8$, *dark green line* $w_{40} = 3.5$, *orange line* $w_{40} = 4.2$, *dark purple line* $w_{40} = 4.9$

important robustness against the SA. Besides, the SAT filter profile moves sideways with respect to the origin in z much less than in the clear aperture case.

5.7 Conclusions

We have developed a mathematic formalism to analyse the SA effect based on the modification of the variance of the axial intensity distribution. In particular, we have considered the case in which the field exiting from the high-NA objective is focused through an interface between two dielectrics. Following, we have defined a merit function, tolerance to the SA, to design an amplitude filter for decreasing the sensibility to SA of the system. Finally, we have measured the experimental axial distribution in a confocal microscope setup, obtaining an important reduction of the SA impact with the SAT filter.

Acknowledgements This work was funded by the Plan Nacional I+D+I (grant FIS2009-9135), Ministerio de Ciencia e Innovación, Spain. Financial support is also acknowledged from Generalitat Valenciana (grant PROMETEO/2009/077), Spain.

References

Booth MJ, Wilson T (2000) Strategies for the compensation of specimen-induced spherical aberration in confocal microscopy of skin. J Microsc 200:68–74

Booth MJ, Neil MAA, Wilson T (1998) Aberration correction for confocal imaging in refractive-index-mismatched media. J Microsc 192:90–98

Braat J (1997) Influence of substrate thickness on optical disk readout. Appl Opt 36:8056–8062

Day D, Gu M (1998) Effects of refractive-index mismatch on three-dimensional optical data-storage density in a two-photon bleaching polymer. Appl Opt 37:6299–6304

Gu M (2000) Advanced optical imaging theory. Springer-Verlag, Berlin

Haeberlé O, Ammar M, Furukawa H, Tenjimbayashi K, Torok P (2003) Point spread function of optical microscopes imaging through stratified media. Opt Express 11:2964–2969

Hell S, Reiner G, Cremer C, Stelzer EHK (1993) Aberrations in confocal fluorescence microscopy induced by mismatches in refractive-index. J Microsc 169:391–405

Ke PC, Gu M (1998) Characterization of trapping force in the presence of spherical aberration. J Mod Opt 45:2159–2168

Lü B, Luo S (2000) Beam propagation factor of hard-edge diffracted cosh-Gaussian beams. Opt Commun 178:275–281

Martinez-Corral M, Saavedra G (2009) The resolution challenge in 3D optical microscopy. Prog Opt 53:1–67

Martínez-Herrero R, Mejías PM (1993) Second-order spatial characterization of hard-edge diffracted beams. Opt Lett 18:1669–1671

Martínez-Herrero R, Mejías PM, Arias M (1995) Parametric characterization of coherent, lowest-order Gaussian beams propagating through hard-edged apertures. Opt Lett 20:124–126

Reihani SNS, Khalesifard HR, Golestanian R (2006) Measuring lateral efficiency of optical traps: the effect of tube length. Opt Commun 259:204–211

Saavedra G, Escobar I, Martínez-Cuenca R, Sánchez-Ortiga E, Martínez-Corral M (2009) Reduction of spherical-aberration impact in microscopy by wavefront coding. Opt Express 17:13810–13818

Schwertner M, Booth MJ, Wilson T (2005) Simple optimization procedure for objective lens correction collar setting. J Microsc 217:184–187

Sheppard CJR (1987) Scanning optical microscopy. In: Barer R, Cosslett VE (eds) Advances in optical and electron microscopy, vol 10. Academic, London, pp 1–98

Sheppard CJR (1995) Aberrations in high aperture optical systems. Optik 101:1–5

Sheppard CJR (1998) Aberrations in high aperture conventional and confocal imaging systems. Appl Opt 27:4782–4786

Sheppard CJR, Cogswell CJ (1991) Effects of aberrating layers and tube length on confocal imaging properties. Optik 87:34–38

Sheppard CJR, Gu M (1993) Imaging by a high aperture optical system. J Mod Opt 40:1631–1651

Stallinga S (2005a) Compact description of substrate-related aberrations in high numerical-aperture optical disk readout. Appl Opt 44:849–858

Stallinga S (2005b) Finite conjugate spherical aberration compensation in high numerical-aperture optical disc readout. Appl Opt 44:7307–7312

Török P, Varga P (1997) Electromagnetic diffraction of light focused through a stratified medium. Appl Opt 36:2305–2312

Török P, Varga P, Laczik Z, Broker GR (1995a) Electromagnetic diffraction of light focused through a planar interface between materials of mismatched refractive-indexes: an integral-representation. J Opt Soc Am A 12:325–332

Török P, Varga P, Németh G (1995b) Analytical solution of the diffraction integrals and interpretation of wave-front distortion when light is focused through a planar interface between materials of mismatched refractive-indexes. J Opt Soc Am A 12:2660–2671

Török P, Hewlett SJ, Varga P (1997) The role of specimen-induced spherical aberration in confocal microscopy. J Microsc 188:158–172

Wilson T, Carlini AR (1988) Three-dimensional imaging in confocal imaging-systems with finite sized detectors. J Microsc 149:51–66

Chapter 6
Improving Image Formation by Pushing the Signal-to-Noise Ratio

Emiliano Ronzitti, Giuseppe Vicidomini, Francesca Cella Zanacchi, and Alberto Diaspro

6.1 Introduction

Imaging technologies are essential requirements in order to achieve significant results in biomedical research areas. They can be optimally characterized according to three fundamentals properties, namely: temporal resolution, spatial resolution, and signal-to-noise ratio (SNR) performances of the imaging process. Currently, imaging research aims to optimize each of these crucial aspects (Hell 2007).

Several important results have been obtained during the last decades in the three-dimensional (3D) spatial resolution framework. Confocal single-photon (Wilson and Sheppard 1984) and two-photon excitation (Diaspro et al. 2005) fluorescence microscopy are widely used for biological imaging investigations. Their major benefit is ascribed to the 3D imaging capability obtained through an optical sectioning by a pinhole (confocal laser scanning microscope, CLSM) and by an excitation confinement (two-photon excitation, 2PE). In the CLSM case, the 3D

E. Ronzitti
LAMBS-IFOM, MicroScoBio, Department of Physics, University of Genoa, Via Dodecaneso 33, 16145 Genoa, Italy
and
SEMM-IFOM-IEO, University of Milan, Via Adamello 16, 20139 Milan, Italy

G. Vicidomini
LAMBS-IFOM, MicroScoBio, Department of Physics, University of Genoa, Via Dodecaneso 33, 16145 Genoa, Italy
and
Department of NanoBiophotonics, Max Planck Institute for Biophysical Chemistry, Am Fassberg, Göttingen, Germany

F. Cella Zanacchi and A. Diaspro (✉)
LAMBS-IFOM, MicroScoBio, Department of Physics, University of Genoa, Via Dodecaneso 33, 16145 Genoa, Italy
and
Nanophysics, Italian Institute of Technology, Via Morego 30, 16163 Genova, Italy
e-mail: alberto.diaspro@iit.it.

A. Diaspro (ed.), *Optical Fluorescence Microscopy*,
DOI 10.1007/978-3-642-15175-0_6,

capability is obtained with relevant improvements in terms of optical resolution, while for the TPE scheme, as the excitation wavelengths are nearly twice as large as in the single-photon case, a minimization of the damaging photophysical interactions is achieved. The combination of these two schemes in a two-photon confocal scanning microscopy has been proposed to join their optical advantages (Gu and Sheppard 1993). Further spatial resolution improvements have been reached with the 4Pi confocal fluorescence microscopy (Hell and Steltzer 1992) pushing to the diffractional limit achievable simply by optics rearrangements. Recently, several techniques based on different interferences and photophysical interaction phenomena have been proposed to break the diffraction limits toward fluorescence nanoscopy (STED, PALM, and SSIM) (Gustafsson 2005; Schermelleh et al. 2008; Westphal et al. 2003; Betzig et al. 2006).

Similarly, in order to follow fast (ms) and very fast (sub ms) biological events, several techniques have been elaborated in the last years (Maddox et al. 2003; Reddy and Saggau 2005) significantly improving the temporal resolution.

In any case, each of the developments in the temporal or in the spatial domain cannot neglect the remarkable noise contribution in the imaging formation. As the noise is always present in the imaging process and it influences and deteriorates the imaging quality, it is crucial to elaborate methodologies to minimize its influence (Sheppard et al. 2006).

In the presence of noise, the overall imaging quality could be evaluated by considering the SNR of the image. This is the case of many live-cell confocal and two-photon applications where low fluorophore concentration and deep specimen imaging are involved and noise contributions are not negligible (Jonkman and Steltzer 2002).

The noise collected in the detected image signal arises from different sources. Noise can originate, on the one hand, from signals due to the optical elements reflection and the light scattered from within the microscope body, and on the other hand, from the out-of-focus signals emanating uniformly from the entire extension of a thick fluorescent object (Sheppard et al. 2006; Sandison and Webb 1994). Several models have been presented in order to characterize this noise effect taking into account (Sheppard et al. 2006; Sandison and Webb 1994) the confocal pin-hole extension and the fluorescence excitation volume (Gu and Sheppard 1993). Another significant contribution to the image noise is named shot-noise or Poissonian noise. It is a multiplicative noise intrinsically linked to the signal acquired arising because of the quantum nature of light (Young 1996): due to the statistical nature of photon production ruled by quantum physics, the probability distribution of photons in an observation temporal window obeys a Poissonian distribution. This statistical behavior induces an uncertainty in the photon detection counting more prominently as the number of expected photons decrease. Finally, the signal acquired is always affected by an electronic noise due to the detection step.

Noise has a degrading effect on the imaging process deteriorating, particularly the higher frequencies as they are transmitted at lower levels. Substantially, it reduces the transfer function bandwidth and thus decreases the practical resolution power of the system.

In order to enhance the SNR in the high-frequency range the interference effects induced by the insertion on the microscope illumination arm of an annular filter can be exploited.

Recently, several works suggest amplitude and phase filters juxtaposed to the lens as useful tools able to tailor the axial and radial shapes of point spread function of an optical system (Haeberl and Simon 2006; Mondal and Diaspro 2008; Martínez-Corral et al. 2002). According to the concept primarily exposed by Toraldo di Francia (1952), several annuli filters have been designed in order to sharpen up the central lobe of the PSF of the objective lens at the expense of an increase of its side-lobes intensity (Sheppard and Hegedus 1988; Cox et al. 1982). The application of such ring filters is limited by the level of side-lobes, relatively high in comparison with the central peak; moreover, since filter can partially limit the illumination light, the intensity on the object is considerably low with respect to the unobstructed case. However, these limits have been partially overcome in confocal and 2PE scanning microscopy as the multiplicative nature of their PSF reduces appreciably the side-lobes intensity, and the scanning laser illumination could be tuned to have adequate intensity (Neil et al. 2000; Caballero et al. 2006).

The annular filter analysis reveals that the overall spatial frequency transfer bandwidth extension is unchanged (Davis et al. 2004). However, within this transfer bandwidth, a redistribution of the spatial frequencies transmission is obtained and a significant improvement of the SNR at high-frequency range is reached (Davis et al. 2004; O'Neill 1956).

6.2 PSF and OTF

The imaging performances of an optical system can be evaluated through the intensity 3D-point spread function (3D-PSF), that is, the intensity response of an optical system illuminated by a point-like source and through the 3D-optical transfer function (3D-OTF), that is the 3D Fourier transformation of the PSF.

The intensity 3D-PSF of an optical system could be determined by Gu and Sheppard (1993)

$$\mathrm{PSF}_{\mathrm{system}}\,(u,v,\phi)=\mathrm{PSF}_{\mathrm{ill}}(u,v,\phi)\times\mathrm{PSF}_{\mathrm{det}}(u,v,\phi) \tag{6.1}$$

where $\mathrm{PSF}_{\mathrm{ill}}$ and $\mathrm{PSF}_{\mathrm{det}}$ represent, respectively, the illumination and the detection PSF, given by

$$\mathrm{PSF}_{\mathrm{ill}}(u,v,\phi)=\left(|h(u,v,\phi)|^{2}\right)^{m} \tag{6.2}$$

$$\mathrm{PSF}_{\mathrm{det}}(u,v,\phi)=D(v)\otimes\left(|h(u,v,\phi)|^{2}\right) \tag{6.3}$$

Here, a point-like source is assumed, m is the number of photons absorbed during the excitation process, and $D(v)$ is the intensity sensitivity of the detector, defined for a circular pinhole as

$$D(v) = \begin{cases} 1 & v < v_{\mathrm{d}} \\ 0 & \text{otherwise} \end{cases} \tag{6.4}$$

where v_{d} is the normalized detector radius,

$$v_{\mathrm{d}} = \frac{2\pi\sqrt{x_{\mathrm{d}}^2 + y_{\mathrm{d}}^2}}{\lambda} \sin \alpha_{\mathrm{d}} \tag{6.5}$$

with α_{d} the angular aperture of the objective in the detection side. $|h(u, v, \phi)|^2$ represents the intensity point spread function of a single lens given by Wolf (1959) and Richards and Wolf (1959),

$$|h(u, v, \phi)|^2 = |\mathbf{E}(u, v, \phi)|^2 \tag{6.6}$$

$\mathbf{E}(u, v, \phi)$ represents the electric field induced in a point $\mathrm{p}(u, v, \phi)$ of the image plane defined according to the diffractional vectorial theory by Richards and Wolf (1959) and Wolf (1959):

$$\mathbf{E}(u, v, \phi) = \begin{pmatrix} -iA(I_0 + I_2 \cos 2\alpha) \\ -iAI_2 \sin 2\alpha \\ -2AI_1 \cos \alpha \end{pmatrix} \tag{6.7}$$

A is a constant, $I_{0,1,2}(u, v)$ are integrals over the aperture angle $0 \leq \vartheta \leq \alpha$, i.e.,

$$\begin{aligned} I_0(u, v) &= \int_0^\alpha \sqrt{\cos \vartheta} \sin \vartheta (1 + \cos \vartheta) J_0\left(\frac{v \sin \vartheta}{\sin \alpha}\right) \mathrm{e}^{iu \cos \vartheta / \sin^2 \alpha} \mathrm{d}\vartheta \\ I_1(u, v) &= \int_0^\alpha \sqrt{\cos \vartheta} \sin^2 \vartheta J_1\left(\frac{v \sin \vartheta}{\sin \alpha}\right) \mathrm{e}^{iu \cos \vartheta / \sin^2 \alpha} \mathrm{d}\vartheta \\ I_2(u, v) &= \int_0^\alpha \sqrt{\cos \vartheta} \sin \vartheta (1 - \cos \vartheta) J_2\left(\frac{v \sin \vartheta}{\sin \alpha}\right) \mathrm{e}^{iu \cos \vartheta / \sin^2 \alpha} \mathrm{d}\vartheta \end{aligned} \tag{6.8}$$

ϕ is the azimuthal angle, λ the wavelength of the light, and α the angular aperture of the lens, and u, v are the radial and axial coordinates given by

$$\begin{aligned} u &= \frac{2\pi}{\lambda} z \sin^2 \alpha \\ v &= \frac{2\pi}{\lambda} \sqrt{x^2 + y^2} \sin^2 \alpha \end{aligned} \tag{6.9}$$

An illumination linearly polarized along the x-direction is assumed here.

The confocal scheme can be modeled considering a single-photon excitation and an ideal point-like pin-hole aperture approximating $D(v)$ as a delta Dirac function. On the contrary, the 2PE scheme adopted is considered with no pin-hole detection approximating $D(v)$ as infinitely extensive with unitary response. With this usual assumption we have defined the confocal and two-photon intensity PSF (Jonkman and Steltzer 2002):

$$\mathrm{PSF_{CLSM}}(u, v, \phi) = \left(|h_{\mathrm{ex}}(u, v, \phi)|^2\right)\left(|h_{\mathrm{em}}(u, v, \phi)|^2\right) \tag{6.10}$$

$$\mathrm{PSF_{2PE}}\,(u, v, \phi) = \left(|h_{\mathrm{ex}}(u, v, \phi)|^2\right)^2 \tag{6.11}$$

where $h_{\mathrm{ex}}\,(u, v, \phi)$ and $h_{\mathrm{em}}\,(u, v, \phi)$ represent the amplitude PSF of the objective lens for the excitation and the emission wavelength, respectively.

6.3 Pupil-Plane Filter Effects

The insertion of a pupil-plane filter juxtaposed to the objective lens in the illumination pathway of light induces a structure modification of the electromagnetic field near the focus of the optical system. In particular, the electric field components given by (6.7) are modified by the pupil-plane function insertion. As a consequence, the intensity distribution of light in the image plane given by (6.6) changes, and the PSF and the OTF of the microscope system are significantly modified. Several pupil-plane filters have been elaborated to properly engineer the optical system response, usually tuning the amplitude and the phase of the incident light.

Assuming an obscuring annular amplitude filter centered on the lens optical axis and completely absorbing the incident light over its extension (Fig. 6.1), the pupil function can be defined by

$$K(\vartheta) = \begin{cases} 0 & \frac{\alpha}{2}\left(1 - \frac{C}{100}\right) < \vartheta < \frac{\alpha}{2}\left(1 + \frac{C}{100}\right) \\ 1 & \text{otherwise} \end{cases} \tag{6.12}$$

where C is a constant and represents the annular covering percentage of the overall angular aperture. In this case, (6.8) is modified by the insertion of the $K(\theta)$ multiplicative factor in the integrand of the $I_{0,1,2}(u,v)$ integrals.

Figures 6.2 and 6.3 show the filter effect on the PSF for a CLSM and a 2PE scheme when an amplitude ring filter is inserted on the illumination pathway of the excitation light assumed linearly polarized in the x-direction (NA 1.4; n = 1.5; 2PE ex 900nm, CLSM ex 400nm; em 500nm).

As observed by the graphs in Figs. 6.2 and 6.3, either for CLSM or for 2PE, the filter insertion results in a double PSF effect: a sharpening up of the central lobe,

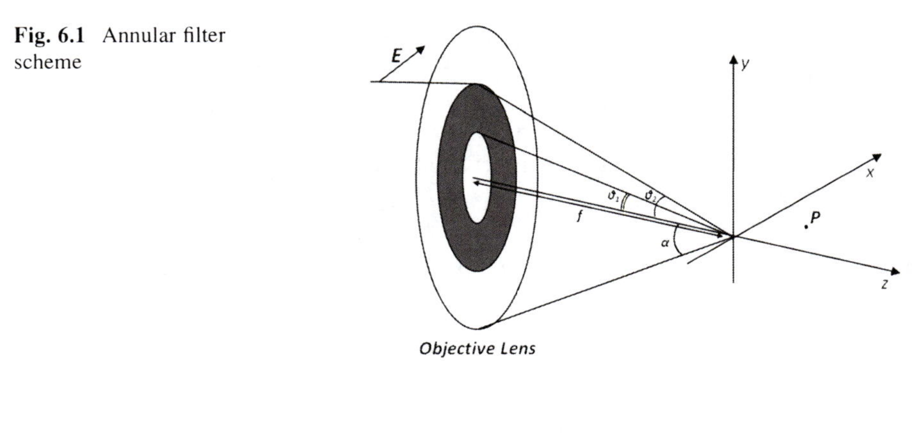

Fig. 6.1 Annular filter scheme

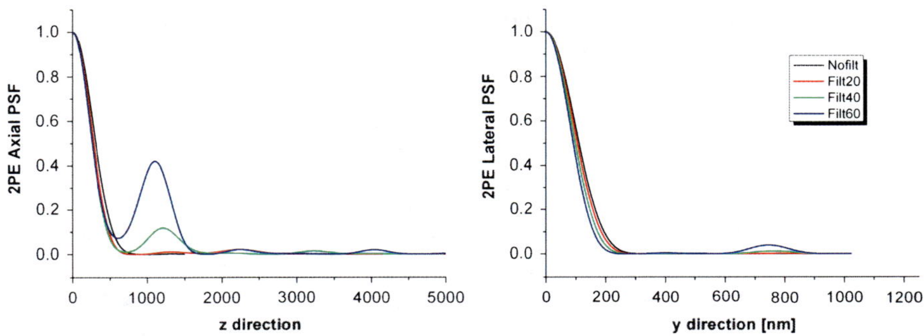

Fig. 6.2 2PE PSF for different covering filters in axial (a) and lateral directions (b)

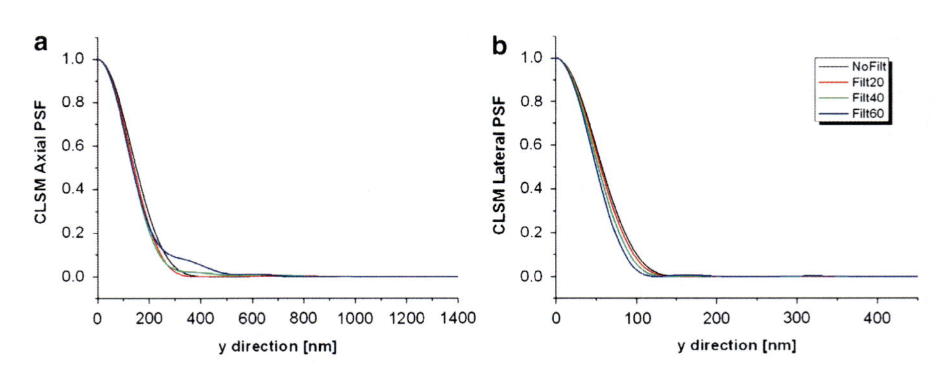

Fig. 6.3 CLSM PSF for different covering filters in axial (a) and lateral directions (b)

especially along the direction perpendicular to the polarization direction of the illumination light (hence, in this case in the *y*-direction) and an increase in the side-lobe levels. Both effects become more prominent as the covering portion *C* increases.

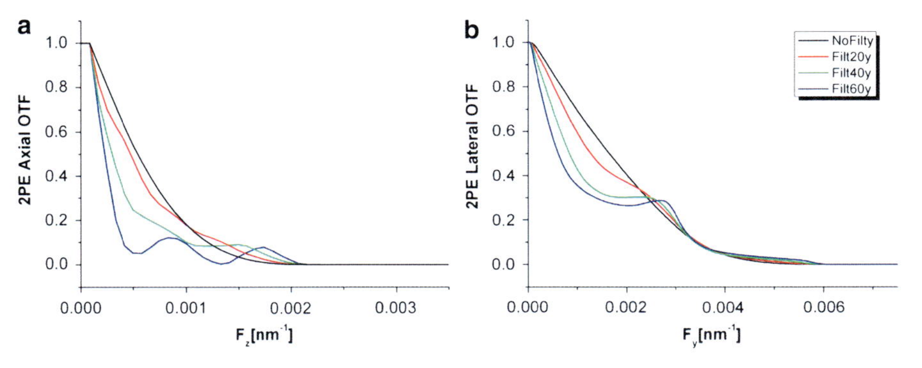

Fig. 6.4 2PE OTF for different covering filters in axial (a) and lateral directions (b)

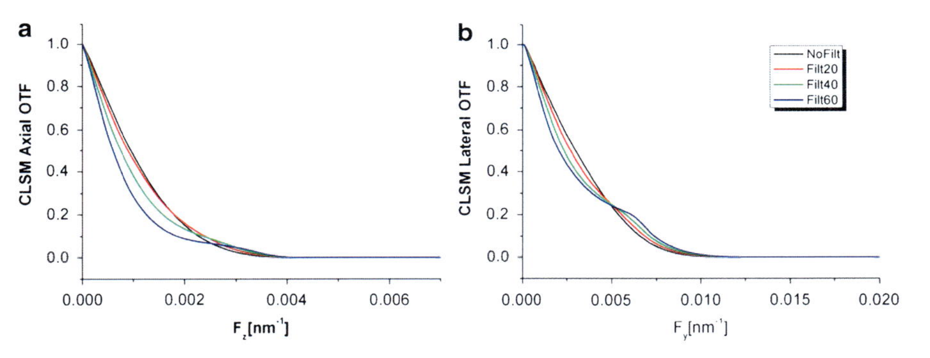

Fig. 6.5 CLSM OTF for different covering filters in axial (a) and lateral directions (b)

The 3D-OTF has been performed by Fourier transforming the 3D-PSF. The axial and the radial profiles of the 3D OTF for 2PE and CLSM scheme have been plotted in Figs. 6.4 and 6.5. The theoretical bandwidth extension is not affected by the pupil filter insertion, hence, no spatial frequencies are gained. Nevertheless, within this transfer bandwidth, a remarkable increase of the high-frequency and a reduction of the low-frequency transmissions are obtained (such an effect is prominent in the axial direction). Therefore, since noise affects particularly the information transmission at high frequencies, such a filter can be employed in some practical situations in order to partially recover the high-frequency information loss and consequently the resolution deteriorated.

This imaging quality improvement can be observed coupling such a filter with an imaging restoration method. Figure 6.6a represents an object phantom formed by a stack of 64 planes at a distance of 60 nm, with a matrix of dots in the middle plane. Each dot is $60 \times 60 \times 300$ nm at a distance of 180 nm. A simulation of the phantom perturbed by shot-noise and imaged by a 2PE is shown for a filter and an unfilter configuration in Figs. 6.6b and 6.6c, respectively. Applying a

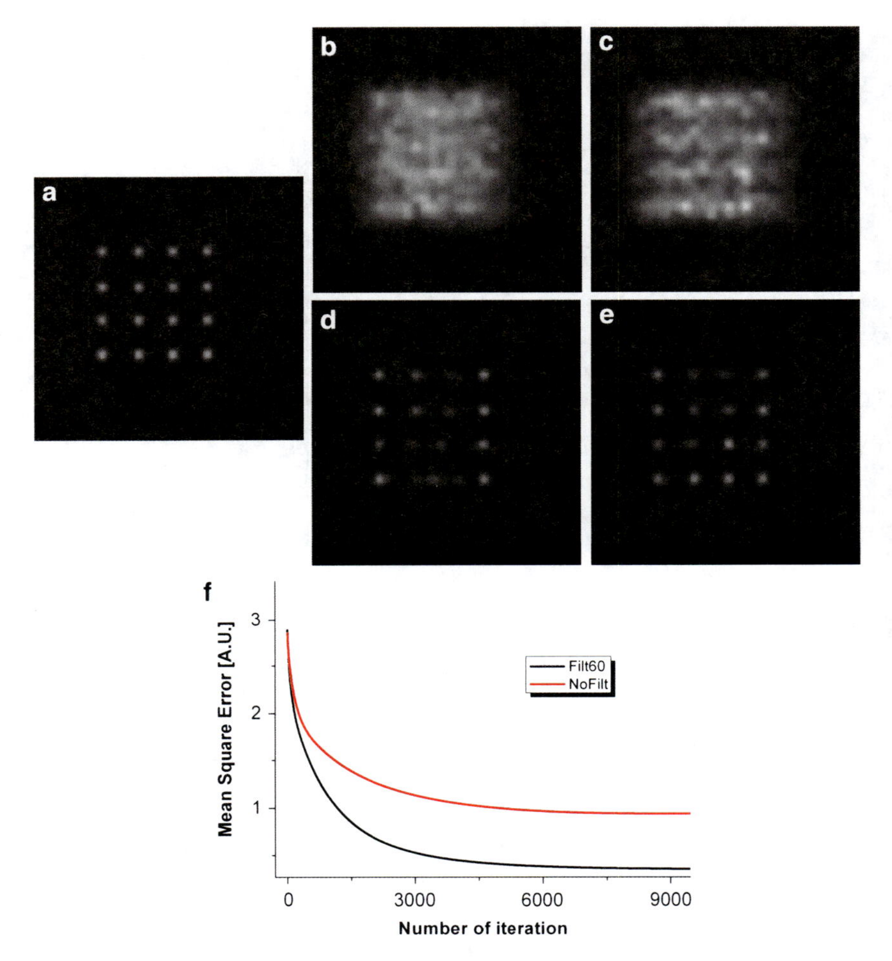

Fig. 6.6 Richardson–Lucy imaging restoration application. Original object (a); object imaged by 2PE (900 nm excitation light, NA 1.4, refractive index 1.5) in the absence of (b) and in the presence of a filter (c) assuming a shot-noise due to 50 photons/voxel; image obtained after RL algorithm methods in the absence of (d) and in the presence of a filter (e). MSE comparison between the original and the restorated image in filter and unfilter schemes (f)

Richardson–Lucy imaging restoration algorithm, a significant recovery of the original object image is possible in both the cases (Fig. 6.6d, unfilter and Fig. 6.6e, filter). As shown in the graph of Fig. 6.6f, the mean square error between the original object image and the restorated image is minimized in the filter configuration. This indicates that in this case, as filter configuration boosts the high-frequency information, it allows an improvement in the imaging reconstruction of the object.

6.4 Conclusion

The insertion of a pupil-plane filter on the light illumination pathway of a CLSM and TPE microscope allows to engineer the PSF response and rearrange the OTF distribution. In particular, an amplitude ring filter permits an enhancement of the high-frequency transmission of the optical system. Although the employment of such filters does not allow strictly a theoretical improvement in the spatial resolution, it improves the SNR in the high-frequency range and hence it can improve the quality of image in several situations affected by remarkable noise contribution.

References

Betzig E, Patterson GH, Sougrat R, Lindwasser OW, Olenych S, Bonifacino JS, Davidson MW, Lippincott-Schwartz J, Hess HF (2006) Imaging intracellular fluorescent proteins at nanometer resolution. Science 313:1642–1645

Caballero MT, Ibanez-Lopez C, Martinez-Corral M (2006) Shaded-mask filtering: novel strategy for improvement of resolution in radial polarization scanning microscopy. Opt Eng 45:0980031

Cox IJ, Sheppard CJR, Wilson T (1982) Reappraisal of arrays of concentric annuli as superresolving filters. J Opt Soc Am 72:1287–1291

Davis BJ, Karl WJ, Swan AK, Unlu MS, Goldberg BB (2004) Capabilities and limitations of pupil-plane filters for superresolution and image enhancement. Opt Express 12:4150–4156

Diaspro A, Chirico G, Collini M (2005) Two photon fluorescence excitation and related techniques in biological microscopy. Q Rev Biophys 38:97–166

Gu M, Sheppard CJR (1993) Effects of a finite-sized pinhole on 3D image formation in confocal two-photon fluorescence microscopy. J Mod Opt 40:2009–2024

Gustafsson MGL (2005) Nonlinear structured-illumination microscopy: wide-field fluorescence imaging with theoretically unlimited resolution. Proc Natl Acad Sci USA 102:13081–13086

Haeberl O, Simon B (2006) Improving lateral resolution in confocal fluorescence microscopy using laterally interfering excitation beams. Opt Commun 259:400–408

Hell SW (2007) Far-field optical nanoscopy. Science 316:1153–1158

Hell SW, Steltzer EHK (1992) Properties of a 4Pi confocal fluorescence microscope. J Opt Soc Am A 9:2159–2166

Jonkman JEN, Steltzer EHK (2002) Resolution and contrast in confocal two-photon microscopy. In: Diaspro A (ed) Confocal and two photons: foundations, applications and advances. Wiley-Liss, Wilmington

Maddox PS, Moree B, Canman JC, Salmon ED (2003) Spinning disk confocal microscope system for rapid high-resolution, multimode, fluorescence speckle microscopy and green fluorescent protein imaging in living cells. Methods Enzymol 360:597–617

Martínez-Corral M, Caballero MT, Stelzer EHK, Swoger J (2002) Tailoring the axial shape of the point spread function using the Toraldo concept. Opt Express 10:98–103

Mondal PP, Diaspro A (2008) Lateral resolution improvement in two photon excitation microscopy by aperture engineering. Opt Commun 281:1855–1859

Neil M, Juskaitis R, Wilson T, Laczik ZJ, Sarafis V (2000) Pupil-plane filters for confocal microscope point spread function engineering. Opt Lett 25:245–247

O'Neill EL (1956) Transfer function for an annular aperture. J Opt Soc Am 46:285–288

Reddy GD, Saggau P (2005) Fast three-dimensional laser scanning scheme using acousto-optic deflectors. J Biomed Opt 10:064038

Richards B, Wolf E (1959) Electromagnetic diffraction in optical systems II. Structure of the image field in an aplanatic system. Proc R Soc Lond A 253:358–379
Sandison DR, Webb WW (1994) Background rejection and signal-to-noise optimization in confocal and alternative fluorescence microscopes. Appl Opt 4:603–615
Schermelleh L, Carlton PM, Haase S, Shao L, Winoto L, Kner P, Burke B, Cardoso MG, Agard DA, Gustafsson MGL, Leonhardt H, Sedat JW (2008) Subdiffraction multicolor imaging of the nuclear periphery with 3D structured illumination microscopy. Science 320:1332–1336
Sheppard CJR, Hegedus ZS (1988) Axial behavior of pupil-plane filters. J Opt Soc Am A 5:643–647
Sheppard CJR, Gan X, Gu M, Roy M (2006) Signal to noise ratio in confocal microscopes. In: Pawley JB (ed) Handbook of biological confocal microscopy
Toraldo di Francia G (1952) Nuovo pupille superrisolvente. Atti Fond Giorgio Ronchi 7:366–372
Westphal V, Kastrup L, Hell SW (2003) Lateral resolution of 28 nm ($\lambda/25$) in far-field fluorescence microscopy. Appl Phys B 77:377–380
Wilson T, Sheppard CJR (1984) Theory and practice of scanning optical microscopy. Academic, London
Wolf E (1959) Electromagnetic diffraction in optical systems I. An integral representation of the image field. Proc R Soc Lond A 253:349–357
Young IT (1996) Quantitative microscopy. IEEE Eng Med Biol 15:59–66

Chapter 7
Site-Specific Labeling of Proteins in Living Cells Using Synthetic Fluorescent Dyes

Gertrude Bunt

7.1 Introduction

In the last decade, fluorescence microscopy has evolved from a classical "retrospective" microscopy approach into an advanced imaging technique that allows the observation of cellular activities in living cells with increased resolution and dimensions. This development was fuelled by the discovery of the *Aequoria victoria* green fluorescent protein (Tsien 1998) that can be genetically fused to almost every protein of interest. Its discovery opened up possibilities and stimulated the development of new and advanced microscopy techniques. Conversely, their introduction created a demand for new spectroscopic sensing properties and novel fluorescent labels, and labeling techniques quickly became indispensable. The developments that followed in the disciplines of optics, molecular biology, and chemistry went hand in hand, and have led to new innovative read-out techniques and fluorescent probes.

A large collection of fluorophores and labeling methods (Bunt and Wouters 2004) are available nowadays that range from chemical to genetic approaches. In the early days, the imaging of fluorescently labeled proteins in cells involved in vitro labeling of purified proteins with synthetic dyes and their subsequent cellular introduction, for example by microinjection. Genetic fusion to members of the green fluorescent protein family then took over as routine method. In the last few years, the use of synthetic fluorophores is regaining popularity by the implementation of in vivo-labeling methods that combine the selectivity of genetic targeting with the versatility of chemical conjugation.

Synthetic probes offer a wide choice of custom-designed spectral and sensing properties and their typical small footprint ensures minimal intrusion of the biological activity of the labeled proteins. Their re-introduction was further complemented by the developments in the synthesis of new fluorescent probes such as

G. Bunt
Molecular and Cellular Systems, Department of Neuro and Sensory Physiology, University Medicine Goettingen, Humboldtallee 23, 37073 Goettingen, Germany
e-mail: gbunt@gwdg.de

A. Diaspro (ed.), *Optical Fluorescence Microscopy*,
DOI 10.1007/978-3-642-15175-0_7,

semiconductor nanocrystals (quantum dots) and small organic dyes with novel fluorescence properties. The orthogonality of mild and biocompatible in vivo site-directed protein-labeling procedures for different dyes allows the design and implementation of advanced optical assays for multi-parameter read-out of complex cellular events.

In this chapter, recent developments in fluorescent probes and a selection of frequently used in vivo-labeling methods and applications will be presented.

7.2 New Fluorescent Labels

The increased focus on fluorescent microscopy as a tool in cell biology has put increasing demands on the fluorophores used for imaging (Fig. 7.1). These range from "better" fluorophores with generally improved spectral properties to more "functional" fluorophores that allow the direct interrogation of a specific cellular condition by a change in its properties.

The number and variation of genetically encoded fluorophores has been dramatically expanded in recent years by the ongoing search for new biological sources and extensive mutational strategies. However, some of their properties remain less than ideal. Fluorescent proteins generally suffer from rather broad and sometimes structured excitation and emission spectra, relatively low quantum yield, and/or low photostability. The generation of fluorescent protein-based biosensors, for example, in the form of FRET biosensors, though steadily giving rise to new functionalities, is not always straightforward and equally suffers from these limitations. The availability of synthetic dyes with specifically optimized fluorescent behavior and custom-tailored functionalities, together with the means for their selective attachment to proteins, will allow the construction of novel bioassays with a high level of sophistication. The following section will describe three examples of the unique properties of synthetic dyes that can be exploited.

7.2.1 Quantum Dots

The first example of improved labels is given by the inorganic "quantum dot" (Qdot) nanocrystals (Bruchez et al. 1998; Bruchez 2005; Chan and Nie 1998; Chan et al. 2002; Medintz et al. 2005) as superior fluorophore alternative. These nanometer-size semiconductor clusters generally consist of a CdSe or CdTe core and a ZnS shell, and are highly fluorescent. Their most important property is their exceptional photostability. Together with their high fluorescence yield, it allows them to generate tens of thousands times more photons per fluorophore as compared to fluorescent proteins and organic fluorophores. These properties give rise to remarkably high signal-to-noise levels that even allow the imaging of single Qdots with conventional fluorescence microscope-camera setups. Their fluorescence

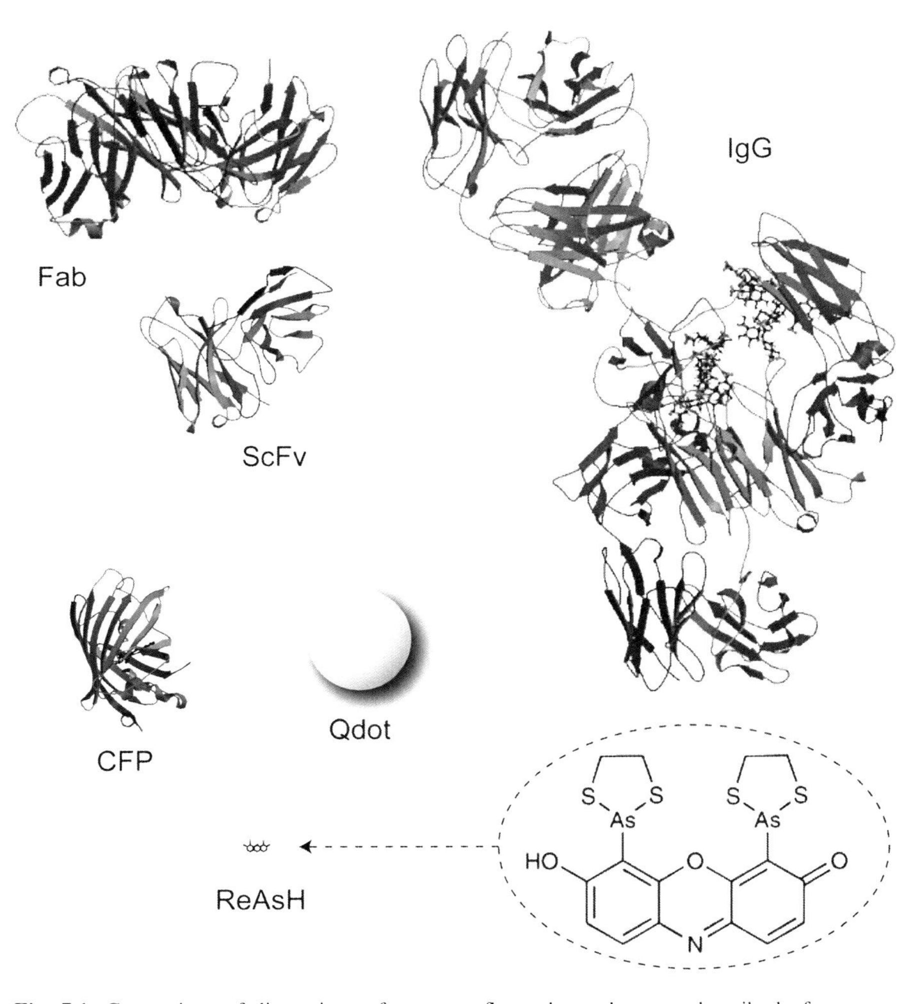

Fig. 7.1 Comparison of dimensions of common fluorophore classes and antibody fragments. Antigen-recognizing fragments Fab and scFv that are derived from the IgG molecule are shown in scale. These binding elements can be labeled with various fluorophores. The core of a Qdot of 5 nm in diameter is shown, without the biocompatible shell. CFP represents the green fluorescent protein family. Organic dyes, e.g., ReAsH, are the smallest labels available. The structures for the proteins shown were obtained from the Brookhaven Protein Database (www.pdb.org/ accession numbers: Fab: 12e8, scFv: 1ap2, IgG: 1igt, CFP: 1cv7)

emission is tunable as the quantum confinement effect that gives rise to their fluorescence properties is highly dependent on their size: their synthesis is tunable between 2 and 10 nm with high homogeneity (±3%), corresponding to an emission maximum range from 525 to 655 nm, respectively. Moreover, the fluorescence emission of the differently sized Qdots is sharp and symmetric around discrete wavelengths. Their excitation spectra, on the other hand, are broad extending from the violet range, where the absorbance is highest, to close to their emission in the visible spectrum. The large (>100 nm) Stokes shift between violet excitation and green to red emission further adds to the high signal-to-noise level. In particular, their single wavelength excitation coupled to their different and easily separable emission windows make Qdots ideal fluorophores for multiplexing, allowing the simultaneous observation of different labeled components.

The introduction of Qdots in cell biology required the development of a biocompatible coating that made them water-soluble, overcame the loss of fluorescence in an aqueous medium, and provided the functional groups for their covalent attachment to proteins. Apart from their use as alternative fluorescent label in immunofluorescence applications, their use with living cells has been successfully pioneered in the last few years (Wu et al. 2003; Derfus et al. 2004; Rosenthal et al. 2002; Dahan et al. 2003; Lidke et al. 2004; Vu et al. 2005). However, due to their relatively large size, which is further increased by the addition of the biocompatible coating, their use is mainly restricted to the outer cell membrane that is easily accessible and to their intracellular uptake into endocytotic compartments. The main limitation to their intracellular use is a convenient means to introduce them into the cytosol. Microinjection has been used with some success (Roberti et al. 2009), but this cannot be used as a routine method for large numbers of cells.

Because of their exceptional properties, the use of Qdots has revolutionized the technique of (single and multi-color) single particle tracking. However, the Qdots, like many other fluorophores, can spontaneously switch on and off their fluorescence emission. This blinking behavior can be a problem when tracking a single Qdot-labeled protein over time as it creates "blind spots." Their blinking behavior can be exploited to separate two Qdots that have approached each other to a distance below the diffraction-limited resolution. This has been used in an elegant study by (Lidke et al. 2005) that in fact preceded the publication of other super-resolution techniques called PALM and STORM, which solved the problem of overlapping positions by repeated and incremental imaging of low concentrations of fluorescent emitters. In spite of limitations to their intracellular use, Qdots are powerful fluorophores with valuable fluorescence properties that are increasingly taking over the classical fluorophores in various applications.

7.2.2 Environmentally Sensitive Dyes

One particularly interesting class of fluorophores possesses sensitivity toward the physicochemical properties of the surrounding solvent, referred to as solvatochromicity. Some of these fluorophores show sensitivity toward the polarity of the environment, whereas others, as for fluorescent rotors, depend on the viscosity of the medium. Solvent-sensitivity is difficult to recreate with GFPs as their chromophores are shielded from the environment by their β-barrel. The use of solvatochromic dyes in cell biology was pioneered by the Toutchkine lab (Toutchkine et al. 2003) who explored their potential for the detection of protein–protein interactions. Here, the challenge was to probe the exclusion of the aqueous medium between two interacting proteins using solvatochromic dyes. The rationale was that the interaction interface between two proteins shows higher viscosity and lower polarity as compared to the solvent-exposed protein surface. This way, proteins strategically labeled with dyes that change fluorescence upon a change in polarity and/or viscosity at the binding site for another protein will report on the binding of this partner protein with an increase or decrease in fluorescence. This principle was used to create a sensor for the interaction between the GTPase Cdc42 and N-WASP, which initiates actin polymerization by the Arp2/3 complex. Attachment of a solvatochromic merocyanin fluorophore to the WASP protein was shown to detect the binding of active, GTP-bound, Cdc42 by an increase in fluorescence of a factor of ±5. This basic principle was further developed to create a live cell imaging biosensor for the activation of endogenous Ccd42 (Nalbant et al. 2004). This sensor was based on the Cdc42-binding domain from N-WASP, the CRIB domain, which was tagged with GFP and labeled with a merocyanin fluorophore. The doubly labeled construct reported on the binding of active Cdc42 by a change in the ratio of merocyanin-to-GFP fluorescence, in which the optically inert GFP serves as a reference for concentration differences of the probe.

A merocyanin variant was also used to probe the conformational change in the S100A4 protein upon binding to calcium (Garrett et al. 2008). The S100A4 protein plays an important role in the modulation of actomyosin cytoskeletal dynamics, and the increased expression of this protein correlates with malignant tumors, which are characterized by invasion and metastasis. The merocyanin was introduced at a site that is shielded in the apo- and is solvent-exposed in the calcium-bound form of S100A4. The sensor for the calcium-binding status of S100A4 was constructed by the co-injection of FITC-labeled and merocyanin-labeled S100A4 proteins at a fixed stoichiometry of 1:2. Here, FITC serves as concentration reference for the solvatochromic effect of the merocyanin-labeled protein since it is not influenced by calcium or conformational changes. The calcium-sensitivity of S100A4 couples cell motility to Ca^{2+}-based signaling reactions. Stimulation of fibroblasts with lysophosphatidic acid, a strong inducer of cell motility, was shown to increase the fraction of calcium-bound S100A4 in areas of increased membrane protrusion early during incubation. These simple ratiometric assays thus allow the direct demonstration and localization of protein interactions or conformational changes in proteins by solvatochromic fluorescent probes.

The environmental sensitivity of this type of fluorophores can also be combined with site-directed labeling. One example is the development of the environmentally sensitive biarsenical dye derivative BarNile (Nakanishi et al. 2004), which was used to probe conformational changes of an expressed calmodulin protein tagged with an internal tetracysteine motif (see Section 7.3.2.1) in living cells. As with the S100A4 protein, conformational changes in calmodulin are related to changes in calcium concentration.

7.2.3 Photochromic Dyes

Another unique property of fluorophores that can be exploited for sophisticated sensing schemes is that of photochromicity, i.e., the change in spectral properties induced by illumination at appropriate wavelengths that are ideally different from the excitation wavelength. The change in spectral properties can be a shift in absorption/emission spectra or a switching on and off of fluorescence yield (photo-activation) with or without an appreciable change in color. This effectively creates a light-induced switching between two different forms. Ideally, the light-induced change is reversible, and switching can be performed many times before the system fatigues.

One of the applications of photo-switching is the suppression of autofluorescence in the same emission range as the fluorescently labeled compound of interest (Marriott et al. 2008). This was achieved in living cells by the use of nitrobenzo-spiropyran (nitroBIPS)-based probes, as well as reversible photo-switchable green fluorescent protein (Dronpa), in demanding tissues like explanted neurons, live *Xenopus* embryos, and zebrafish larvae. Switching of the nitroBIPS probe is very fast; the photo-isomerization of the non-fluorescent spiro state to the fluorescent merocyanin state is induced by illumination with 365 nm light (or 720 nm two-photon excitation), the back-conversion by illumination with 543 nm creates the 612 nm emission.

The basis of background suppression is the modulation of photochromic properties by the repetitive back and forth switching between nonfluorescent and fluorescent states coupled to detection in exact synchrony (Marriott et al. 2008). This produces a stimulus-coupled change that is distinct from the native sources. This optical lock-in detection method therefore isolates the signal of molecules that switch on and off in response to the optical switching signal, allowing the specific determination of the fluorescence of the probe, even if the photochromic dye contributes to less than 0.1% of total fluorescence. This detection scheme is extremely sensitive and generates essentially background-free images of astoundingly high contrast.

This lock-in detection principle based on photochromic dyes has also been exploited for sensitive FRET measurements. When the photochromic dye is used as acceptor for a donor fluorophore, then FRET can be detected with high sensitivity from the increase in donor fluorescence that correlates with the periodic switching off of acceptor absorption.

The basic principle of photochromic FRET was shown for a fusion construct of a lucifer yellow donor fluorophore with a diheteroarylethene acceptor (Giordano et al. 2002). The open form of the diheteroarylethene acceptor is colorless. UV-induced photoisomerization creates the closed colored and FRET-competent form. The reaction can be reversed by illumination with visible light. Consequently, photoswitching creates a FRET-ON (for the merocyanin state) and FRET-OFF (for the spiro state) condition. The switching of the acceptor was faithfully followed by the fluorescence variations of the lucifer yellow donor group in this dual fluorophore compound, proving the operation principle of this FRET sensing scheme. Complete switching is not required as less than 10% variation is sufficient for sensitive lock-in detection.

The sensitivity limit of the detection of photochromic FRET in cells was investigated using the GFP–nitroBIPS FRET pair in living cells (Mao et al. 2008). A fusion protein between GFP and alkylguaninetransferase allowed the site-directed labeling with benzylguanine–nitroBIPS in living cells (see Sect 7.3.2.2), bringing together the donor and photochromic acceptor fluorophore of the FRET pair in the same molecule. Titration of the amount of GFP–nitroBIPS with unlabeled donor allowed the variation of observed FRET efficiency in the cells. It should be noted, however, that this procedure dilutes high FRET complexes with non-FRET donors, but does not vary the FRET efficiency inside the complex. Nevertheless, the apparent FRET efficiency judged by lock-in detection of the GFP changes varied linearly with the fractional labeling of the GFP with the nitro-BIPS acceptor, and could be unambiguously estimated down to a few percent of donor–acceptor-tagged complexes, i.e., an apparent FRET efficiency of lower than 1%. This result makes photochromic FRET the most sensitive detection method for protein interactions and conformational changes available to date.

In recent years, super-resolution fluorescence microscopy techniques (STED, PALM, STORM, and F-PALM) (Hofmann et al. 2005; Betzig et al. 2006; Hess et al. 2006; Rust et al. 2006) have been developed that exploit the properties of photo-switchable fluorescent probes. The imaging process is based on the determination of the localization of individual probes with high precision by fitting a Gaussian profile. In these techniques, the fluorescence emission of individual fluorophores is modulated in time such that only an optically resolvable subset of fluorophores is activated and imaged. This implies cycles of imaging, where activation and deactivation of the probe result in images of individual fluorophores that do not overlap. This way, the background noise is minimized and the photon count of the fluorophore maximized resulting in a resolution accuracy of about 10–20 nm.

Both photo-switchable fluorescent proteins and organic dyes are available and have been used for these methods (Fernández-Suárez and Ting 2008). However, the organic switchable dyes are normally brighter as they exhibit higher extinction coefficients, thereby allowing higher numbers of photons to be collected, resulting in larger contrast ratios. Furthermore, the speed of switching, the reversibility, and extent of the photochromic effect in fluorescent proteins are modest when compared to their chemical counterparts, and the number of available colors is very limited. Synthetic photo-switchable dyes include rhodamines, diarylethenes, as well as switchable cyanine dyes. Of these, the rhodamines have a high potential for live cell imaging (Folling et al. 2007) as they are membrane-permeable and allow the imaging of intracellular proteins. Lately, a family of bright photo-switchable probes with distinct colors based on the pairing of cyanine dyes was created (Bates et al. 2007). It was found that Cy3 functions as an activator as it facilitates the switching of other cyanines as Cy5, Cy5.5, and Cy7. Alexa405 and Cy2 can also activate the switching of Cy5. The pairing of an activator dye with switching dyes of different colors allowed multicolor STORM imaging (Bates et al. 2007). Combined with the broad range of available colors, this pairing of organic dyes allows a high level of experimental freedom in high-resolution microscopy and exemplifies the value of custom-tailored properties that are given by synthetic probes.

7.3 Site-Specific Chemical Labeling in Living Cells

A number of biocompatible site-specific labeling chemistries have been developed to introduce the new dyes into cell biology. The utility of a new chemical-labeling method critically depends on a number of general requirements. These are first of all a high selectivity and sensitivity. Obviously, fluorescent labeling of proteins should not interfere with their specific function for in vivo labeling, and should therefore be carried out under mild biocompatible-labeling conditions. Moreover, the labeling approaches should preferentially allow the introduction of several probes with different characteristics, such as color, in an orthogonal manner so as to allow multiplexing. Under such optimal conditions, proteins can be fitted with a range of probes, in particular those that possess innovative spectral and sensing properties.

An important prerequisite to obtain a highly specific labeling of proteins is that the fluorescence of nonbound probes is eliminated. The easiest implementation of in vivo fluorophore conjugation reactions is therefore the labeling of extracellular epitopes, where the unbound dye is simply washed out. In the case of intracellular labeling, this can be achieved by using probes that can be washed out of the cell or by using fluorophores at sub-saturating conditions. Alternatively, labeling strategies can be used with fluorogenic probes, i.e., dyes that gain fluorescence only upon binding to the protein of interest.

The choice for chemical labeling over a genetic fluorescent protein-based approach is determined by the biological question or the particular microscopy technique used. For example, chemical labeling allows the specific labeling of the cell surface versus the total cellular pool, a condition often required in single-molecule studies. Furthermore, the variability and spectral quality of synthetic dyes allow the attachment of fluorophores with new properties desired for certain advanced microscopy techniques, as is the case for the use of photoswitchable dyes in super-resolution microscopy. These days, one has the possibility to choose from a diverse range of labeling methods, each with their own advantages and disadvantages. The following sections provide an overview of the most popular strategies published ordered by their extracellular or intracellular use.

7.3.1 Extracellular Chemical In Vivo-Labeling Techniques

7.3.1.1 Biotinylated Proteins as Chemical Handle for Labeling

The highest reported affinity for a pair of biological components is that for the binding of biotin and (strept)avidin ($K_d = 10^{-15}$ M). Due to this quality, this pair of proteins has gained popularity as versatile labeling platform. Indeed, a large number of fluorescent streptavidin conjugates are commercially available nowadays. Additionally, the little nonspecific binding of streptavidin and the advantage of its significantly smaller size in comparison to IgG antibodies contributed to their widespread use for protein labeling and recognition purposes.

A robust labeling method for cell surface proteins with excellent labeling specificity that is based on the biotinylation of target proteins and the subsequent binding of streptavidin was developed by the Ting laboratory (Chen et al. 2005; Howarth and Ting 2008). Here, a 15-amino acid acceptor peptide (AP), derived from a naturally biotinated protein of bacterial origin, is fused to the target protein. An amino-reactive biotin is subsequently ligated to the lysine chain of the AP by the *Escherichia coli* enzyme biotin ligase (BirA), upon which the biotinylated protein is bound by fluorescently labeled streptavidin in a second step.

It was found that biotin can be substituted by ketone derivates as the BirA ligase accepts ketones and ligates them to the AP with similar kinetics and high substrate specificity as biotin (Chen et al. 2005). This allows the direct fluorescent labeling of proteins by covalent binding as the ketone can be specifically conjugated to hydrazide- or hydroxylamine-functionalized fluorophores in a following step. The versatility of this labeling method was further increased by the development, using two generations of phage display selection from 15-mer peptide libraries, of a new 15-amino acid substrate that is biotinylated by the yeast biotin ligase (Chen et al. 2007). This yeast acceptor peptide (yAP) allows the specific labeling of AP and yAP fusion proteins co-expressed in the same cell, as the yAP is not recognized and biotinylated by BirA, and conversely, the AP is not biotinylated by yBL. The applicability of orthogonal labeling with this method was shown for differently colored Qdots.

Since both streptavidin and avidin are tetramers, their use in fluorescent labeling can potentially induce cross-linking – i.e., tetramerization – of the biotin conjugate, which can interfere with protein function and molecular interactions. To circumvent this problem, a monovalent streptavidin was generated using a chimeric streptavidin tetramer with only one single functional biotin-binding site (Howarth et al. 2006).

Furthermore, it should be noted that the use of (strept)avidin–biotin schemes for in vivo labeling is limited to extracellular approaches as the intracellular presence of streptavidin can lead to toxic biotin depletion, which interferes with the function of biotin in the cytoplasm and mitochondria.

7.3.1.2 Labeling of Carrier Protein Moieties – ACP and PCP

Another enzyme-based labeling method for extracellular labeling of proteins is the post-translational modification of carrier proteins. This labeling method is based on the transfer of 4'-phosphopantetheine from CoA by phosphopantetheine transferase (PPtase) to the serine side chain of an 80-amino acid sequence derived from either peptide carrier protein (PCP) or acyl carrier protein (ACP) to form a specific covalent attachment (George et al. 2004; Yin et al. 2004). As the sulfhydryl group in the phosphopantetheine is not required for its recognition or coupling by the enzyme, it can be derivatized with a fluorescent dye to effect covalent fluorescent labeling of the carrier peptide. Fluorescent labeling of the free sulfhydryl group in CoA can be achieved by a simple reaction with maleimide-carrying fluorophores, allowing the facile construction of a wide range of fluorescent substrates by the user. A variety of labeled substrates is also commercially available.

Dual color labeling can be achieved by the use of PPtases of different origins as the Sfp PPtase of *Bacillus subtilis* and the AcpS of *E. coli* have different substrate specificities; whereas Sfp PPtase modifies both the PCP and ACP domains, the AcpS only modifies the ACP domain. Consequently, PCP or ACP tags can be differentially labeled with different fluorophore–CoA conjugates by the incubation of the cells with AcpS to label the ACP-tagged protein, followed by Sfp-catalyzed labeling of the PCP-tagged protein (Vivero-Pol et al. 2005).

Recently, short carrier tags (S6 and A1) of 12 amino acids each were selected from a phage-displayed peptide library as efficient substrates, while maintaining their orthogonality for the two PPtases (Zhou et al. 2007). The PPtase carrier protein method is restricted to labeling of cell surface proteins, because the probes are not membrane-permeable. On the other hand, this way selective labeling and imaging of receptors on the plasma membrane can be achieved without the interference of a high background resulting from the intracellular pool. This is essential in single-molecule studies to efficiently and properly detect the individual fluorescent signals arising from the plasma membrane, as was shown for the investigation of the motility of individual neurokinin-1 receptors (Prummer et al. 2006). Moreover, this is exemplified by the selective FRET measurements on interactions between cell surface-located neurokinin-1 receptors, which were simultaneously labeled with donors and acceptors using ACP tags (Meyer et al. 2006). The other advantage is that the optimal donor-to-acceptor ratio for the FRET detection method of choice can be controlled precisely in the creation of stochastic FRET pairs, and that this ratio is maintained homogeneously throughout the cell population. The independence on the absolute and relative expression level of donor/acceptor pairs allows for very precise FRET measurements. The relative simple chemistry of the probes so that one can easily attach fluorophores with the optimal desired spectral properties, and for which several probe variants can even be obtained commercially these days, makes this method straightforward applicable.

7.3.1.3 Sortagging: Sortase-Mediated Transpeptidation

A third enzymatic extracellular-labeling method is based on the proteolytic cleavage and subsequent amide linkage of a C-terminal LPXTG peptide by the *Staphylococcus aureus* sortase (Popp et al. 2007). Bacterial sortases mediate the covalent attachment of proteins to the bacterial cell wall. Each sortase from different bacterial species recognizes its own specific sequence. In case of the *S. aureus* sortase, the motif is proteolytically cleaved between threonine and glycine, leaving a C-terminal-LPXT that is instantly amide-linked by the enzyme to a pentaglycine nucleophile, which is normally present in the cell wall. To achieve in vivo labeling of target proteins expressed on the cell surface, the peptide tag LPETGG was genetically fused to the C-terminus of osteoclast differentiation factor (Tanaka et al. 2008). The addition of sortase enzyme and a fluorescently labeled N-terminal triglycine probe was shown to result in the site-specific labeling of the protein. An advantage of this method is that the fluorescent probes can be easily made by standard solid-phase peptide synthesis, i.e., can be directly ordered from a variety of peptide synthesis services. However, the method suffers from the disadvantage that it is restricted to the labeling of the C-terminus of proteins as the LPXTG motif needs a flexible and unstructured region close to the C-terminus to allow the transpeptidation reaction.

7.3.2 *Intracellular Chemical In Vivo-Labeling Techniques*

7.3.2.1 Biarsenical-EDT2-Labeling

The biarsenical–tetracysteine system is perhaps the best known site-specific in vivo-labeling technique (Adams et al. 2002; Griffin et al. 1998, 2000) (Fig. 7.2). In this method, a tetracysteine motif CCXXCC is genetically introduced in the expressed protein. This motif is specifically recognized by membrane-permeable fluorophores that are substituted with two arsenite groups with a precisely defined spacing. These biarsenical compounds were developed to be non-fluorescent when present free in solution. Upon binding to the tetracysteine motif, the compounds gain fluorescence as a result of structural stabilization exerted upon binding. Several biarsenical fluorophore variants have been generated. Derivatization by the addition of two arsenite groups to the aromatic rings of fluoresceine and resorufin resulted in the compounds called FlAsh (green emission) and ReAsh (red emission), respectively (Adams et al. 2002). In addition to these, the blue variant ChoXAsH (Adams et al. 2002) and a biarsenical derivative of Nile Red (Nakanishi et al. 2001) have been synthesized.

The tetracysteine motif is small and can be introduced *de novo* or by modification of endogenous cysteine-rich sequences in proteins. Optimization of the tetracysteine context (Martin et al. 2005) by directed evolution has led to improved binding affinity and quantum yield, improving signal levels, and reduced cytotoxicity.

Nonetheless, the presence of background staining and the need for the reducing environment of the cytoplasm and nucleus for proper labeling set limits to the applicability of this method. In addition, the poor photostability and the sensitivity of fluorescein derivatives for pH changes in the physiological range represent inherent limitations. The generation of fluoro-substituted versions, F2FlAsH and F4FlAsH, exhibited significant improvements in photostability and pH-dependence over the original FlAsH (Spagnuolo et al. 2006). Moreover, F2FlAsH possesses

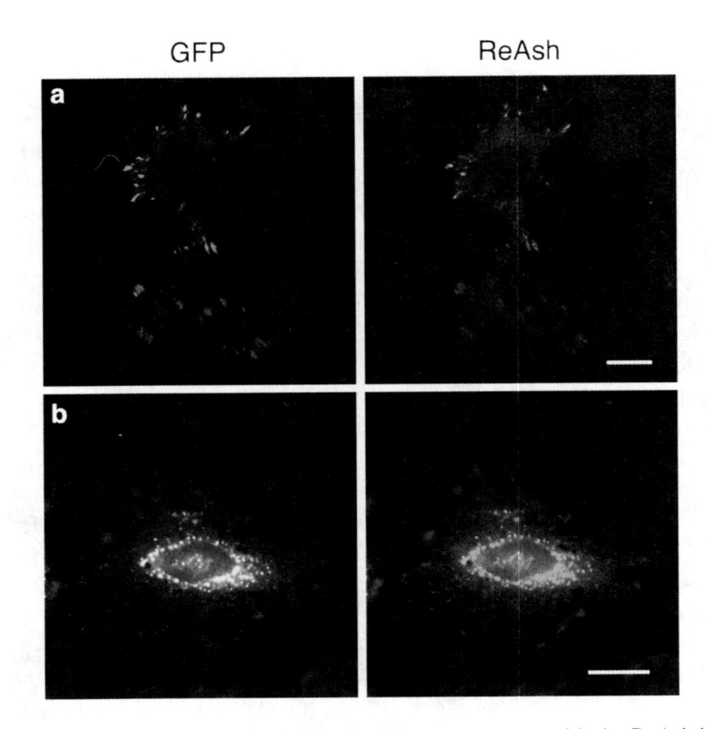

Fig. 7.2 Tetracysteine labeling of focal adhesion kinase constructs with the ReAsh biarsenical-EDT2 dye. GFP was fused to the N-terminus of the kinase as a reference for the recognition by ReAsh. ReAsh labeling of (**a**) the tetracysteine motif FLNCCPGCCMEP (×2) at its C-terminus and (**b**) the motif EAAAREACCRECCARA inserted into the kinase is shown. The bright fluorescent structures at the cell periphery are focal adhesions where the kinase resides. Note the complete overlap of the signals of both fluorophores indicating correct and complete labeling. Bars: 10 μm

improved signal levels in the form of higher absorbance, quantum yield, and larger Stokes shift. The emission of F4FlAsH lies intermediate to that of FlAsH and ReAsH, expanding the color palette of biarsenical compounds. In addition, the two new probes form an excellent FRET pair with a substantially larger critical distance for FRET (R_0) than obtained so far with the biarsenical dyes. This is important for the generation of controlled FRET pairs as described above for ACP labeling.

Generally speaking, multicolor labeling requires orthogonal-labeling sequences to allow the simultaneous labeling of proteins with different spectral variants. If only one labeling sequence is available, temporally separated protein pools can be labeled by sequential conjugation steps. Sequential labeling with FlAsh and ReAsH was used in a pulse chase determination of the age and fate of tetracysteine-labeled proteins to study the turnover rate of connexin-43 in gap junctions (Gaietta et al. 2002) and the dendritic synthesis and trafficking of AMPA receptors (Ju et al. 2004). The absence of orthogonal-labeling peptides inhibited the wider adoption of this method for other biological questions that require multicolor labeling.

Recently, a long awaited orthogonal biarsenical fluorophore was introduced. The new red biarsenical–tetracysteine fluorophore (AsCy3) allows the selective labeling of multiple proteins with different fluorophores using the same biarsenical-labeling chemistry (Cao et al. 2007). Here, the conserved interatomic distance between the two arsenic moieties of 6 Å was increased to 14.5 Å. Coupled with the identification of a complementary high-affinity-binding sequence CCKAEAACC (Cy3Tag), with wider spacing of the two cysteine pairs, this fluorophore permits its simultaneous application with the original biarsenic compounds in cells. Moreover, AsCy3 acts as an efficient FRET acceptor for FlAsH (R_0 of 6.5 nm), as its absorption overlaps with the emission of FlAsH. This is important as it extends the use of FRET from the investigation of homotypic interactions to interactions between two different proteins. AsCy3 possesses superior photostability and exhibits a minimal environmental sensitivity compared to the existing biarsenical probes FlAsH and ReAsH, which are important attributes for single-molecule detection. These newly obtained properties are expected to increase the use of biarsenical-EDT2 labeling for the functional studies of proteins in living cells, making full use of the gained orthogonality.

7.3.2.2 O6-Alkylguanine-DNA Alkyltransferase Labeling (AGT/SNAP Tag)

This covalent in vivo-labeling method is based on fusion proteins with the human DNA repair protein hAGT; O6-alkylguanine-DNA alkyltransferase (Keppler et al. 2003, 2004a, b). This enzyme transfers the alkyl group from its substrate O6-alkylguanine-DNA to one of its cysteine residues. As this covalent linkage still occurs with fluorescently labeled alkyl derivatives, it leads to self-labeling of the enzyme tag that is fused to the protein of interest. Fluorescent O6-benzylguanine derivatives are available across the visible spectrum (Keppler et al. 2006), in cell-permeable and -impermeable forms. Furthermore, affinity and bifunctional probes were designed (Juillerat et al. 2005). The relatively uncomplicated synthesis method makes this a popular method for microscopy techniques that rely on the use of fluorophores with special and customized properties. Moreover, labeling is short, labeling conditions are very mild, and the reaction can be performed in any cellular compartment. An early disadvantage of the method, however, was the background of endogenous alkyltransferase that required the use of AGT knockout cells. The development of an inhibitor of endogenous AGT and the genetic engineering of fast and highly active hAGT mutants that are not inhibited by this inhibitor solved this problem and resulted in specific labeling and the general application of this method (Juillerat et al. 2005; Gronemeyer et al. 2006). The tag based on the mutant with the highest activity (^{N}AGT) is called SNAP-tag. To allow multicolor labeling, an additional AGT mutant was engineered for labeling with derivates that are unreactive toward the SNAP-tag. Together with the SNAP-tag, this new CLIP-tag allows dual-color labeling of proteins in a single cell (Gautier et al. 2008).

To conclude, the current specificity of this labeling method and the broad range of available substrates are attractive features of this approach, but the size of the AGT tag (21 kDa) remains an intrinsic disadvantage.

7.3.2.3 HaloTag: Enzyme–Ligand Interaction Self-Labeling

A second enzyme-mediated self-labeling reaction is based on the engineered activity-dead haloalkane dehalogenase enzyme (HaloTag) (Los et al. 2008; Los and Wood 2007; Zhang et al. 2006). This GFP-size (33 kDa) monomeric tag covalently couples a haloalkane substrate to an aspartate residue near its active site by trapping of this intermediate in its enzymatic processing. The covalent bond formation between the protein tag and the chloroalkane linker is highly specific, occurs rapidly (within 15 min), and is essentially irreversible. The probe chemistry is relatively simple and many variants of the probes are commercially available. Do-it-yourself linkers for labeling to fluorophores of choice are also available to allow the construction of custom-tailored ligands. The ligands can be obtained in cell-permeable and -impermeable forms. However, the relatively large size of the HaloTag indicates that its advantage will lie in those applications where the spectral properties cannot be met by the visible fluorescent proteins, where selective labeling at the membrane is required, or for pulse-labeling experiments.

7.4 In Vivo Labeling of Endogenous Proteins

The ultimate goal of fluorescent observation of cellular events is the live labeling of endogenous proteins. Traditionally, labeling of endogenous proteins has been achieved by immunohistochemical approaches on fixed cells. Antibodies bind antigens with high affinity and high specificity and are unquestionably an ideal tool for fluorescent labeling of endogenous proteins. However, their use in live cells is restricted as they are sensitive to the reducing cellular environment, leading to inefficient folding and assembly of their variable heavy and light chain. This prohibits the expression of unmodified antibodies in the cellular interior. Therefore, their intracellular applications mostly relied on their technically demanding introduction in cells by microinjection.

Recently, smaller recombinant antibody fragments, such as the classic monovalent antibody fragments (Fab, scFv) (Fig. 7.1), and engineered variants thereof are emerging (Holliger and Hudson 2005). The epitope-recognition domains of IgGs have been dissected into either monovalent (Fab, scFv, and single variable V_H and V_L domains) or bivalent fragments by genetic engineering. One recombinant antibody fragment that is rapidly gaining popularity is the single-chain antibody (scFv) as it retains the specific, monovalent, and antigen-binding affinity of the parent IgG while showing improved penetration and kinetics. An scFv represents the minimal epitope recognition part of an IgG (±28 kDa) in which the V_H and V_L domains are linked with a flexible polypeptide linker such as to prevent their dissociation.

One of the first reported use of single-chain antibodies for in vivo protein labeling was by Farinas and Verkman (1999), who targeted a single-chain antibody to specified sites in the cell to trap cell-permeable hapten–fluorophore conjugates. A phOx–fluorescein conjugate was applied to measure the pH value in the Golgi lumen in live cells. In this case, a ±5-fold fluorescence increase of the bound hapten–fluorophore conjugate over the unbound conjugate was observed. This fluorogenic effect improved the specificity and sensitivity of the approach as the increased fluorescence allowed imaging with little contribution from the free probe.

The fluorogen concept was extended to a broad new class of protein–dye reporters by Szent-Gyorgyi et al., that are also based on single-chain antibodies, and are called fluorogen-activating proteins (FAPs) (Szent-Gyorgyi et al. 2008). Their basic principle aimed at the optimization of the fluorogenic effect of the binding of fluorophores by antibody fragments. The two structurally unrelated dyes thiazole orange (TO) and malachite green (MG) were selected for this purpose as they are known fluorogens. Strong fluorescence activation was observed when TO intercalated with DNA (550-fold) and when MG bound to a specific RNA aptamer (2360-fold). A library of human scFvs was screened in yeast for enhancing the fluorescence of TO and MG. Eight unique FAPs that elicit dramatic fluorescence enhancement upon binding of TO and MG derivatives were isolated. The principle behind the enhanced fluorescence is that of the "molecular rotors" in which the rapid rotation around a single bond in the fluorophore leads to a loss of fluorescence. This rotation within the chromophore is constrained upon scFv binding, and the fluorescence is recovered.

In addition, the scFv-mediated fluorophore targeting by single-chain antibody represents a valuable variant of the site-specific chemical-labeling approaches discussed so far. However, in the above-described form it is unfortunately not directly applicable to endogenous proteins as it relies on the introduction of fluorophore-recognizing antibody fragments.

The general problem remains that scFvs contain internal disulfide linkages so that functional expression is relatively poor in reducing environments. This restricts their use to mainly the cell surface and secretory pathway. Even though it has been shown that functional scFvs can be expressed in a disulfide-free format, i.e., lacking cysteins (Proba et al. 1998), the real solution to this problem came with the switch to monovalent antibody fragments originating from the camel.

Rothbauer et al. were the first to target and trace endogenous antigens in live cells using antibody fragments from *Camelidae* sp. (Rothbauer et al. 2006, 2008). Camelid-Ig, and also shark Ig-NARs, are unusual in that they comprise a homodimeric pair of only heavy chains and lack light chains (Holliger and Hudson 2005). In camelids, the high-affinity single V-like domain is called VhH. It represents the smallest intact antigen-binding fragment available. These small VhH fragments are also referred to as nanobody and retain the targeting specificity of whole mAbs and can be produced as stable in vitro targeting reagents. VhHs possess long surface loops that even enable the recognition of epitopes in cavities of targets, such as enzyme active sites. Rothbauer et al. generated fluorescent nanobodies, called chromobodies, in which heavy-chain antibodies from *Camelidae* sp. were fused with fluorescent proteins and expressed in living cells (Rothbauer et al. 2008). Chromobodies recognizing endogenous cytoplasmic and nuclear antigens and fused to mRFP were used to follow antigens in different subcellular compartments throughout S phase and mitosis, a compelling demonstration of the biocompatibility of this approach. Chromobodies will enable new functionalities in the imaging field, as they can potentially target any antigenic structure in living cells.

7.5 Conclusion

The wealth of new spectral and sensing properties offered by synthetic fluorophores – as exemplified in this chapter by the near-ideal fluorescence properties of quantum dots, the versatility of environmental sensitivity in the generation of powerful imaging biosensors, and the exquisite contrast and FRET sensitivity offered by the light-controlled modulation of fluorescence spectral properties in organic dyes – is made available to cell biological questions by the concurrent development in site-specific protein labeling. The introduction of mild bioconjugation chemistries with varying degrees of orthogonality for the labeling with multiple fluorophores offers the cell biologist a versatile toolset that allow a high level of sophistication in the investigation of the working of living cells. The use of genetically engineered antibody-derived recognition modules extend fluorescent labeling to endogenous proteins and are expected to have a huge impact in the imaging field.

Important to note is that no single labeling approach is ideally suited for all aspects encountered in different experimental settings, and that therefore the choice for a particular approach should be carefully evaluated on a per-case basis. The decisive arguments in favor of a certain method can amongst others be given by the availability of fluorophore substrates with the right properties, or if not available commercially, the ease of synthesis of this substrate in the laboratory, the permeability of the membrane for the labeling substrate if intracellular targets are to be labeled, and the size of the recognition tag on the protein of interest.

In summary, the recent combined developments in the generation of synthetic fluorophores, detection strategies for their specific functionalities, and the means for their site-specific labeling of proteins, have generated a unique window of opportunity for modern cell biology to visualize molecular events in living cells with tools custom-tailored for the task.

Acknowledgments I thank Ekatarina Papucheva for the images on the FAK-ReAsh constructs, and Fred S. Wouters for critical reading of the manuscript.

References

Adams SR, Campbell RE, Gross LA, Martin BR, Walkup GK, Yao Y, Llopis J, Tsien RY (2002) New biarsenical ligands and tetracysteine motifs for protein labeling in vitro and in vivo: synthesis and biological applications. J Am Chem Soc 124:6063–6076

Bates M, Huang B, Dempsey GT, Zhuang X (2007) Multicolor super-resolution imaging with photo-switchable fluorescent probes. Science 317:1749–1753

Betzig E, Patterson GH, Sougrat R, Lindwasser OW, Olenych S, Bonifacino JS, Davidson MW, Lippincott-Schwartz J, Hess HF (2006) Imaging intracellular fluorescent proteins at nanometer resolution. Science 313:1642–1645

Bruchez MP (2005) Turning all the lights on: quantum dots in cellular assays. Curr Opin Chem Biol 9:533–537

Bruchez M Jr, Moronne M, Gin P, Weiss S, Alivisatos AP (1998) Semiconductor nanocrystals as fluorescent biological labels. Science 281:2013–2016

Bunt G, Wouters FS (2004) Visualization of molecular activities inside living cells with fluorescent labels. Int Rev Cytol 237:205–277

Cao H, Xiong Y, Wang T, Chen B, Squier TC, Mayer MU (2007) A red Cy3-based biarsenical fluorescent probe targeted to a complementary binding peptide. J Am Chem Soc 129: 8672–8673

Chan WC, Nie S (1998) Quantum dot bioconjugates for ultrasensitive nonisotopic detection. Science 281:2016–2018

Chan WC, Maxwell DJ, Gao X, Bailey RE, Han M, Nie S (2002) Luminescent quantum dots for multiplexed biological detection and imaging. Curr Opin Biotechnol 13:40–46

Chen I, Howarth M, Lin W, Ting AY (2005) Site-specific labeling of cell surface proteins with biophysical probes using biotin ligase. Nat Methods 2:99–104

Chen I, Choi YA, Ting AY (2007) Phage display evolution of a peptide substrate for yeast biotin ligase and application to two-color quantum dot labeling of cell surface proteins. J Am Chem Soc 129:6619–6625

Dahan M, Levi S, Luccardini C, Rostaing P, Riveau B, Triller A (2003) Diffusion dynamics of glycine receptors revealed by single-quantum dot tracking. Science 302:442–445

Derfus AM, Chan WCW, Bhatia SN (2004) Intracellular delivery of quantum dots for live cell labeling and organelle tracking. Adv Mater 16:961–966

Farinas J, Verkman AS (1999) Receptor-mediated targeting of fluorescent probes in living cells. J Biol Chem 274:7603–7606

Fernández-Suárez M, Ting AY (2008) Fluorescent probes for super-resolution imaging in living cells. Nat Rev Mol Cell Biol 9:929–943

Folling J, Belov V, Kunetsky R, Medda R, Schonle A, Egner A, Eggeling C, Bossi M, Hell SW (2007) Photochromic rhodamines provide nanoscopy with optical sectioning. Angew Chem Int Ed Engl 46:6266–6270

Gaietta G, Deerinck TJ, Adams SR, Bouwer J, Tour O, Laird DW, Sosinsky GE, Tsien RY, Ellisman MH (2002) Multicolor and electron microscopic imaging of connexin trafficking. Science 296:503–507

Garrett SC, Hodgson L, Rybin A, Toutchkine A, Hahn KM, Lawrence DS, Bresnick AR (2008) A biosensor of S100A4 metastasis factor activation: inhibitor screening and cellular activation dynamics. Biochemistry 47:986–996

Gautier A, Juillerat A, Heinis C, Correa IR Jr, Kindermann M, Beaufils F, Johnsson K (2008) An engineered protein tag for multiprotein labeling in living cells. Chem Biol 15:128–136

George N, Pick H, Vogel H, Johnsson N, Johnsson K (2004) Specific labeling of cell surface proteins with chemically diverse compounds. J Am Chem Soc 126:8896–8897

Giordano L, Jovin TM, Irie M, Jares-Erijman EA (2002) Diheteroarylethenes as thermally stable photoswitchable acceptors in photochromic fluorescence resonance energy transfer (pcFRET). J Am Chem Soc 124:7481–7489

Griffin BA, Adams SR, Tsien RY (1998) Specific covalent labeling of recombinant protein molecules inside live cells. Science 281:269–272

Griffin BA, Adams SR, Jones J, Tsien RY (2000) Fluorescent labeling of recombinant proteins in living cells with FlAsH. Methods Enzymol 327:565–578

Gronemeyer T, Chidley C, Juillerat A, Heinis C, Johnsson K (2006) Directed evolution of O6-alkylguanine-DNA alkyltransferase for applications in protein labeling. Protein Eng Des Sel 19:309–316

Hess ST, Girirajan TP, Mason MD (2006) Ultra-high resolution imaging by fluorescence photoactivation localization microscopy. Biophys J 91:4258–4272

Hofmann M, Eggeling C, Jakobs S, Hell SW (2005) Breaking the diffraction barrier in fluorescence microscopy at low light intensities by using reversibly photoswitchable proteins. Proc Natl Acad Sci U S A 102:17565–17569

Holliger P, Hudson PJ (2005) Engineered antibody fragments and the rise of single domains. Nat Biotechnol 23:1126–1136

Howarth M, Ting AY (2008) Imaging proteins in live mammalian cells with biotin ligase and monovalent streptavidin. Nat Protoc 3:534–545

Howarth M, Chinnapen DJ, Gerrow K, Dorrestein PC, Grandy MR, Kelleher NL, El-Husseini A, Ting AY (2006) A monovalent streptavidin with a single femtomolar biotin binding site. Nat Methods 3:267–273

Ju W, Morishita W, Tsui J, Gaietta G, Deerinck TJ, Adams SR, Garner CC, Tsien RY, Ellisman MH, Malenka RC (2004) Activity-dependent regulation of dendritic synthesis and trafficking of AMPA receptors. Nat Neurosci 7:244–253

Juillerat A, Heinis C, Sielaff I, Barnikow J, Jaccard H, Kunz B, Terskikh A, Johnsson K (2005) Engineering substrate specificity of O6-alkylguanine-DNA alkyltransferase for specific protein labeling in living cells. Chembiochem 6:1263–1269

Keppler A, Gendreizig S, Gronemeyer T, Pick H, Vogel H, Johnsson K (2003) A general method for the covalent labeling of fusion proteins with small molecules in vivo. Nat Biotechnol 21:86–89

Keppler A, Kindermann M, Gendreizig S, Pick H, Vogel H, Johnsson K (2004a) Labeling of fusion proteins of O6-alkylguanine-DNA alkyltransferase with small molecules in vivo and in vitro. Methods 32:437–444

Keppler A, Pick H, Arrivoli C, Vogel H, Johnsson K (2004b) Labeling of fusion proteins with synthetic fluorophores in live cells. Proc Natl Acad Sci U S A 101:9955–9959

Keppler A, Arrivoli C, Sironi L, Ellenberg J (2006) Fluorophores for live cell imaging of AGT fusion proteins across the visible spectrum. Biotechniques 41:167–170, 172, 174–175

Lidke DS, Nagy P, Heintzmann R, Arndt-Jovin DJ, Post JN, Grecco HE, Jares-Erijman EA, Jovin TM (2004) Quantum dot ligands provide new insights into erbB/HER receptor-mediated signal transduction. Nat Biotechnol 22:198–203

Lidke KA, Rieger B, Jovin TM, Heintzmann R (2005) Superresolution by localization of quantum dots using blinking statistics. Opt Express 13:7052–7062

Los GV, Wood K (2007) The HaloTag: a novel technology for cell imaging and protein analysis. Methods Mol Biol 356:195–208

Los GV, Encell LP, McDougall MG, Hartzell DD, Karassina N, Zimprich C, Wood MG, Learish R, Ohana RF, Urh M et al (2008) HaloTag: a novel protein labeling technology for cell imaging and protein analysis. ACS Chem Biol 3:373–382

Mao S, Benninger RK, Yan Y, Petchprayoon C, Jackson D, Easley CJ, Piston DW, Marriott G (2008) Optical lock-in detection of FRET using synthetic and genetically encoded optical switches. Biophys J 94:4515–4524

Marriott G, Mao S, Sakata T, Ran J, Jackson DK, Petchprayoon C, Gomez TJ, Warp E, Tulyathan O, Aaron HL et al (2008) Optical lock-in detection imaging microscopy for contrast-enhanced imaging in living cells. Proc Natl Acad Sci U S A 105:17789–17794

Martin BR, Giepmans BN, Adams SR, Tsien RY (2005) Mammalian cell-based optimization of the biarsenical-binding tetracysteine motif for improved fluorescence and affinity. Nat Biotechnol 23:1308–1314

Medintz IL, Uyeda HT, Goldman ER, Mattousi H (2005) Quantum dot bioconjugates for imaging, labelling and sensing. Nat Materials 4:435–446

Meyer BH, Segura JM, Martinez KL, Hovius R, George N, Johnsson K, Vogel H (2006) FRET imaging reveals that functional neurokinin-1 receptors are monomeric and reside in membrane microdomains of live cells. Proc Natl Acad Sci U S A 103:2138–2143

Nakanishi J, Nakajima T, Sato M, Ozawa T, Tohda K, Umezawa Y (2001) Imaging of conformational changes of proteins with a new environment-sensitive fluorescent probe designed for site-specific labeling of recombinant proteins in live cells. Anal Chem 73:2920–2928

Nakanishi J, Maeda M, Umezawa Y (2004) A new protein conformation indicator based on biarsenical fluorescein with an extended benzoic acid moiety. Anal Sci 20:273–278

Nalbant P, Hodgson L, Kraynov V, Toutchkine A, Hahn KM (2004) Activation of endogenous Cdc42 visualized in living cells. Science 305:1615–1619

Popp MW, Antos JM, Grotenbreg GM, Spooner E, Ploegh HL (2007) Sortagging: a versatile method for protein labeling. Nat Chem Biol 3:707–708

Proba K, Worn A, Honegger A, Pluckthun A (1998) Antibody scFv fragments without disulfide bonds made by molecular evolution. J Mol Biol 275:245–253

Prummer M, Meyer BH, Franzini R, Segura JM, George N, Johnsson K, Vogel H (2006) Post-translational covalent labeling reveals heterogeneous mobility of individual G protein-coupled receptors in living cells. Chembiochem 7:908–911

Roberti MJ, Morgan M, Menéndez G, Pietrasanta LI, Jovin TM, Jares-Erijman EA (2009) Quantum dots as ultrasensitive nanoactuators and sensors of amyloid aggregation in live cells. J Am Chem Soc 131:8102–8107

Rosenthal SJ, Tomlinson I, Adkins EM, Schroeter S, Adams S, Swafford L, McBride J, Wang Y, DeFelice LJ, Blakely RD (2002) Targeting cell surface receptors with ligand-conjugated nanocrystals. J Am Chem Soc 124:4586–4594

Rothbauer U, Zolghadr K, Tillib S, Nowak D, Schermelleh L, Gahl A, Backmann N, Conrath K, Muyldermans S, Cardoso MC et al (2006) Targeting and tracing antigens in live cells with fluorescent nanobodies. Nat Methods 3:887–889

Rothbauer U, Zolghadr K, Muyldermans S, Schepers A, Cardoso MC, Leonhardt H (2008) A versatile nanotrap for biochemical and functional studies with fluorescent fusion proteins. Mol Cell Proteomics 7:282–289

Rust MJ, Bates M, Zhuang X (2006) Sub-diffraction-limit imaging by stochastic optical reconstruction microscopy (STORM). Nat Methods 3:793–795

Spagnuolo CC, Vermeij RJ, Jares-Erijman EA (2006) Improved photostable FRET-competent biarsenical-tetracysteine probes based on fluorinated fluoresceins. J Am Chem Soc 128: 12040–12041

Szent-Gyorgyi C, Schmidt BF, Creeger Y, Fisher GW, Zakel KL, Adler S, Fitzpatrick JA, Woolford CA, Yan Q, Vasilev KV et al (2008) Fluorogen-activating single-chain antibodies for imaging cell surface proteins. Nat Biotechnol 26:235–240

Tanaka T, Yamamoto T, Tsukiji S, Nagamune T (2008) Site-specific protein modification on living cells catalyzed by sortase. Chembiochem 9:802–807

Toutchkine A, Kraynov V, Hahn K (2003) Solvent-sensitive dyes to report protein conformational changes in living cells. J Am Chem Soc 125:4132–4145

Tsien RY (1998) The green fluorescent protein. Annu Rev Biochem 67:509–544

Vivero-Pol L, George N, Krumm H, Johnsson K, Johnsson N (2005) Multicolor imaging of cell surface proteins. J Am Chem Soc 127:12770–12771

Vu TQ, Maddipatti R, Blute TA, Nehilla BJ, Nusblat L, Desai TA (2005) Peptide-conjugated quantum dots activate neuronal receptors and initiate downstream signaling of neurite growth. Nano Lett 5:603–607

Wu X, Liu H, Liu J, Haley KN, Treadway JA, Larson JP, Ge N, Peale F, Bruchez MP (2003) Immunofluorescent labeling of cancer marker Her2 and other cellular targets with semiconductor quantum dots. Nat Biotechnol 21:41–46

Yin J, Liu F, Li X, Walsh CT (2004) Labeling proteins with small molecules by site-specific posttranslational modification. J Am Chem Soc 126:7754–7755

Zhang Y, So MK, Loening AM, Yao H, Gambhir SS, Rao J (2006) HaloTag protein-mediated site-specific conjugation of bioluminescent proteins to quantum dots. Angew Chem Int Ed Engl 45:4936–4940

Zhou Z, Cironi P, Lin AJ, Xu Y, Hrvatin S, Golan DE, Silver PA, Walsh CT, Yin J (2007) Genetically encoded short peptide tags for orthogonal protein labeling by Sfp and AcpS phosphopantetheinyl transferases. ACS Chem Biol 2:337–346

Chapter 8
Imaging Molecular Physiology in Cells Using FRET-Based Fluorescent Nanosensors

Fred S. Wouters

8.1 Analytical Fluorescence Microscopy

The utility of fluorescence microscopy in cell biology (Netterwald 2008) stems from its noninvasiveness, coupled to the high sensitivity and multiparameter nature of the detection. Sensitive detectors and cameras are capable of detecting and counting single fluorescence photons, allowing low expression levels of fluorescently marked components in the living cell.

Fluorescence detection is thus ideally suited for the investigation of living cells. Localization and kinetic information of labeled proteins can be quantitatively obtained by a number of techniques that use photobleaching, or photoactivation or photoconversion of photochromic fluorophores (Lippincott-Schwartz et al. 2003), i.e., that change their emission properties upon illumination at a different wavelength, or directly investigate motility by the analysis of the transient occupation of single emitters in the small volume of the focal plane of a high numerical-aperture objective. The latter techniques that can be grouped under the name of fluorescence (Böhmer and Enderlein 2003) or image correlation spectroscopy (Kolin and Wiseman 2007) have seen a tremendous development in the last recent years. Some of these techniques are explained in more detail in a different chapter of this book.

The many parameters of fluorescence provide a wealth of information on the fluorescent species under investigation and permit the differentiation between – and identification of – discrete fluorophores. Of the many parameters, the most often used in biological imaging is the emission spectrum, i.e., the range of energy emitted in the form of fluorescent photons. This allows the selection of different

F.S. Wouters (✉)
Laboratory for Molecular and Cellular Systems, Department of Neuro- and Sensory Physiology, Centre for Physiology and Pathophysiology, University Medicine Göttingen, Humboldtallee 23, 37075 Göttingen, Germany
and
DFG Research Center for Molecular Physiology of the Brain and Excellence Cluster EXC171 for Microscopy on the Nanometer Scale, Göttingen, Germany
e-mail: fred.wouters@gwdg.de

A. Diaspro (ed.), *Optical Fluorescence Microscopy*,
DOI 10.1007/978-3-642-15175-0_8,

fluorophores by emission color, by which multiple fluorescently labeled components can be followed in the same cell by judicious choice of excitation and detection wavelengths. The relatively new semiconductor nanocrystal "quantum dot" fluorophores optimally capitalize on emission separation for simultaneous imaging in that the different spectral variants can be simultaneously excited in the deep blue range, but their emission, which is narrow and symmetric, can be easily differentiated (Bruchez 2005). Another fluorescence parameter that is not used so often is the difference in excitation spectrum, i.e., the energies that can be absorbed and lead to fluorescence. One example here is the mixing of fluorophores of comparable emission wavelengths, but differing Stoke's shift: a long (~100 nm) Stoke's shift variant of the green fluorescent protein (GFP), Turbo-Sapphire (Zapata-Hommer and Griesbeck 2003), and a far-red emitting (~180 nm Stoke's shift) fluorescent protein, mKeima (Kogure et al. 2008) that shares its excitation optimum with cyan fluorescent protein (CFP), are available. Other fluorescence parameters (Lakowicz 2006) are used in more sophisticated spectrofluoroscopic measurements; they include polarization, photostability, and fluorescence lifetime. Polarization refers to the persistence of direction in the emission upon polarized excitation, i.e., using light with one preferred vibration direction. This directional relationship is reduced when the fluorophore rotates between excitation and emission of a photon, providing information on the rotational velocity of a fluorophore and, by this, on protein complex formation or the viscosity of the medium. Photostability refers to the chemical reactivity of the fluorophore in the excited state. These photochemical reactions typically render the fluorophore nonfluorescent in a process called photobleaching. The fluorescence lifetime is the average delay time between excitation and emission of a photon, i.e., the average duration of the fluorophore excited state. For this reason, there is a clear link with the previous two parameters; a longer lifetime will lead to a greater loss of polarization and will leave the fluorophore more vulnerable to photobleaching. The lifetime is directly related to the quantum yield, i.e., the efficiency of fluorescence generation expressed as the number of photons emitted per absorbed photon. The quantum yield (Q) is the ratio of the observed (τ_{obs}) over the radiative lifetime ($\tau_{radiative}$). The latter is the lifetime that a fluorophore would possess when all excited state energy was used to generate fluorescent photons (depopulation with the fluorescence emission rate k_f): as the observed and radiative lifetimes would be the same, the quantum yield would be unity:

$$Q = k_f \Big/ \sum k_{depopulation} = \tau_{obs} / \tau_{radiative} . \quad (8.1)$$

The radiative lifetime is a theoretical fluorophore-specific constant as nonradiative depopulation routes to the ground state are accompanied by energy losses, i.e., by interactions with the environment. The lifetime can thus be described as the inverse of the sum of depopulation rate constants:

$$\tau = 1 \Big/ \sum k_{depopulation} . \quad (8.2)$$

8.1.1 Förster Resonance Energy Transfer

As the radiative lifetime is a constant, the fluorescence lifetime is a measure for the concentration-normalized "molecular" brightness of the fluorophore.

All of the above parameters can be combined in spectroscopic measurements to extract information on the identity and environment of the fluorophore. Perhaps the most informative spectroscopic measurement for the status and environment of the fluorophore is that of Förster resonance energy transfer (FRET) (Förster 1948, 1965; Jares-Erijman and Jovin 2003; Clegg 1996). FRET is a photophysical phenomenon by which energy is transmitted nonradiatively between the excited states of two fluorophores. There are a number of criteria that have to be fulfilled before FRET coupling takes place. The coupling is mediated by coulombic interactions between the electromagnetic oscillations of the transition dipoles of the fluorophores. These resonance interactions are extremely sensitive to the separation distance of the fluorophores, giving rise to a sixth-order dependence on distance and on the orientation between the transition dipoles (described by a geometric term κ^2), i.e., they prefer a more parallel (optimally collinear) orientation and do not exhibit coupling at a perpendicular orientation. A fluorophore that has absorbed a photon can emit a photon or its emission dipole can couple to the absorption dipole of a second fluorophore. When resonance occurs, the excitation energy of the donating fluorophore is transferred to populate the excited state of the accepting fluorophore without the formation of a fluorescent photon from the donor. The probability of coupling not only depends on distance and orientation considerations, but also on the suitability of the dipole pair to resonate per se. That is, the selection of suitable FRET pairs has to take into account that the donor possesses a high quantum yield (Q_D), so that sufficient excited state energy is available for transfer, and the acceptor has to possess a high molar extinction coefficient (ε_A), so that it can absorb the donated energy with high efficiency. Furthermore, the emission spectrum of the donor and the absorption spectrum of the acceptor fluorophore have to overlap, indicating that the energies that are donated can be accommodated in the acceptor excited state. Although these criteria refer to basic energetic conditions, their use in describing the radiative properties of fluorophores can lead to the erroneous assumption that the acceptor would absorb a donor-generated photon in FRET. However, the probability of this process at reasonable fluorophore concentrations is exceedingly low as the emitted photon is radiated in an arbitrary direction and thus mostly does not encounter an acceptor fluorophore in its path. Furthermore, at interfluorophore distances where this process could occur, FRET dominates and prevents the generation of donor emission photons.

The extreme distance and orientation dependence of FRET is the reason for its rising popularity during the last decade as it permits the evaluation of protein–protein interactions, protein modifications, and conformational changes (Bunt and Wouters 2004; Wouters et al. 2001). The highly nonlinear sixth-order distance relationship gives rise to a very steep distance dependence centered around the distance where the

FRET efficiency (E) is 50%, i.e., where 50% of the donor fluorophores participate in FRET. This point, also called the critical or Förster distance R_0, is determined by the spectral properties of the FRET pair that determine its coupling (see above) and lie between 5 and 7 nm for dye pairs that are typically used in the life sciences:

$$E = \frac{R_0^6}{R_0^6 + R^6}\,, \qquad R_0^6 = \frac{\kappa^2}{n^4} Q_\mathrm{D} \int f_\mathrm{D}(\lambda)\varepsilon_\mathrm{A}(\lambda)\lambda^4 \mathrm{d}\lambda\,, \tag{8.3}$$

where n is the refractive index and the integral represents the spectral overlap. It is fortunate that this distance range corresponds to the typical size of proteins so that maximal sensitivity is in this important biological distance scale, and it guarantees that a direct physical interaction between two proteins can be concluded from FRET measurements. Distance considerations dominate the case of intermolecular protein–protein interactions. A second, very popular biological FRET measurement uses genetic fusions of two suitable fluorescent proteins that incorporate a sensing domain in its polypeptide design; conformational change of this sensing domain leads to FRET changes in the embedded FRET pair. These constructs are used as sensors for biological activity in that the conformational change is coupled to the biological read-out. Often, a domain of a signaling protein that undergoes a conformational change upon its activation is taken as surrogate measurement for the activity of the endogenous proteins in the signaling pathway. Alternatively, the conformational change is designed to be brought about by the binding of a protein or analyte, its posttranslational modification, or an in-chain interaction between a domain that is changed and a domain that recognizes this change. These FRET sensors enjoy high popularity because of their "plug-and-play" nature, but suffer from a lack in predictable bandwidth (Bunt and Wouters 2004). That is, in these sensors, the donor and acceptor moieties cannot achieve complete separation, and for this reason, conformational FRET sensitivity often dominates over distance effects. However, maximization of orientation differences between the active and inactive states of the sensor can practically only be achieved by trial-and-error variation of the linkage between the fluorophores.

8.1.2 FRET Consequences

The occurrence of FRET has a number of consequences that can be exploited for its detection and quantification.

FRET coupling reduces the emission yield of the donor fluorophore. This quenched donor emission can be judged from the reduction of the number of donor fluorescence photons per unit time, which is proportional to the efficiency of FRET. The amount of quenching can also be determined very easily by the comparison of donor fluorescence intensities before (F_{DA}, the donor fluorescence in the presence of the acceptor) and after (F_D) selective acceptor photobleaching, which creates a population of donor molecules that no longer undergo FRET and exhibit the full, unquenched emission intensities. Such a measurement is technically simple, especially when implemented in confocal scanning microscopes, but is of course only suitable for fixed cells. Since the quantum yield can be directly measured from the fluorescence lifetime (τ_{DA} for observed, FRET-affected lifetime, τ_D is the zero-FRET reference lifetime, typically derived from an independent measurement), this quenching can also be judged from the reduction in donor fluorescence lifetime, which is proportional to the efficiency of FRET. FRET thus increases the emission probability ($\tau_{DA} < \tau_D$) of the donor in the photophysically coupled pair.

$$E = 1 - \frac{F_{DA}}{F_D} = 1 - \frac{\tau_{DA}}{\tau_D}. \tag{8.4}$$

As a consequence of the shortening of the donor fluorescence lifetime, its photobleaching rate is reduced. This can be exploited in a very simple, yet slow and destructive, measurement that requires only a stable light source and repetitive image acquisition: the exponential decay in fluorescence emission $I(t)$ due to bleaching at constant illumination can be observed in the image stack, and the preexponential factor, the bleaching constant τ_{bleach}, can be fitted pixel-by-pixel that is inversely correlated to the fluorescence lifetime and from which the FRET efficiency can be calculated (Jovin and Arndt-Jovin 1989). A reference τ_{bleach} can be obtained from an independent experiment, omitting the acceptor, or by selective acceptor photobleaching in a region of the same cell to exclude sample preparation variation.

$$\begin{aligned} I(t) &= I_0 e^{\left(\frac{-t}{\tau_{bleach}}\right)} \\ E &= 1 - \frac{\tau_{bleach,\ reference}}{\tau_{bleach,\ observed}} \end{aligned} \tag{8.5}$$

for background corrected images.

FRET coupling causes the population of the excited state of the acceptor, which results in its emission. As a consequence, acceptor photons are generated when the donor is specifically excited. This is called sensitized acceptor emission. It should be noted, however, that the acceptor photons cannot be unequivocally assigned to FRET because some acceptor is directly excited at donor wavelengths and some donor emission contaminates (bleeds-through in) the red-shifted acceptor emission channel. These spectral contaminations can be corrected using a number of reference measurements, but this is explained in more detail elsewhere (van Rheenen et al. 2004). Importantly, the impact of contaminations depends on the relative concentration of donor and acceptor molecules in each pixel of an image. In the case of intramolecular FRET biosensors, where the donor and acceptor molecules are contained in the same polypeptide chain, the contamination is constant and corrections are not required. This advantage has significantly contributed to the popularity of this kind of biosensors. In contrast to the quenched donor emission, there is no easy way to relate (spectrally purified) sensitized emission to the FRET efficiency. In practice, sensitized emission measurements are related to the quenched donor emission or the spectral changes are otherwise normalized to the sensor concentrations. These different methods are explained in more detail elsewhere (Elder et al. 2009; Wouters and Bunt 2009). The fluorescence lifetime again can be used to detect this increase in fluorescence yield. The fluorescence lifetime of the acceptor itself does not change due to FRET, i.e., no difference can be detected in the delay time between direct acceptor excitation and emission. However, the coupled FRET pair can be regarded as a new fluorophore with donor absorption and (donor and) acceptor emission properties, and its lifetime describes the delay between donor excitation and sensitized acceptor emission: its duration therefore includes an additional delay caused by the transfer of energy from the reduced lifetime donor that adds to the pure lifetime of the acceptor. In FRET, the lifetime of sensitized emission thus increases (Esposito et al. 2008; Harpur et al. 2001; Jose et al. 2007). It should be noted, however, that the same limitations apply as for the detection of spectral intensities as directly excited acceptor and bleed-through donor emission will lower the FRET-increased lifetime accordingly.

From these consequences, it follows that the donor and acceptor effects should be detected separately in order to judge FRET. However, FRET measurements also have to satisfy the opposite demand; they have to possess a large spectral overlap between donor emission and acceptor excitation for optimal FRET coupling and are therefore intrinsically limited. For this reason, we have recently developed a nonfluorescent chromoprotein derivative of the yellow fluorescent protein as efficient FRET partner for a GFP donor (Ganesan et al. 2006). As donor lifetime measurements collect information of FRET in the reverse direction of the transfer of energy, no fluorescence is required on the part of the acceptor. Our mutant loses excited state energy much more rapidly than the duration of the excited state by fast internal conversion, i.e., environmental nonradiative interactions, and consequently is fully FRET competent, but does not contaminate the donor emission channel with otherwise inseparable yellow fluorescence. We realized that, apart from liberating a large part of the visible spectrum and allowing the collection of the complete, rather than only the nonacceptor-overlapping, part of the donor emission spectrum, this FRET couple also benefits from the removal of less obvious limitations by the use of a nonfluorescent acceptor: at high donor excitation probability, as for instance, occurring in confocal scanning microscopes, the same excited donor will attempt to donate its energy to the same acceptor in its FRET radius. This acceptor, however, requires its fluorescence lifetime to depopulate its excited state before it is competent again to accept another quantum of energy from the donor. It can therefore occur that a donor unsuccessfully tries to couple with a still-occupied acceptor, and no transfer will be possible (Hänninen et al. 1996; Beutler et al. 2008). The exceedingly short lifetime of the nonfluorescent acceptor chromoprotein prevents this "FRET frustration" effect. The second effect is also related to the very short acceptor lifetime. As explained above, the susceptibility of a fluorophore to photobleaching photochemical reactions depends on its lifetime and is also a function of the chemical composition of the fluorophore itself. The FRET-induced population of the acceptor excited state can lead to its photobleaching just as radiative excitation does. Unfortunately, most FRET pairs in use already have an unfavorable photostability relationship between the donor and the acceptor, and FRET can thus cause the destruction of the acceptor in proportion to its efficiency. This process has been exploited for a very sensitive FRET assay (Mekler 1994; Mekler et al. 1997). Because the dark acceptor only spends very little time in its excited state, it does not appreciably photobleach. Dark acceptors are thus ideal "sinks" for FRET.

As described above, the fluorescence lifetime carries information on the emission yield of the donor (reduced lifetime) and sensitized acceptor (increased lifetime) and possesses a decisive advantage over spectral emission intensity measurements as, in contrast to these, it does not depend on the concentration of fluorophores. Intensity measurements require normalization to fluorophore concentration to isolate emission yield changes from concentration differences. Furthermore, lifetimes are intrinsically quantitative and can provide information on the presence of multiple interacting (or conformational) species, their relative concentration, and the FRET efficiency in the complex. See Fig. 8.1 for an example of the application of FRET/FLIM to a biological question.

8.1.3 Lifetime Detection for FRET

However, in contrast to intensity measurements in different spectral ranges, lifetime measurements are more complicated and require specialized equipment. The different implementations of fluorescence lifetime imaging (FLIM) equipment and their comparative advantages are given elsewhere (Esposito et al. 2007b; Esposito and Wouters 2004). Functionally, all measurements are equivalent as they determine the lifetime-delay-induced distortion in the emission upon excitation with a temporally encoded pattern. When this pattern is a train of Dirac-type, i.e., approaching infinitely short pulses, the excited state is populated instantaneously, and the stochastic nature of fluorescence emission will cause an exponential decay with a characteristic preexponential factor, which is equal to the lifetime. Multiple lifetime emitters will introduce several weighted (fraction α_i) exponential decays:

$$I(t) = I_0 \sum_i \alpha_i e^{\frac{-t}{\tau_i}}. \tag{8.6}$$

Only when the lifetime is essentially zero will the excited state depopulate instantaneously and faithfully reproduce the pulse pattern. The measurement thus consists of following the time pattern of exponential decay and is therefore said

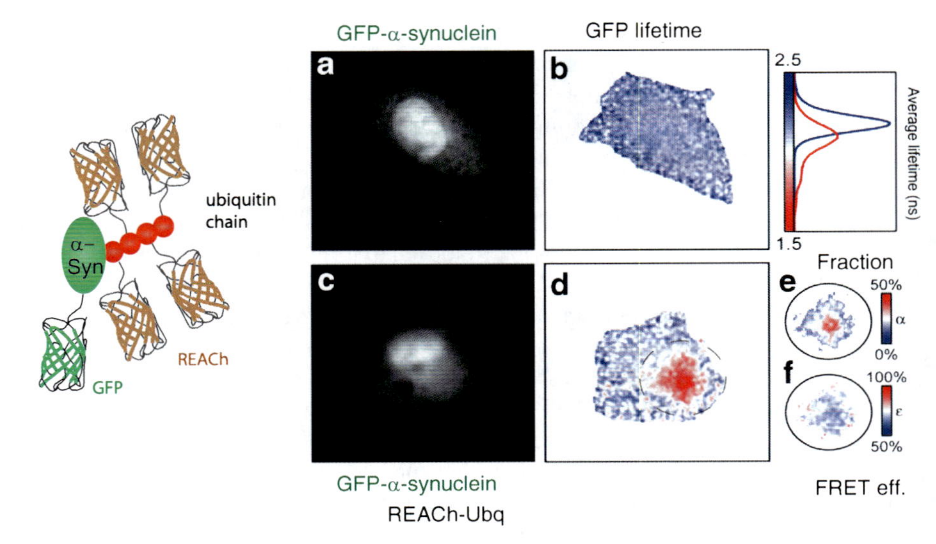

Fig. 8.1 Lifetime imaging of ubiquitinated α-synuclein. Aggregation of the α-synuclein protein is causative for neuronal death in Parkinson's disease. Dysfunctional proteins are often targeted for proteasomal degradation by the covalent chain-like polymeric attachment of ubiquitin proteins. *Left panel*: ubiquitination of α-synuclein in cells was visualized by FRET between the donor GFP, fused to α-synuclein, and ubiquitin carrying a nonfluorescent, FRET competent, YFP-based acceptor (REACh) for GFP (Esposito et al. 2007a; Esposito 2006). The GFP signal of α-synuclein in cells that do not (**a**) or do (**c**) coexpress REACh-ubiquitin, and the corresponding fluorescence lifetime image of the GFP signal (**b** and **d**, respectively), is shown. The distribution of lifetimes is shown in the histogram. Lower lifetimes corresponding to higher FRET efficiencies in the cell that coexpresses donor and acceptor (*red trace*) compared to the cell that only expresses the donor (*blue trace*) show the presence of an ubiquitinated aggregate of α-synuclein. (**e, f**) Fraction (**e**) and FRET efficiency in the complex (**f**) in the indicated region in (**d**), as calculated according to (Esposito et al. 2005a)

to be performed in the time-domain. Most implementations of this method are based on fast point detectors and scanning microscopy. They either map the arrival time of the emitted photons in time-correlated single-photon counting (TCSPC) (Ghiggino et al. 1992) or collect photons in preset-timed collection bins (gating), to create an emission histogram from which the lifetime is also fitted (Buurman et al. 1992). Alternatively, a high-frequency repetitive wave time pattern (square wave–sine wave, circular frequency ω) is used as temporal pattern and the delay-induced phase shift ($\boldsymbol{\phi}$) and demodulation (m) of the harmonic information is retrieved by various cross-correlation and lock-in techniques.

$$\begin{aligned}\phi &= \arctan(\omega\tau),\\ m &= \left[1+(\omega\tau)^2\right]^{-1/2}.\end{aligned} \tag{8.7}$$

As these techniques can be performed on the entire field-of-view by image-intensifier gated cameras (Lakowicz and Berndt 1991) and new directly modulated solid-state cameras (Esposito et al. 2005b, 2006; Mitchell et al. 2002a, b), the so-called frequency-domain measurements are usually implemented on wide-field microscopes and are typically faster in acquisition than time-domain methods. It should be said, however, that both methods can be implemented in either type of microscope and that their functional equivalence even allows the use of each other's specific analysis solutions (Digman et al. 2008) by simple data transformation.

8.2 Designing FRET-Based Biosensors

The process of biosensor design for the study of the molecular physiology of the cell involves a number of decision steps. A given physiological reaction or pathway has to be represented by an identifier event that is either part of the event or a direct consequence of it. Selectivity issues dominate this choice, as the read-out should be interpretable in terms of the selected event. Biological knowledge therefore is the major guiding factor at this stage. There is a current effort that aims to shift the function of biosensors from the isolated detection of one event to the simultaneous detection of multiple events. This multiplexing delivers not only more, but also deeper information as cellular physiological responses rarely rely on one event.

Once (single or multiple) selectivity has been achieved, the objective is to introduce in the cell the fluorescently labeled components that comprise the biosensor. Besides the choice of donor/acceptor fluorophores and detection method, which is mainly set by the instrumentation capabilities of the laboratory and the level of detail, sensitivity, and speed that needs to be reached, a major decision step is whether to construct a report- or actuator-type sensor (Bunt and Wouters 2004).

8.2.1 Reporters

Reporter sensors fulfill – as much as possible – a bystander role and serve as surrogate markers for the physiological event. The best example is the intramolecular donor – sensing domain – acceptor design sensor. One of the earliest examples of this type of sensor includes protease sensors that consist of a donor/acceptor-flanked protease substrate recognition peptide sequence (Heim and Tsien 1996; Jones et al. 2000; Mitra et al. 1996; Xu et al. 1998). Activity of the respective protease will result in cleavage of the substrate and a loss of FRET in the construct. Another prototypic example is the Cameleon sensor for intracellular calcium levels (Miyawaki et al. 1999). Here, the donor/acceptor moieties flank calmodulin and its binding peptide (M13). Calcium binding will induce a conformational change by the interaction of M13 and calmodulin that is measured by a change in FRET. These sensors are designed to minimally interfere with the event under investigation even when expressed at high concentrations. However, apart from a buffering function that these sensors could have, their interference with physiology can also stem from the fact that the domains are often "borrowed" from signaling proteins that might still retain part of their function or protein-binding capacity. For the example of the Cameleon, the choice for calmodulin has a drawback as this protein engages in multiple protein interactions for different pathways in cells. These interactions can reduce the sensitivity toward calcium and can potentially interfere with unrelated events. Another potential problem is that these domains, removed from their original physiological setting, do not faithfully reproduce the behavior of the parental proteins. A major issue with phosphoaminoacid-recognizing domains like SH2 domains in constructs with a respective kinase substrate peptide is the reversibility of the reaction (Sato et al. 2002). Access to a modulatory phosphatase might not be given either due to differences in localization/compartmentalization, or by the fact that the phosphorylated and bound peptide in the FRET sensor is sterically not accessible.

One major problem with the design of single-chain intramolecular sensors is the connection between the fluorophores and the sensing domain(s). The conformational change in the sensing domain must be "felt" by the fused fluorophores, which argues for a stiff and short linker sequence between the individual units. On the other hand, the recognition between a modified substrate and a recognition domain in the same construct would require a very flexible linker, and the folding of the domain and fluorophores must not be sterically hindered by too little spatial "freedom." The effect of a flexible linker of varying length was investigated in a CFP–YFP construct where a GGSGGS sequence was inserted in one to nine copies, and the resulting FRET efficiency evaluated by spectral change and lifetime (Evers et al. 2006). It was found that simple distance calculations from these insertions failed as the flexible linkers adopted a remarkably compact structure where the end-to-end distance was six to seven times smaller than the linker length. This meant that high FRET was observed even with a very long random coil linker. Under the realization that flexible linkers can also bring the donor and acceptor together at long linker distances, the same group investigated the effect of enhancing the dimerization potential between the donor and acceptor fluorescent protein (Vinkenborg et al. 2007). The current generation of fluorescent protein has almost all been optimized for monomerization by mutation of the dimerization interface (Zacharias et al. 2002; Campbell et al. 2002). Even though the affinity for dimerization is in the millimolar range, the local concentration of fluorescent proteins in biosensors or in interacting fluorescently labeled proteins can be high enough to support a substantial interaction. Introduction of two mutations S208F and V224L in the barrel of the fluorescent proteins increases their interaction (Vinkenborg et al. 2007). Under these conditions, the FRET efficiency will be set by this interaction, independent of the length of the intervening linker region (or any sensing domain that allows the approach of these dimerization interfaces in the fluorescent proteins). Fortuitously, this interaction was found to depend on the integrity of the linker chain. Cleavage of a protease recognition site caused a loss of FRET, indicating that the interaction, though efficient in linking the two fluorescent proteins, is transient in nature. The upshot of this finding is that it provides a rational method of increasing the detection bandwidth of intramolecular FRET biosensors and producing a truly modular system with reproducible high FRET efficiency limits. The applicability of this promising approach to other events, also conformational changes, awaits verification.

8.2.2 Actuators

One can also treat the problem of the functional interaction with the molecular process under investigation in intramolecular FRET sensors differently by embracing the complexity of the system; rather than attempting to reduce it to a single biosensor construct, actuator-type sensors can be employed. Actuators are fully functional components that have been fluorescently tagged. They will participate in the signaling reactions, ideally as well as their unlabeled endogenous counterparts, and faithfully report on the reaction in all its complexity. These sensors can thus produce detailed insight into molecular processes that include modulatory effects. The major issue with these sensors is that their functional equivalence has to be guaranteed, which is not always straightforward given that our information on the biological function of the parental protein is likely incomplete, and that their expression level should not upset the function of the network in which they operate. Most implementations of this type of sensors detect those protein–protein interactions that are meaningful intermediates in a complex signaling pathway. Other actuator designs probe changes in the conformation of proteins as they engage in protein interactions or changes in the molecular architecture of multiprotein complexes (Bunt and Wouters 2004).

8.2.3 Multispecificity Detectors

A radical departure from the aim of detecting a single molecular event with the highest possible selectivity is required to understand the biological complexity in typical cellular physiological responses that are themselves multifactorial in nature. Multiplexing multiple sensors in the parallel detection of related events is the first step toward unraveling complexity as temporally and/or spatially coordinated changes could reveal their codependence. FRET allows the investigation of isolated biochemical events with high precision and specificity, and current technology on the detection and labeling side in principle already allows the multiplexing of two FRET assays.

An example is the combination of two different proteolytical recognition sites in a single sensor for two different caspase activities (Wu et al. 2006). Here, two intramolecular FRET sensors were combined in a single construct in a

CFP–DEVD–YFP–VEID–mRFP design. In this sensor, the DEVD sequence is recognized and cleaved by caspase 3, and the VEID sequence by caspase 6. In this case, the two FRET pairs were combined by sharing a common fluorophore, YFP, as both acceptor for CFP and donor for mRFP, allowing the interrogation and comparison of both proteolytic activities in a single cell. The benefit of such an approach is, of course, that the intrinsically heterogeneous response of cells to apoptosis induction can now be investigated at a hitherto unattained level of precision: new behavioral populations can be created from the combined information. It was for instance shown that cells that expressed caspase 3 activity, but did not up-regulate caspase 6 activity similarly, increased within the first 2 h of treatment with the apoptosis-inducing agent staurosporin. Different combinations of fluorescent proteins and otherwise labeled proteins, and different schemes are of course possible and have been published.

Another recent example uses two separate FRET assays, with four fluorescent proteins in the same cell; a CFP–Venus [Venus is a variant of YFP (Nagai et al. 2002)]-based intramolecular Cameleon calcium sensor and a TagRFP–mPlum [mPlum is an mRFP variant (Shaner et al. 2004)]-based intermolecular FRET assay for Ras activity (Grant et al. 2008). In this the Cameleon signal is read out by ratiometric imaging and the binding of the TagRFP-labeled Raf-RBD to GTP-containing, active, mPlum-labeled Ras is judged from the reduced lifetime of Tag-Red (Merzlyak et al. 2007) by FLIM. Exposure of cells that expressed both individual FRET sensors with epidermal growth factor demonstrated a transient calcium rise, followed by a sustained activation of Ras.

The following directions are available for expanding the information range in FRET sensors. They provide different viewpoints on the function of components:

8.2.3.1 Many Assays, Few Directions

A *functional* viewpoint answers the question "what does the component *do*?" It aims at the identification of its output characteristics and serves its classification. Such an approach provides a unique tool for the identification of biological activities at high resolution. With dimensional expansion of multiplexing, i.e., massively parallel detection, the measurements are obviously no longer feasible inside single cells, but have to be performed on biochemical preparations or on large numbers of cells. An example is the fingerprinting of protease activities by the use of large combinatorial FRET-based substrate libraries (Sun et al. 2007). Here, different proteases were tested for their reactivities toward different systematically varied substrates, in order to obtain patterns of proteolytical preference for peptide sequences that allowed the determination of functional "familial" relationships between different proteases. Such an unbiased approach therefore carries a lot of valuable information. It can be easily appreciated that similar multipoint identification approaches for cellular activities and responses could uncover information and relationships that are impossible to obtain otherwise.

8.2.3.2 Few Assays, Many Directions

A *structural* footprint answers the question "how does the component *behave* (under different conditions)?" It aims at the description of the component's input response. The required increase in information content within one assay can be obtained by variation on its design. An example from the field of protein folding exemplifies this approach. The Barstar enzyme, which carryied a single tryptophane residue, was modified by site-directed introduction of single cysteins, whose thiol groups can be specifically labeled with fluorescent dyes, or in this case, with a trinitrobenzoic acid group. These (10) derivatized cysteins form a quenching FRET acceptor for the intrinsically fluorescent donor tryptophane residue in many different possible geometric combinations. Using these proteins, the geometric changes inside the protein upon (un)folding could be addressed at high resolution, as subnanometer resolution between the different positions in the protein could be monitored at high (ms) temporal resolution (Sinha and Udgaonkar 2007). The unique information gained from this multiplexed assay is that the protein folds in an uncooperative manner, with gradual transition between extended and collapsed forms. No other structural biological tool could have provided this information with similar spatiotemporal resolution. The other angle of multipoint analysis would be to have multiple labels on proteins, and this type of analysis is expected to increase in importance as more sophisticated measurements becomes available, allowing the simultaneous acquisition of multiple fluorescent labels and fluorescence parameters.

8.2.3.3 Many Assays, Many Directions

An *organizational* viewpoint (or map) answers the question "what is the *relation* of the component to other components and events?" It aims at the description of input/output connections and serves the understanding of system hierarchy. When describing a more variable set of conditions than a single biochemical activity or the behavior of a single protein, the approach needs to be expanded on both sides in order to obtain a scalable assay with a controllable level of precision. A recent elegant example of such biological question and multisite detection solution is provided by the exploration of (changes in the) nanoscape of membrane-associated proteins in microdomains. Remarkably, the possibility of organized domains in the plasma membrane has been anticipated already in the seminal paper by Singer and Nicholson that essentially established the opposite as a general conclusion, i.e., that membrane proteins can freely diffuse in the membrane lipid bilayer (Singer and Nicholson 1972). Since then, the membrane microdomain concept has gained a lot of popularity as it served to explain heterogeneity in membrane signaling responses, but has also attracted a lot of controversy that mainly relate to the fact that these domains are below the detection resolution of most investigation methods. One of the driving principles behind trimeric G-protein signaling is the dissociation of the Gα from the G$\beta\gamma$ subunits after activation. Under the assumption that different lipid modifications would direct these subunits to different lipid microdomains, the location of G-protein subunits in different microdomains was investigated by testing a large number (38) of different FRET pairs composed of (the lipid anchors of, and also complete) heterotrimeric G-protein constructs and lipid anchors of proteins that signify different lipid microdomains. From these many interactions, different microdomain clustering behaviors could be identified for the G-proteins, different types of microdomains were identified, and changes in localization/partitioning of G-proteins for the active and inactive state could be identified (Abankwa and Vogel 2007). Display of the aggregated information of the individual FRET assays in tensor-plots provides an immediate insight into the nanoscale organization of protein distributions and interactions that is, in principle, fully scalable to any number of components. Again, the technical limitation to the number of fluorophores that can be simultaneously observed in cells, here just two for one FRET pair, required the observation of large numbers of cells in a cytometer in order to extract generalities in the behavior of the pairs. Even at the current state of state of the art FRET technology, the information content per cell could be doubled. In any case, this paper (Abankwa and Vogel 2007) shows the road ahead, as the new multicomponent groups that were identified as single biologically relevant states are important basic building blocks of the cellular signaling circuitry.

8.2.4 Coincidence Detectors

However, the biochemical network comprises of a large number of interconnected components and responses can be encoded in subtle changes at multiple sites in the network (Ma'ayan and Iyengar 2006). The challenge for modern system- and mechanism-oriented analytical cell biology is therefore to learn the organizational rules, which cannot be derived from the observation of a single or a few, unconnected, component(s). Once a minimum of two FRET pairs can be encoded within a cell, like is the case for the first examples of expanding multiplexing FRET detection, the assays can be coupled (Fig. 8.2). The progression from an intrinsically one-dimensional observation like a single FRET pair to the two-dimensional observation of two (and more) FRET pairs is logical and feasible, as the design of the dual caspase sensor example above illustrates. Rather than keeping the two observations unlinked, however, designs can be adopted that probe the third dimension, i.e., that link the information contained in both FRET pairs in a manner that reveals connection points in the biochemical network (Bunt and Wouters 2004). Three variations are imaginable:

1. Two interactions converging on a single connection, i.e., a *node*, can represent a decision point when, for instance, two proteins compete for binding to the same protein, or when both interactions are not mutually exclusive, indicate a coincidence detector where two events have to occur before the signal is propagated. There are numerous examples of this kind of signaling modules, for instance,

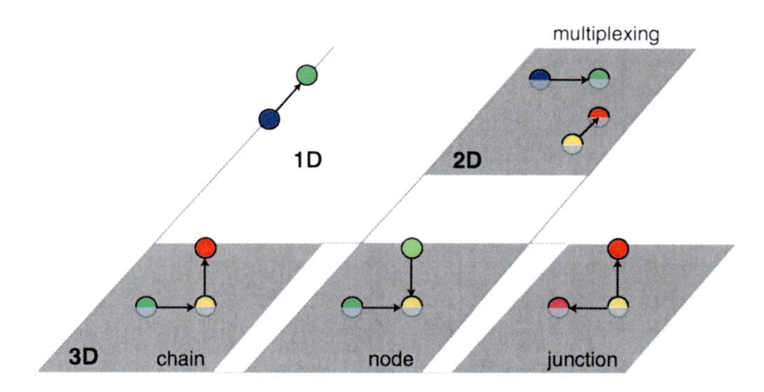

Fig. 8.2 Expansion of FRET measurements. A graphical overview of the development of FRET measurements from the current single-event assays (one-dimensional) to their parallelization for the detection of multiple, uncoupled events (two-dimensional), and their connection for the detection of coupled events (three-dimensional) in hierarchically sequential (chain), converging (node), or diverging (junction) schemes is illustrated

in the form of scaffolding proteins that bind multiple adapter and/or effector proteins in order to diversify/adapt the possible responses downstream of the scaffolding protein in a combinatorial manner. Moreover, information weighing is an important regulatory aspect at nodes. This kind of interaction can, for instance, be detected by the use of a shared acceptor fluorophore on the common binding partner. The two binding proteins would then have to carry fluorophores that are suitable donors for this shared acceptor. This is most practically implemented by the use of two fluorophores with different Stoke's shift, but similar emission spectrum, which allows them to be addressed by selective excitation. Alternatively, detection schemes involving photoactivatable or photoswitchable fluorophores are thinkable.

2. Situations where a single protein has the choice to bind to two different interaction partners, i.e., in a *junction*, are also often encountered in signaling. One example is the heterodimerization of many membrane receptors, where the interactions bring together subunits with different signaling preferences. The many interaction choices of trimeric G-proteins to GPCRs (see above) and the adapter molecules that are recruited to activated tyrosine receptor kinases are also examples of a junctional organizational scheme. The choice of binding to one of many possible receptors effectively encodes the identity of the signaling pathway downstream of this receptor. How this "promiscuity" in binding leads to specified signaling outcomes is a difficult and intensely investigated question. Analytical fluorescence microscopy plays an important role in this investigation by virtue of its quantitative discriminatory power. It is clear, however, that in order to derive an understanding of the process from the (large amounts of) information obtained from such studies, a systematic interpretation aided by computer modeling will play an increasingly important role.

3. A *chain* of two consecutive events can be probed by the same fluorophore arrangement as in the dual caspase example, where, for instance, CFP and YFP report on the first interaction, and YFP and mRFP on the second interaction of the YFP-labeled component. Different fluorophore combinations are of course possible. The difference in this case is that the YFP-labeled component now actually bridges the two individual steps. In this case, changes in FRET that are transmitted to the intermediate fluorophore are sensed also in the second interaction, and these changes carry additional information. This design of assay is unique in that it permits true coincidence detection. The most interesting case in signaling would be when the second reaction is conditional on the first. Many signaling pathways use such conditional gates, for instance, in the form of protein–protein interactions that are conditional on the phosphorylation of specific residues in one or both of the interacting proteins. The two coupled events cannot only be analyzed separately from each other, but the co-occurrence on the same molecules in time and space can be extracted as the donor properties of the YFP (for FRET with mRFP) and its acceptor properties (for FRET with CFP) are coupled. The transfer of energy down the complete chain of FRET fluorophores is indicative of the coupled occurrence of both events. One of the first papers to investigate the possibility for such assays also used a dual protease design with the same fluorophores (Galperin et al. 2004). In this case, both protease substrates were for the same protease factor Xa. Treatment of the construct with factor Xa uncoupled all fluorophores. Spectral analysis of the three emission spectra permitted the identification of both single transfer steps as well as the long-range coupled two-step transfer between the terminal fluorophores in the chain. The FRET efficiencies of both steps were determined by sequential acceptor photobleaching in fixed cells: bleaching of mRFP in a region of the cell unquenched its YFP donor in the same region, and bleaching of this unquenched YFP led to the unquenching of its CFP donor.

Another example of the use of fluorophore chain FRET arrangements in a single construct is for the detection of conformational changes in a protein. The incorporation of two FRET pairs, in principle, allows not only the probing of conformational changes in two parts of the same protein, but also the sensitivity toward a single conformational change can be tuned and enhanced by the use of a bridging fluorophore, for instance, in the class of intramolecular FRET biosensors. For the sake of argument, the example here again uses a CFP–YFP–mRFP arrangement. If a conformational sensing domain is placed between CFP and YFP, the FRET efficiency might not be very high and might also not be easily improved as this would require the manipulation of the transition dipole orientations. However, an mRFP can be positioned immediately adjacent to the YFP and the coupling between the YFP and mRFP can be optimized. Once optimized, this dual fluorophore module can also be used to increase the FRET transfer between CFP and YFP, and it can do this in a variety of sensors that vary in the sensing domain between CFP and YFP. The reason for the increased FRET transfer in the first half of the chain is the same as for the nonfluorescent acceptor in that the excited state energy that arrives in YFP is rapidly transferred to mRFP. Another advantage of such a chain FRET biosensor is that there is less spectral contamination between CFP and mRFP and that the sensor can be used in situations where there is a yellow fluorescent contamination in the cell. One particular consequence is that the FRET sensor might even be combined with the detection of a yellow fluorescently labeled additional component, at least when the localizations do not overlap too much. The yellow component from the sensor could then be subtracted from the additional component by its correlated CFP or mRFP signal. An example of such a sensor for PKA activity is CRY-AKAR (*c*yan-*r*ed-*y*ellow *A*-*k*inase *a*ctivation *r*eporter) that flanges the Forkhead-associated domain 1 and an LRRATLVD PKA phosphorylation substrate peptide between Cerulean [a variant of CFP (Rizzo et al. 2004)] and mCherry [an mRFP variant (Shaner et al. 2004, 2005)] and couples mVenus immediately adjacent to mCherry (Allen and Zhang 2008). Activation of the sensor causes binding of the Forkhead domain to the peptide, which brings Cerulean in a favorable position to undergo FRET with the mVenus–mCherry unit. The red/cyan ratio provides a very robust measurement of PKA activity. The enhancement effect can be seen from the FRET efficiencies of the individual steps: 19% for cyan to yellow, 36% from yellow to red, but overall still 18% from cyan to red. Furthermore, the measurement was shown to be insensitive to the incubation of cells with L-sepiapterin, a GTP cyclohydrolase I inhibitor with CFP-like excitation and broad, predominantly yellow emission. It should be noted that many pharmacological inhibitors/activators used in cell biology for the perturbation of pathways are intrinsically fluorescent. The same is true for many components in drug screening. This approach can circumvent these issues of spectral contamination.

The most obvious application of a chain FRET arrangement is the investigation of (changes in) the composition of multiprotein complexes. Here, the two sides of the FRET chain represent a physical interaction between the first and second, and the second and third protein. Under the realization that the two FRET steps are linked and influence each other's FRET coupling, conditional binding events can be investigated. The sequential photobleaching approach used by Galperin et al. to estimate the FRET efficiencies in a CFP–YFP–mRFP construct does not provide information on this coupling as the chain is degraded back-to-front by the photobleaching of the mRFP terminal acceptor, effectively collapsing the situation into a typical CFP–YFP sensor. The FRET then measured by photobleaching of the YFP is no longer under influence of the FRET "pull" by mRFP in the second FRET step.

In an elegant variation of the sequential photobleaching approach, Fazekas et al. (2008) combine an acceptor photobleaching step with a donor photobleaching step. Using antibodies against interacting membrane proteins that were labeled with XFITC (green emission), Cy3 (red emission) or Cy5 (far-red emission), multiple protein interactions could be investigated. The approach is based on the fact that donor photobleaching (XFITC) is slowed down by FRET (see Sect. 1.2), but the acceptor (Cy3) is not destructed in this measurement. Furthermore, acceptor photobleaching (Cy5) unquenches donor fluorescence (Cy3). In both instances, Cy3 survives the FRET measurement in both directions. This method was therefore dubbed "two-sided FRET." In contrast to the other (one-sided, "retrograde") sequential photobleaching method, the donor photobleaching step for the determination of XFITC-Cy3 FRET can be performed in the presence of the FRET "pull" between Cy3 and Cy5. Correlation of the FRET efficiencies thus obtained provides information on the coupling between both types of protein interactions in the complex: no correlation between both FRET sides indicates constant intramolecular distances. This could be demonstrated using three antibodies against the same transmembrane protein, ErbB2, a family member of the EGFR family that is often upregulated in cancer. When high FRET efficiencies correlate, the binding between the three proteins is cooperative. When high FRET efficiencies in one direction correlate with low FRET efficiencies in the other direction, then the protein interaction probed in the first direction inhibits the interaction probed in the second direction. The latter could be shown for ErbB2 homodimerization, which anticorrelated with the interaction between the cell adhesion receptor β1-integrin and ErbB2, uncovering a mechanism for the modulation of ErbB2 homodimerization by its interaction with β1-integrin. Furthermore, a modest positive correlation was found between the interaction of β1-integrin and ErbB2, and β1-integrin with another cell adhesion molecule, CD44.

It should be noted, however, that a correlation between the two FRET sides in this method also does not imply the presence of a ternary complex as both individual interactions can occur independently from each other in the same pixel. New detection techniques and paradigms are required to detect the proper molecular coincidence of two protein interactions, i.e., a trimer, or the direct coupling between biochemical events like a binding that occurs downstream of a conformational change or modification.

One way of detecting such "Boolean AND" interactions that relies on the conventional intensity-based FRET detection techniques makes use of an elegant biotechnological solution. It combines the process of bimolecular fluorescence complementation (BiFC) with FRET (Shyu et al. 2008). In BiFC, two halves of the polypeptide of YFP are expressed as fusion protein with interacting proteins. When the host proteins interact, the two halves are brought into close proximity and this suffices for the reconstitution of a folded, fluorescent YFP molecule. The information for the correct folding solution, contained in the primary sequence, is apparently sufficiently strong to drive and guide the folding of the protein also when the chain is interrupted, but the two halves are spatially constrained by the interacting proteins. As this interaction between two proteins generates YFP, this molecule can be used as acceptor for the detection of the binding of a third protein, labeled with the donor CFP. FRET thus occurs exclusively upon the formation of a ternary complex. Failure of ternary complex formation can still be investigated in more detail as the formation of fluorescent YFP by BiFC can be used to identify the missing protein interaction. This method was used for the detection of a ternary transcription regulation protein complex consisting of bFos, bJun (by BiFC), and NFAT-1 (by FRET). It was shown that a mutant form of NFAT-1 did allow the formation of a bFos–bJun heterodimer (YFP was formed by BiFC), but that it did not assemble with the heterodimer to form the ternary complex, as FRET was significantly reduced. A new interaction was also observed by this method between the Fos–Jun heterodimer and the NFκ-B subunit p65, demonstrating cross talk between these two families of transcription factors.

8.3 Conclusion

FRET microscopy, especially when based on FLIM, offers a quantitative view of functional mechanisms of cellular machines by its discriminating power on the molecular scale. The continued development of new fluorophores and analytical and optical approaches fuels this discovery process. However, an understanding of the functioning of cells demands a shift in focus from isolated events to a wider view that encompasses multiple aspects of the system under investigation simultaneously. Knowledge on the correlations, connections, and hierarchy of multiple events is required to reach an understanding of the intricate cellular biochemical network that gives rise to adequate cellular physiological responses to stimuli. This asks for an increase in the "dimensionality" of FRET (and other quantitative biophysical optical) assays, which will be reached in two steps: the parallelization of assays in or between cells – providing functional, structural, and organizational viewpoints of activities – and the creation of unique coincidence detecting assays that report exclusively on coupled events, i.e., where both events originate from (act on) the same molecule in time and space. The first examples of such next generation optical assays light the way toward an integrative view of the working of complex cellular systems.

Acknowledgments Additional support is acknowledged from the "FLI-Cam" project in the Biophotonics program of the Federal Ministry of Science and Education (BMBF). I thank Alessandro Esposito for the figure of ubiquitinated α-synuclein.

References

Abankwa D, Vogel H (2007) A FRET map of membrane anchors suggests distinct microdomains of heterotrimeric G proteins. J Cell Sci 120:2953–2962

Allen MD, Zhang J (2008) A tunable FRET circuit for engineering fluorescent biosensors. Angew Chem Int Ed Engl 47:500–502

Beutler M, Makrogianneli K, Vermeij RJ, Keppler M, Ng T, Jovin TM, Heintzmann R (2008) satFRET: estimation of Forster resonance energy transfer by acceptor saturation. Eur Biophys J 38:69–82

Böhmer M, Enderlein J (2003) Fluorescence spectroscopy of single molecules under ambient conditions: methodology and technology. Chemphyschem 4:793–808

Bruchez MP (2005) Turning all the lights on: quantum dots in cellular assays. Curr Opin Chem Biol 9:533–537

Bunt G, Wouters FS (2004) Visualization of molecular activities inside living cells with fluorescent labels. Int Rev Cytol 237:205–277

Buurman EP, Sanders R, Draaijer A, Gerritsen HC, Vanveen JJF, Houpt PM, Levine YK (1992) Fluorescence lifetime imaging using a confocal laser scanning microscope. Scanning 14: 155–159

Campbell RE, Tour O, Palmer AE, Steinbach PA, Baird GS, Zacharias DA, Tsien RY (2002) A monomeric red fluorescent protein. Proc Natl Acad Sci USA 99:7877–7882

Clegg RM (1996) Fluorescence resonance energy transfer. In: Wang XF, Herman B (eds) Fluorescence imaging spectroscopy and microscopy. Wiley, London

Digman MA, Caiolfa VR, Zamai M, Gratton E (2008) The phasor approach to fluorescence lifetime imaging analysis. Biophys J 94:L14–L16

Elder AD, Domin A, Kaminski Schierle GS, Lindon C, Pines J, Esposito A, Kaminski CF (2009) A quantitative protocol for dynamic measurements of protein interactions by Forster resonance energy transfer-sensitized fluorescence emission. J R Soc Interface 6:S59–S81

Esposito A (2006) Molecular and cellular quantitative microscopy. Thesis, Utrecht University, The Netherlands and Göttingen University, Germany

Esposito A, Wouters FS (2004) Fluorescence lifetime imaging microscopy. In: Bonifacino JS, Dasso M, Harford JB, Lippincott-Schwartz J, Yamada KM (eds) Current protocols in cell biology. Wiley, London

Esposito A, Gerritsen HC, Wouters FS (2005a) Fluorescence lifetime heterogeneity resolution in the frequency domain by lifetime moments analysis. Biophys J 89:4286–4299

Esposito A, Oggier T, Gerritsen HC, Lustenberger F, Wouters FS (2005b) All-solid-state lock-in imaging for wide-field fluorescence lifetime sensing. Opt Express 13:9812–9821

Esposito A, Gerritsen HC, Oggier T, Lustenberger F, Wouters FS (2006) Innovating lifetime microscopy: a compact and simple tool for life sciences, screening, and diagnostics. J Biomed Opt 11:34016

Esposito A, Dohm CP, Bahr M, Wouters FS (2007a) Unsupervised fluorescence lifetime imaging microscopy for high content and high throughput screening. Mol Cell Proteomics 6:1446–1454

Esposito A, Gerritsen HC, Wouters FS (2007b) Optimizing frequency-domain fluorescence lifetime sensing for high-throughput applications: photon economy and acquisition speed. J Opt Soc Am A Opt Image Sci Vis 24:3261–3273

Esposito A, Gralle M, Dani MA, Lange D, Wouters FS (2008) pHlameleons: a family of FRET-based protein sensors for quantitative pH imaging. Biochemistry 47(49):13115–13126

Evers TH, van Dongen EM, Faesen AC, Meijer EW, Merkx M (2006) Quantitative understanding of the energy transfer between fluorescent proteins connected via flexible peptide linkers. Biochemistry 45:13183–13192

Fazekas Z, Petras M, Fabian A, Palyi-Krekk Z, Nagy P, Damjanovich S, Vereb G, Szollosi J (2008) Two-sided fluorescence resonance energy transfer for assessing molecular interactions of up to three distinct species in confocal microscopy. Cytom A 73:209–219

Förster T (1948) Zwischenmolekulare Energiewanderung und Fluoreszenz. Ann Phys 2:55–75

Förster T (1965) Delocalized excitation and excitation transfer. In: Sinanoglu O (ed) Modern quantum chemistry – Istanbul lectures, Part III. Academic, New York

Galperin E, Verkhusha VV, Sorkin A (2004) Three-chromophore FRET microscopy to analyze multiprotein interactions in living cells. Nat Methods 1:209–217

Ganesan S, Ameer-Beg SM, Ng TT, Vojnovic B, Wouters FS (2006) A dark yellow fluorescent protein (YFP)-based resonance energy-accepting chromoprotein (REACh) for Forster resonance energy transfer with GFP. Proc Natl Acad Sci USA 103:4089–4094

Ghiggino KP, Harris MR, Spizzirri PG (1992) Fluorescence lifetime measurements using a novel fiberoptic laser scanning confocal microscope. Rev Sci Instrum 63:2999–3002

Grant DM, Zhang W, McGhee EJ, Bunney TD, Talbot CB, Kumar S, Munro I, Dunsby C, Neil MA, Katan M, French PM (2008) Multiplexed FRET to image multiple signaling events in live cells. Biophys J 95:L69–L71

Hänninen PE, Lehtelä L, Hell SW (1996) Two- and multiphoton excitation of conjugate dyes with continuous wave lasers. Opt Commun 130:29–33

Harpur AG, Wouters FS, Bastiaens PI (2001) Imaging FRET between spectrally similar GFP molecules in single cells. Nat Biotechnol 19:167–169

Heim R, Tsien RY (1996) Engineering green fluorescent protein for improved brightness, longer wavelengths and fluorescence resonance energy transfer. Curr Biol 6:178–182

Jares-Erijman EA, Jovin TM (2003) FRET imaging. Nat Biotechnol 21:1387–1395

Jones J, Heim R, Hare E, Stack J, Pollok BA (2000) Development and application of a GFP-FRET intracellular caspase assay for drug screening. J Biomol Screen 5:307–318

Jose M, Nair DK, Reissner C, Hartig R, Zuschratter W (2007) Photophysics of Clomeleon by FLIM: discriminating excited state reactions along neuronal development. Biophys J 92: 2237–2254

Jovin TM, Arndt-Jovin DJ (1989) Luminescence digital imaging microscopy. Annu Rev Biophys Biophys Chem 18:271–308

Kogure T, Kawano H, Abe Y, Miyawaki A (2008) Fluorescence imaging using a fluorescent protein with a large Stokes shift. Methods 45:223–226

Kolin DL, Wiseman PW (2007) Advances in image correlation spectroscopy: measuring number densities, aggregation states, and dynamics of fluorescently labeled macromolecules in cells. Cell Biochem Biophys 49:141–164

Lakowicz JR (2006) Principles of fluorescence spectroscopy. Springer, New York

Lakowicz JR, Berndt KW (1991) Lifetime-selective fluorescence imaging using an Rf phase-sensitive camera. Rev Sci Instrum 62:1727–1734

Lippincott-Schwartz J, Altan-Bonnet N, Patterson GH (2003) Photobleaching and photoactivation: following protein dynamics in living cells. Nat Cell Biol (Suppl): S7–14

Ma'ayan A, Iyengar R (2006) From components to regulatory motifs in signalling networks. Brief Funct Genomic Proteomic 5:57–61

Mekler VM (1994) A photochemical technique to enhance sensitivity of detection of fluorescence resonance energy transfer. Photochem Photobiol 59:615–620

Mekler VM, Averbakh AZ, Sudarikov AB, Kharitonova OV (1997) Fluorescence energy transfer-sensitized photobleaching of a fluorescent label as a tool to study donor-acceptor distance distributions and dynamics in protein assemblies: studies of a complex of biotinylated IgM with streptavidin and aggregates of concanavalin A. J Photochem Photobiol B 40:278–287

Merzlyak EM, Goedhart J, Shcherbo D, Bulina ME, Shcheglov AS, Fradkov AF, Gaintzeva A, Lukyanov KA, Lukyanov S, Gadella TW, Chudakov DM (2007) Bright monomeric red fluorescent protein with an extended fluorescence lifetime. Nat Methods 4:555–557

Mitchell AC, Wall JE, Murray JG, Morgan CG (2002a) Direct modulation of the effective sensitivity of a CCD detector: a new approach to time-resolved fluorescence imaging. J Microsc 206:225–232

Mitchell AC, Wall JE, Murray JG, Morgan CG (2002b) Measurement of nanosecond time-resolved fluorescence with a directly gated interline CCD camera. J Microsc 206:233–238

Mitra RD, Silva CM, Youvan DC (1996) Fluorescence resonance energy transfer between blue-emitting and red-shifted excitation derivatives of the green fluorescent protein. Gene 173: 13–17

Miyawaki A, Griesbeck O, Heim R, Tsien RY (1999) Dynamic and quantitative Ca2+ measurements using improved cameleons. Proc Natl Acad Sci USA 96:2135–2140

Nagai T, Ibata K, Park ES, Kubota M, Mikoshiba K, Miyawaki A (2002) A variant of yellow fluorescent protein with fast and efficient maturation for cell-biological applications. Nat Biotechnol 20:87–90

Netterwald J (2008) Emerging trends in cell biological research. Drug Discovery Dev 10:39

Rizzo MA, Springer GH, Granada B, Piston DW (2004) An improved cyan fluorescent protein variant useful for FRET. Nat Biotechnol 22:445–449

Sato M, Ozawa T, Inukai K, Asano T, Umezawa Y (2002) Fluorescent indicators for imaging protein phosphorylation in single living cells. Nat Biotechnol 20:287–294

Shaner NC, Campbell RE, Steinbach PA, Giepmans BN, Palmer AE, Tsien RY (2004) Improved monomeric red, orange and yellow fluorescent proteins derived from *Discosoma* sp. red fluorescent protein. Nat Biotechnol 22:1567–1572

Shaner NC, Steinbach PA, Tsien RY (2005) A guide to choosing fluorescent proteins. Nat Methods 2:905–909

Shyu YJ, Suarez CD, Hu CD (2008) Visualization of AP-1 NF-kappaB ternary complexes in living cells by using a BiFC-based FRET. Proc Natl Acad Sci USA 105:151–156

Singer SJ, Nicholson GL (1972) The fluid mosiac model of the structure of cell membranes. Science 175:720–731

Sinha KK, Udgaonkar JB (2007) Dissecting the non-specific and specific components of the initial folding reaction of barstar by multi-site FRET measurements. J Mol Biol 370:385–405

Sun H, Panicker RC, Yao SQ (2007) Activity based fingerprinting of proteases using FRET peptides. Biopolymers 88:141–149

van Rheenen J, Langeslag M, Jalink K (2004) Correcting confocal acquisition to optimize imaging of fluorescence resonance energy transfer by sensitized emission. Biophys J 86:2517–2529

Vinkenborg JL, Evers TH, Reulen SW, Meijer EW, Merkx M (2007) Enhanced sensitivity of FRET-based protease sensors by redesign of the GFP dimerization interface. Chembiochem 8: 1119–1121

Wouters FS, Bunt G (2009) Molecular resolution of cellular biochemistry and physiology by FRET/FLIM. In: Diaspro A (ed) Nanoscopy and multidimensional optical fluorescence microscopy. Taylor & Francis, Boca Raton

Wouters FS, Verveer PJ, Bastiaens PI (2001) Imaging biochemistry inside cells. Trends Cell Biol 11:203–211

Wu X, Simone J, Hewgill D, Siegel R, Lipsky PE, He L (2006) Measurement of two caspase activities simultaneously in living cells by a novel dual FRET fluorescent indicator probe. Cytom A 69:477–486

Xu X, Gerard AL, Huang BC, Anderson DC, Payan DG, Luo Y (1998) Detection of programmed cell death using fluorescence energy transfer. Nucleic Acids Res 26:2034–2035

Zacharias DA, Violin JD, Newton AC, Tsien RY (2002) Partitioning of lipid-modified monomeric GFPs into membrane microdomains of live cells. Science 296:913–916

Zapata-Hommer O, Griesbeck O (2003) Efficiently folding and circularly permuted variants of the Sapphire mutant of GFP. BMC Biotechnol 3:5

Chapter 9
Measuring Molecular Dynamics by FRAP, FCS, and SPT

Kevin Braeckmans, Hendrik Deschout, Jo Demeester, and Stefaan C. De Smedt

9.1 Introduction

In many research areas, it is important to obtain quantitative information on dynamic properties of molecules and nanoparticulate matter in biomaterials. In cell biology, for example, it has become clear that the cell organization is a highly dynamic process and that advanced microscopy techniques can aid in unraveling cellular molecular dynamics. Microscopy techniques are also useful in the drug delivery field, where substantial efforts are being made to develop smart nanomedicines for delivering therapeutics to specific target tissues in the body. Having a detailed understanding of the stability and transport of such nanomedicines in the blood circulation, extracellular matrix, and target cells is crucial for further improvement of their structure and composition (De Smedt et al. 2005; Remaut et al. 2007b). Several complementary advanced fluorescence microscopy techniques have been developed for studying the mobility of molecules and particles on the micro- and nanoscale. In this chapter, the most important techniques are discussed: fluorescence recovery after photobleaching (FRAP), fluorescence correlation spectroscopy (FCS), and single particle tracking (SPT).

9.2 Fluorescence Recovery After Photobleaching

FRAP is a well-known fluorescence microscopy technique that has been around since the 1970s (Axelrod et al. 1976; Peters et al. 1974). FRAP allows to measure the diffusion of fluorescently labeled molecules or particles on a micrometer scale. A typical FRAP experiment consists of three distinct phases, as is depicted in Fig. 9.1. First, with a low-intensity excitation beam, the fluorescence signal is

K. Braeckmans (✉), H. Deschout, J. Demeester, and S.C. De Smedt
Laboratory of General Biochemistry and Physical Pharmacy, Ghent University, Harelbekestraat 72, 9000 Ghent, Belgium
e-mail: Kevin.Braeckmans@UGent.be

A. Diaspro (ed.), *Optical Fluorescence Microscopy*,
DOI 10.1007/978-3-642-15175-0_9,

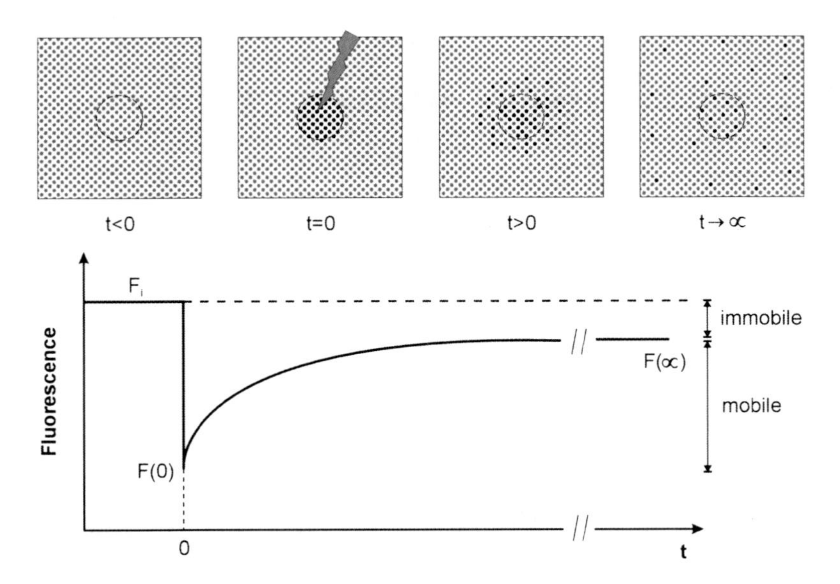

Fig. 9.1 Schematic representation of a typical FRAP experiment. With an intense laser beam, the fluorescent molecules are quickly photobleached inside a particular area. After photobleaching, diffusional exchange of the bleached and non-bleached molecules will occur, resulting in a gradual recovery of the fluorescence inside the photobleached area. With a suitable mathematical model, it is possible to extract the diffusion coefficient and the local (im)mobile fraction of labeled molecules. Reprinted from De Smedt et al. (2005) with permission from Elsevier

measured coming from the region of interest in the fluorescently labeled sample. Next, with a high-power excitation beam, the fluorescent molecules are quickly photobleached in a particular area, typically in the order of a few micrometer up to tens of micrometers in diameter. The photobleaching step creates a local concentration gradient of fluorescent molecules, which will cause net diffusion of the photobleached molecules out of the photobleached area and of intact fluorescent molecules from the surroundings into the photobleached area. The diffusion process after photobleaching is again monitored with a low-intensity light beam. By plotting the fluorescence intensity of the photobleached area as a function of time, where $t = 0$ is the time point immediately after photobleaching, one obtains a typical FRAP curve as shown in Fig. 9.1. By fitting of a suitable FRAP model to the recovery curve, it is possible to extract the diffusion coefficient D of the fluorescently labeled molecules in that area. Moreover, if a fraction of the molecules are immobile inside the photobleached area, they cannot be replaced by intact fluorophores and the fluorescence intensity inside this area will not fully recover. Therefore, from the asymptotic fluorescence intensity value, it is possible to calculate the local (im)mobile fraction as well.

Although all based on the same principle, a multitude of different FRAP analysis methods have been reported. Until the mid 1990s, FRAP was mostly carried out on custom-built instruments where a focused stationary laser beam was used to photobleach a spot in the sample. An attenuated stationary laser beam was used to record the fluorescence recovery in the photobleached spot (Axelrod et al. 1976; Kao et al. 1993; Lopez et al. 1988). In order to obtain spatial rather than just temporal information of the recovery process, a digital camera on an epi-fluorescence microscope was used in combination with Fourier analysis (Berk et al. 1993; Tsay and Jacobson 1991). As these early experiments required a dedicated custom-built microscope setup, until then FRAP experiments were limited to a few specialized laboratories only. In the second half of the 1990s, however, a gradual revival of the FRAP technique occurred due to the advent of user-friendly confocal laser scanning microscopes (CLSMs) equipped with an acousto-optic modulater (AOM) or acousto-optic tunable filter (AOTF). An AOM or AOTF can act as a very fast laser shutter, allowing the intensity of the scanning laser beam to be modulated on a pixel-by-pixel basis, giving the possibility to photobleach any kind of user-defined area in the sample. At first, a stationary laser beam was still used for bleaching but a scanning beam for recording the fluorescence recovery images (Blonk et al. 1993; Cutts et al. 1995). This was followed by other CLSM-based methods making use of the scanning beam for photobleaching small line segments. This approach, however, required extensive numerical computations for calculating the diffusion coefficient (Kubitscheck et al. 1998; Wedekind et al. 1994, 1996). Therefore, despite the simplicity to carry out a FRAP experiment on a CLSM, we saw the need in the early 2000s to develop CLSM-based FRAP methods that are accurate but straightforward to perform and interpret by the nonspecialist. One method makes use of a circular bleach area (Braeckmans et al. 2003), whereas another one is based on the photobleaching of a line profile (Braeckmans et al. 2007). The former is especially useful for diffusion measurements in 3D extended samples, such as extracellular matrices, whereas the latter is more appropriate for measurements in smaller objects, such as living cells. Both require the use of a low-numerical aperture (NA) objective lens to eliminate diffusion along the optical axis in thick specimen, which is not taken into account by the models because of the mathematical complexity. Therefore, we have more recently developed a multiphoton FRAP method which is compatible with high-NA objective lenses and takes diffusion along the optical axis into account (Mazza et al. 2008). A high-NA objective lens is a clear asset in optical microscopy since it provides for a better imaging resolution.

In the drug delivery field, FRAP was used to study the mobility of macromolecules and nanomedicines in extracellular matrices, such as mucus (Braeckmans et al. 2003; Olmsted et al. 2001; Saltzman et al. 1994), (tumor) cell interstitium (Brown et al. 2000, 2004; Papadopoulos et al. 2004; Pluen et al. 2001; Ramanujan et al. 2002), and vitreous (Braeckmans et al. 2003; Peeters et al. 2005). For example, by studying the mobility of differently sized macromolecules in lung mucus and bovine vitreous, we have found that the hyaluronic acid network in the interfibrillar spaces of vitreous poses an extra-sterical hindrance on the diffusing molecules as a function of their size, which is not the case for lung mucus (Braeckmans et al. 2003). We have also shown using FRAP that attaching hydrophilic polyethylene glycol (PEG) chains at the surface of polystyrene nanospheres can circumvent the binding to fibrillar structures in the vitreous and increases their mobility (Peeters et al. 2005). Furthermore, FRAP has been used to study the mobility of nucleotide acids in living cells (Cui et al. 2005; Lukacs et al. 2000; Politz et al. 1998). In related pharmaceutical research, FRAP has been used for studying gel systems for time-controlled drug release as well (Alvarez-Mancenido et al. 2006; Burke et al. 2000; Censi et al. 2009; De Smedt et al. 1994, 1997; van de Manakker et al. 2009).

9.3 Fluorescence Correlation Spectroscopy

FCS is a powerful complementary technique to FRAP that was developed during the same period (Ehrenberg and Rigler 1974; Elson and Magde 1974; Magde et al. 1972, 1974, 1978). As illustrated in Fig. 9.2, an FCS experiment is based on

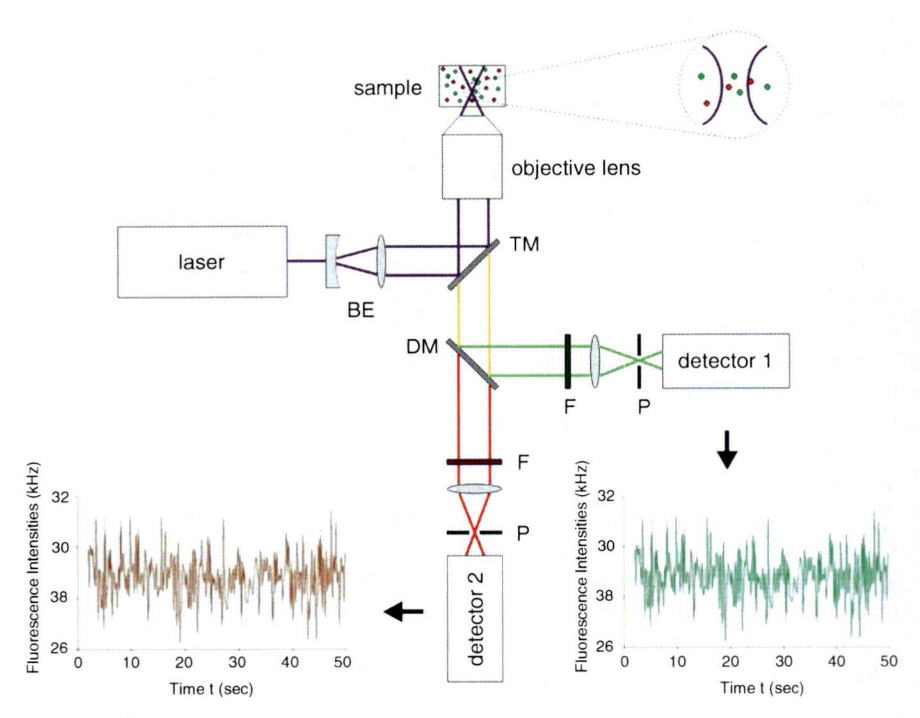

Fig. 9.2 Schematic representation of an FCS setup. A laser beam passes through a beam expander (BE) and is focused by the objective lens to a diffraction limited volume in the sample after being reflected by the trichroic mirror (TM). The fluorescence light generated by the sample is again collected by the objective lens and is sent to the detectors. A dichroic mirror (DM) in combination with suitable fluorescence emission filters (F) can split up the signal into different spectral channels (e.g., *green* and *red*). The confocal pinhole (P) in front of each detector makes sure that only light coming from the focused spot effectively reaches the detectors. Reprinted from Remaut et al. (2007a), with permission from Elsevier

a CLSM type of instrument such that only light from the focal spot can reach the detector. However, rather than being scanned across the sample for obtaining an image, here the focused laser beam is held stationary at one particular location of interest in the sample. The fluorescence intensity is monitored with very high sensitivity and temporal resolution using avalanche photo diode detectors. When a fluorescent molecule enters the focused spot, also called the confocal detection volume, the fluorescence intensity will increase. Conversely, when a molecule leaves the detection volume, the fluorescence intensity will decrease. Thus, the raw output from an FCS experiment is a fluorescence time trace with fluctuations originating from fluorescent molecules moving in and out of the detection volume. Rapid diffusion will result in rapid fluctuations and vice versa. By applying autocorrelation analysis to the fluorescence fluctuation profiles, it is possible to extract the diffusion coefficient as well as the average concentration of molecules in the detection volume (Chen et al. 1999; Elson 2001; Gosch and Rigler 2005; Haustein and Schwille 2004; Krichevsky and Bonnet 2002; Levin and Carson 2004). Apart from diffusion, the fluorescence fluctuations might also arise from other kinds of processes, such as a change in quantum yield, a change in conformation and chemical reactions of the molecules. Therefore, FCS can be used to study photodynamics of fluorophores, as well as binding equilibria and kinetics of enzymes, proteins, and nucleic acids (Levin and Carson 2004; Schwille 2001; Thompson et al. 2002a).

By labeling two different species with fluorophores of a different color, it is possible to quantify their interaction by monitoring the fluorescence time traces in different channels. When the two components are interacting, they will both move together through the detection volume, giving rise to synchronized fluorescence fluctuations in both channels. Conversely, the fluctuations will not be correlated when there is no interaction between the two components as both will move independently from one another through the detection volume. Quantitative information on association and dissociation can thus be obtained by performing cross-correlation analysis (Bacia et al. 2002; Bacia and Schwille 2003; Schwille et al. 1997; Schwille 2001). Our group has used the dual-color FCS technique to study the association and dissociation of antisense oligonucleotides and cationic polymers in buffer and of antisense oligonucleotides and cationic liposomes in living cells (Remaut et al. 2007a). We have also shown that dual-color FCS is a suitable method for studying the integrity of liposomal siRNA formulations in biological fluids, such as full human serum (Buyens et al. 2008).

9.4 Single Particle Tracking

The quantitative information obtained from a FRAP experiment is derived from the diffusion of many hundreds or even thousands of molecules in and around the photobleaching region. The same holds true for information derived from an FCS experiment. While the fluorescence fluctuations are originating from single molecules, the mathematical analysis is performed on a time trace comprising many fluctuations so that the results resemble the ensemble average of many hundreds or thousands of molecules. SPT, however, is a fluorescence microscopy technique capable of visualizing the movement of individual molecules or nanoparticles directly (Saxton and Jacobson 1997; Suh et al. 2005). This is possible by making use of a fast and sensitive camera, such as an electron-multiplying CCD camera, and efficient widefield laser excitation. Although the molecules and nanoparticles are typically below the resolution of the fluorescence microscope, they can be seen as dots of light with a Gaussian-like intensity distribution. The size of these dots is directly related to the resolution of the objective lens being used and is typically around 500 nm diameter for visible light. Therefore, it will be possible to see individual molecules/nanoparticles as long as they are separated by a distance of at least 500 nm. Despite the limited optical resolution, the position of the molecules can be calculated with subresolution precision, typically in the order of tens of nanometers (Kubitscheck et al. 2000; Thompson et al. 2002b). By using suitable image processing algorithms, it is possible to accurately find the center of the dots of light (Anthony et al. 2006; Cheezum et al. 2001; Sbalzarini and Koumoutsakos 2005). Having found the positions of all particles in the SPT movie, their individual trajectories can be calculated, an example of which is given in Fig. 9.3. This is usually done using a nearest neighbor algorithm, although similar particle features can be used in addition to increase the tracking accuracy (Anderson et al. 1992; Lakadamyali et al. 2003). Finally, the trajectories can be analyzed to determine the mode of motion and the corresponding motion parameters. For example, using mean square displacement analysis, it is possible to distinguish free from anomalous diffusion or directed transport (Levi and Gratton 2007; Saxton and Jacobson 1997; Suh et al. 2005). Instead of analyzing the average

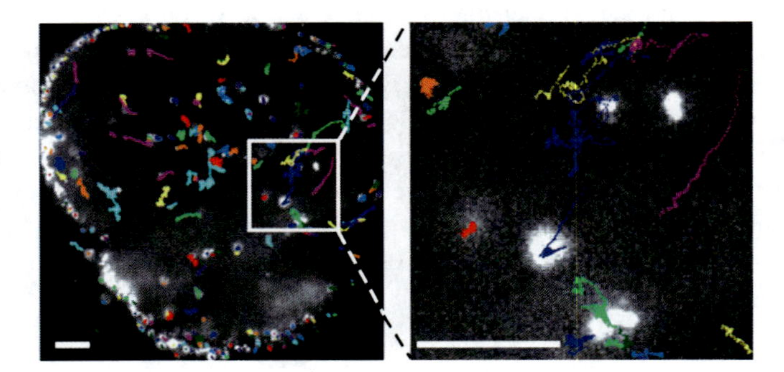

Fig. 9.3 Cytoplasmic transport in living retinal pigment epithelium cells of 100-nm fluorescent polysterene nanospheres coated with a cationic polymer. Trajectories have been calculated from a 60-s SPT movie acquired at 14.4 frames per second. The *inset* shows a variety of trajectories with different modes of motion: restricted motion (*red* and *orange*), directed motion typical for microtubule transport (*pink trajectory to the right*), heterogeneous movement with alternating periods of stalled and directed motion (*dark blue trajectories in the middle* and the *green trajectory at the bottom*). The scale bar is 5 μm. Reprinted from Braeckmans et al. (2009), with permission from John Wiley and Sons and GIT Publishers

displacements along a trajectory, others have been looking at the entire distribution of (square) displacements (Anderson et al. 1992; Apgar et al. 2000; Hellriegel et al. 2004, 2005; Schutz et al. 1997; Sonnleitner et al. 1999; Vrljic et al. 2002). As more information remains available, this type of analysis also allows to study multicomponent mobilities (Hellriegel et al. 2004; Schutz et al. 1997). For example, a molecule may diffuse within a compartment for a certain period of time before "hopping" to the next compartment. If sampled at a sufficiently high rate, the trajectory of such a molecule will show restricted diffusion on a small time scale and free diffusion (hopping between compartments) on a longer time scale (Ritchie et al. 2005). This kind of movement has been observed, for example, for membrane incorporated molecules in cells (Fujiwara et al. 2002; Kusumi et al. 1993). SPT can provide detailed information on the interaction of the molecules of interest and their local environment. For example, by fluorescent labeling of the actin and tubulin networks of living cells, Bausinger et al. (2006) have shown directly the interaction of polyplexes with both cytoskeletons and the different modes of motion in the various stages of intracellular delivery of nanomedicines.

9.5 Conclusion

Advanced fluorescence microscopy methods have been used extensively in the past two decades to gain further insight in the dynamics of molecules in their biological environment. FRAP is very well suited to assess the average diffusion properties of an ensemble of molecules in a micrometer sized area. It also allows to calculate the fraction of particles that are immobile inside the bleached area. FCS can do the same for very fast processes in a submicron detection area. Dual-color FCS with cross-correlation analysis is especially suited for studying intermolecular interactions. SPT allows visualizing the movement of individual fluorescently labeled molecules or particles and can give detailed information on their interaction with the local environment. The combination of such advanced microscopy techniques provides for a powerful set of tools for studying molecular dynamics and will continue to play an important role in cell biology, pharmaceutical, and material research.

References

Alvarez-Mancenido F, Braeckmans K, De Smedt SC, Demeester J, Landin M, Martinez-Pacheco R (2006) Characterization of diffusion of macromolecules in konjac glucomannan solutions and gels by fluorescence recovery after photobleaching technique. Int J Pharm 316:37–46

Anderson CM, Georgiou GN, Morrison IEG, Stevenson GVW, Cherry RJ (1992) Tracking of cell-surface receptors by fluorescence digital imaging microscopy using a charge-coupled device camera. Low-density-lipoprotein and influenza-virus receptor mobility at 4-degrees-C. J Cell Sci 101:415–425

Anthony S, Zhang LF, Granick S (2006) Methods to track single-molecule trajectories. Langmuir 22:5266–5272

Apgar J, Tseng Y, Fedorov E, Herwig MB, Almo SC, Wirtz D (2000) Multiple-particle tracking measurements of heterogeneities in solutions of actin filaments and actin bundles. Biophys J 79:1095–1106

Axelrod D, Koppel DE, Schlessinger J, Elson E, Webb WW (1976) Mobility measurement by analysis of fluorescence photobleaching recovery kinetics. Biophys J 16:1055–1069

Bacia K, Schwille P (2003) A dynamic view of cellular processes by in vivo fluorescence auto- and cross-correlation spectroscopy. Methods 29:74–85

Bacia K, Majoul IV, Schwille P (2002) Probing the endocytic pathway in live cells using dual-color fluorescence cross-correlation analysis. Biophys J 83:1184–1193

Bausinger R, von Gersdorff K, Braeckmans K, Ogris M, Wagner E, Brauchle C, Zumbusch A (2006) The transport of nanosized gene carriers unraveled by live-cell imaging. Angew Chem Int Ed 45:1568–1572

Berk DA, Yuan F, Leunig M, Jain RK (1993) Fluorescence photobleaching with spatial Fourier-analysis – measurement of diffusion in light-scattering media. Biophys J 65:2428–2436

Blonk JCG, Don A, Van Aalst H, Birmingham JJ (1993) Fluorescence photobleaching recovery in the confocal scanning light-microscope. J Microsc 169:363–374

Braeckmans K, Peeters L, Sanders NN, De Smedt SC, Demeester J (2003) Three-dimensional fluorescence recovery after photobleaching with the confocal scanning laser microscope. Biophys J 85:2240–2252

Braeckmans K, Remaut K, Demeester J, De Smedt SC (2007) Line-FRAP with the confocal laser scanning microscope for diffusion measurements in 3-D samples. Biophys J 92:2172–2183

Braeckmans K, Vercauteren D, Demeester J, De Smedt SC (2009) Measuring molecular dynamics. Imaging Microsc 11(2):26–28

Brown E, Pluen A, Compton C, Boucher Y, Jain RK (2000) Measurement of diffusion coefficients in spontaneous human tumors. FASEB J 14:A167

Brown EB, Boucher Y, Nasser S, Jain RK (2004) Measurement of macromolecular diffusion coefficients in human tumors. Microvasc Res 67:231–236

Burke MD, Park JO, Srinivasarao M, Khan SA (2000) Diffusion of macromolecules in polymer solutions and gels: a laser scanning confocal microscopy study. Macromolecules 33:7500–7507

Buyens K, Lucas B, Raemdonck K, Braeckmans K, Vercammen J, Hendrix J, Engelborghs Y, De Smedt SC, Sanders NN (2008) A fast and sensitive method for measuring the integrity of siRNA-carrier complexes in full human serum. J Control Release 126:67–76

Censi R, Vermonden T, van Steenbergen MJ, Deschout H, Braeckmans K, De Smedt SC, van Nostrum CF, di Martino P, Hennink WE (2009) Photopolymerized thermosensitive hydrogels for tailorable diffusion-controlled protein delivery. J Control Release 140:230–236

Cheezum MK, Walker WF, Guilford WH (2001) Quantitative comparison of algorithms for tracking single fluorescent particles. Biophys J 81:2378–2388

Chen Y, Muller JD, Berland KM, Gratton E (1999) Fluorescence fluctuation spectroscopy. Methods 19:234–252

Cui ZQ, Zhang ZP, Zhang XE, Wen JK, Zhou YF, Xie WH (2005) Visualizing the dynamic behavior of poliovirus plus-strand RNA in living host cells. Nucleic Acids Res 33:3245–3252

Cutts LS, Robberts PA, Adler J, Davies MC, Melia CD (1995) Determination of localized diffusion-coefficients in gels using confocal scanning laser microscopy. J Microsc 180: 131–139

De Smedt SC, Lauwers A, Demeester J, Engelborghs Y, Demey G, Du M (1994) Structural information on hyaluronic-acid solutions as studied by probe diffusion experiments. Macromolecules 27:141–146

De Smedt SC, Meyvis TKL, Demeester J, Van Oostveldt P, Blonk JCG, Hennink WE (1997) Diffusion of macromolecules in dextran methacrylate solutions and gels as studied by confocal scanning laser microscopy. Macromolecules 30:4863–4870

De Smedt SC, Remaut K, Lucas B, Braeckmans K, Sanders NN, Demeester J (2005) Studying biophysical barriers to DNA delivery by advanced light microscopy. Adv Drug Deliv Rev 57:191–210

Ehrenberg M, Rigler R (1974) Rotational Brownian-motion and fluorescence intensity fluctuations. Chem Phys 4:390–401

Elson EL (2001) Fluorescence correlation spectroscopy measures molecular transport in cells. Traffic 2:789–796

Elson EL, Magde D (1974) Fluorescence correlation spectroscopy. 1. Conceptual basis and theory. Biopolymers 13:1–27

Fujiwara T, Ritchie K, Murakoshi H, Jacobson K, Kusumi A (2002) Phospholipids undergo hop diffusion in compartmentalized cell membrane. J Cell Biol 157:1071–1081

Gosch M, Rigler R (2005) Fluorescence correlation spectroscopy of molecular motions and kinetics. Adv Drug Deliv Rev 57:169–190

Haustein E, Schwille P (2004) Single-molecule spectroscopic methods. Curr Opin Struct Biol 14:531–540

Hellriegel C, Kirstein J, Brauchle C, Latour V, Pigot T, Olivier R, Lacombe S, Brown R, Guieu V, Payrastre C, Izquierdo A, Mocho P (2004) Diffusion of single streptocyanine molecules in the nanoporous network of sol-gel glasses. J Phys Chem B 108:14699–14709

Hellriegel C, Kirstein J, Brauchle C (2005) Tracking of single molecules as a powerful method to characterize diffusivity of organic species in mesoporous materials. New J Phys 7:1–14

Kao HP, Abney JR, Verkman AS (1993) Determinants of the translational mobility of a small solute in cell cytoplasm. J Cell Biol 120:175–184

Krichevsky O, Bonnet G (2002) Fluorescence correlation spectroscopy: the technique and its applications. Rep Prog Phys 65:251–297

Kubitscheck U, Wedekind P, Peters R (1998) Three-dimensional diffusion measurements by scanning microphotolysis. J Microsc 192:126–138

Kubitscheck U, Kuckmann O, Kues T, Peters R (2000) Imaging and tracking of single GFP molecules in solution. Biophys J 78:2170–2179

Kusumi A, Sako Y, Yamamoto M (1993) Confined lateral diffusion of membrane-receptors as studied by single-particle tracking (nanovid microscopy) – effects of calcium-induced differentiation in cultured epithelial-cells. Biophys J 65:2021–2040

Lakadamyali M, Rust MJ, Babcock HP, Zhuang XW (2003) Visualizing infection of individual influenza viruses. Proc Natl Acad Sci U S A 100:9280–9285

Levi V, Gratton E (2007) Exploring dynamics in living cells by tracking single particles. Cell Biochem Biophys 48:1–15

Levin MK, Carson JH (2004) Fluorescence correlation spectroscopy and quantitative cell biology. Differentiation 72:1–10

Lopez A, Dupou L, Altibelli A, Trotard J, Tocanne JF (1988) Fluorescence recovery after photobleaching (FRAP) experiments under conditions of uniform disk illumination – critical comparison of analytical solutions, and a new mathematical method for calculation of the diffusion coefficient D. Biophys J 53:963–970

Lukacs GL, Haggie P, Seksek O, Lechardeur D, Freedman N, Verkman AS (2000) Size-dependent DNA mobility in cytoplasm and nucleus. J Biol Chem 275:1625–1629

Magde D, Webb WW, Elson E (1972) Thermodynamic fluctuations in a reacting system – measurement by fluorescence correlation spectroscopy. Phys Rev Lett 29:705–708

Magde D, Elson EL, Webb WW (1974) Fluorescence correlation spectroscopy. 2. Experimental realization. Biopolymers 13:29–61

Magde D, Webb WW, Elson EL (1978) Fluorescence correlation spectroscopy. 3. Uniform translation and laminar-flow. Biopolymers 17:361–376

Mazza D, Braeckmans K (with equal contribution), Cella F, Testa I, Vercauteren D, Demeester J, De Smedt SC, Diaspro A (2008) A new FRAP/FRAPa method for three-dimensional diffusion measurements based on multiphoton excitation microscopy. Biophys J 95:3457–3469

Olmsted SS, Padgett JL, Yudin AI, Whaley KJ, Moench TR, Cone RA (2001) Diffusion of macromolecules and virus-like particles in human cervical mucus. Biophys J 81: 1930–1937

Papadopoulos MC, Binder DK, Verkman AS (2004) Enhanced macromolecular diffusion in brain extracellular space in mouse models of vasogenic edema measured by cortical surface photobleaching. FASEB J 18:425–427

Peeters L, Sanders NN, Braeckmans K, Boussery K, de Voorde JV, De Smedt SC, Demeester J (2005) Vitreous: a barrier to nonviral ocular gene therapy. Investig Ophthalmol Vis Sci 46: 3553–3561

Peters R, Peters J, Tews KH, Bahr W (1974) Microfluorimetric study of translational diffusion in erythrocyte-membranes. Biochim Biophys Acta 367:282–294

Pluen A, Boucher Y, Ramanujan S, McKee TD, Gohongi T, di Tomaso E, Brown EB, Izumi Y, Campbell RB, Berk DA, Jain RK (2001) Role of tumor-host interactions in interstitial diffusion of macromolecules: cranial vs. subcutaneous tumors. Proc Natl Acad Sci U S A 98:4628–4633

Politz JC, Browne ES, Wolf DE, Pederson T (1998) Intranuclear diffusion and hybridization state of oligonucleotides measured by fluorescence correlation spectroscopy in living cells. Proc Natl Acad Sci U S A 95:6043–6048

Ramanujan S, Pluen A, McKee TD, Brown EB, Boucher Y, Jain RK (2002) Diffusion and convection in collagen gels: implications for transport in the tumor interstitium. Biophys J 83:1650–1660

Remaut K, Lucas B, Raemdonck K, Braeckmans K, Demeester J, De Smedt SC (2007a) Can we better understand the intracellular behavior of DNA nanoparticles by fluorescence correlation spectroscopy? J Control Release 121:49–63

Remaut K, Sanders NN, De Geest BG, Braeckmans K, Demeester J, De Smedt SC (2007b) Nucleic acid delivery: where material sciences and bio-sciences meet. Mater Sci Eng R Rep 58:117–161

Ritchie K, Shan XY, Kondo J, Iwasawa K, Fujiwara T, Kusumi A (2005) Detection of non-Brownian diffusion in the cell membrane in single molecule tracking. Biophys J 88:2266–2277

Saltzman WM, Radomsky ML, Whaley KJ, Cone RA (1994) Antibody diffusion in human cervical-mucus. Biophys J 66:508–515

Saxton MJ, Jacobson K (1997) Single-particle tracking: applications to membrane dynamics. Annu Rev Biophys Biomol Struct 26:373–399

Sbalzarini IF, Koumoutsakos P (2005) Feature point tracking and trajectory analysis for video imaging in cell biology. J Struct Biol 151:182–195

Schutz GJ, Schindler H, Schmidt T (1997) Single-molecule microscopy on model membranes reveals anomalous diffusion. Biophys J 73:1073–1080

Schwille P (2001) Fluorescence correlation spectroscopy and its potential for intracellular applications. Cell Biochem Biophys 34:383–408

Schwille P, MeyerAlmes FJ, Rigler R (1997) Dual-color fluorescence cross-correlation spectroscopy for multicomponent diffusional analysis in solution. Biophys J 72:1878–1886

Sonnleitner A, Schutz GJ, Schmidt T (1999) Free Brownian motion of individual lipid molecules in biomembranes. Biophys J 77:2638–2642

Suh J, Dawson M, Hanes J (2005) Real-time multiple-particle tracking: applications to drug and gene delivery. Adv Drug Deliv Rev 57:63–78

Thompson NL, Lieto AM, Allen NW (2002a) Recent advances in fluorescence correlation spectroscopy. Curr Opin Struct Biol 12:634–641

Thompson RE, Larson DR, Webb WW (2002b) Precise nanometer localization analysis for individual fluorescent probes. Biophys J 82:2775–2783

Tsay TT, Jacobson KA (1991) Spatial Fourier-analysis of video photobleaching measurements – principles and optimization. Biophys J 60:360–368

van de Manakker F, Braeckmans K, el Morabit N, De Smedt SC, van Nostrum CF, Hennink WE (2009) Protein-release behavior of self-assembled PEG-beta-cyclodextrin/PEG-cholesterol hydrogels. Adv Funct Mater 19:2992–3001

Vrljic M, Nishimura SY, Brasselet S, Moerner WE, McConnell HM (2002) Translational diffusion of individual class II MHC membrane proteins in cells. Biophys J 83:2681–2692

Wedekind P, Kubitscheck U, Peters R (1994) Scanning microphotolysis – a new photobleaching technique based on fast intensity modulation of a scanned laser-beam and confocal imaging. J Microsc 176:23–33

Wedekind P, Kubitscheck U, Heinrich O, Peters R (1996) Line-scanning microphotolysis for diffraction-limited measurements of lateral diffusion. Biophys J 71:1621–1632

Chapter 10
In Vitro–In Vivo Fluctuation Spectroscopies

M. Collini, L. D'Alfonso, M. Caccia, L. Sironi, M. Panzica, G. Chirico, I. Rivolta, B. Lettiero, and G. Miserocchi

10.1 Introduction

Up to the middle of last century, fluctuations were seen as a source of experimental uncertainty. During the Second World War, the technology for the fast analysis of the spectral content of the electromagnetic radiation was developed due to the effort to enhance the performance of the radar systems. The same technology, transposed in the visible range of the electromagnetic radiation, allowed to obtain for the first time quantitative information on the system structure and dynamics from the noise superimposed on its scattered light (Berne and Pecora 2000). Although initially these experiments were based on the computation of the Fourier spectrum of the acquired scattering signal, the modern developments are based on the computation of the autocorrelation function (ACF) of the scattered intensity, $I(t)$:

$$G(\tau) = \langle I(t+\tau)I(t)\rangle_t = \int_0^T \frac{I(t+\tau)I(t)}{T}\,\mathrm{d}t\,. \qquad (10.1)$$

Photon correlation spectroscopy (PCS) is now able to investigate through the computation of (10.1) dynamic processes as fast as 20 ns (Hobel and Ricka 1994; Chirico and Gardella 1999) being mostly limited by the number of photons collected per sampling time.

The principles of fluorescence correlation spectroscopy (FCS) have been introduced by the Webb group at the Cornell University in the 1970s and experimentally proved on drug–DNA kinetics measurements (Magde et al. 1972). The major difference between PCS and FCS techniques lies in the fact that PCS is based on the phase

M. Collini, L. D'Alfonso, M. Caccia, L. Sironi, M. Panzica, and G. Chirico (✉)
Dipartimento di Fisica G.Occhialini, Università di Milano Bicocca, Piazza della Scienza 3, 20126 Milan, Italy
e-mail: giuseppe.chirico@mib.infn.it

I. Rivolta, B. Lettiero, and G. Miserocchi
Dipartimento di Medicina Sperimentale, Università di Milano Bicocca, Via Cadore 48, 20052 Monza, Italy

A. Diaspro (ed.), *Optical Fluorescence Microscopy*,
DOI 10.1007/978-3-642-15175-0_10,

fluctuations of a coherent signal (scattering) induced by the motion of the scattering centers within the observation volume, whereas FCS is based on the amplitude fluctuations of an incoherent signal (fluorescence) due to the motion of fluorescent particles through the observation volume. FCS is a rapidly developing field in itself, and it has been already recognized as an essential tool for the in vitro characterization of absolute concentrations, molecular interactions, and kinetic processes, such as diffusion and chemical reactions (Kohl et al. 2005; Haustein and Schwille 2003; Gosch and Rigler 2005; Enderlein et al. 2004). In vivo (cellular) applications of FCS have also been reported (Chen et al. 2002; Weidemann et al. 2003).

In this chapter, we will focus on autocorrelation and cross-correlation methods applied to the study of the dynamics of in vitro and in vivo systems. In particular, we discuss the case of the photodynamics of dyes or fluorescent proteins trapped in gel matrices and the diffusion of nanoparticles in the cellular matrix and within the cells. The following derivation has been made for confocal one photon excitation (OPE). However no substantial difference appears in the case of two photon excitation (TPE).

10.2 Fluctuation Spectroscopy: General Principles

10.2.1 Average Fluctuations of the Fluorescence Signal

The description of the fluorescence fluctuations arising from the emission of few or single molecules diffusing through the observation volume is better derived, within the FCS method (Muller et al. 2003; Bismuto et al. 2001; Berland et al. 1995), by means of the fluorescence fluctuation ACF:

$$g(\tau) = \frac{\langle \delta F(t+\tau)\delta F(t)\rangle_t}{\langle F(t)\rangle_t^2} = \frac{\langle F(t+\tau)F(t)\rangle_t - \langle F(t)\rangle_t^2}{\langle F(t)\rangle_t^2}\,. \tag{10.2}$$

Since the fluorescence signal, $F(t)$, is proportional to the number of molecules within the observation volume, N, the zero lag extrapolation of the ACF, $g(\tau \to 0)$, has a simple and important meaning:

$$g(\tau \to 0) = \frac{\left\langle \delta F(t)^2 \right\rangle_t}{\langle F(t)\rangle_t^2} = \gamma \frac{\langle \delta N^2\rangle_t}{\langle N\rangle_t^2}\,. \tag{10.3}$$

In fact, from (10.3), by assuming a Poisson distribution for N, we find that:

$$g_0 = g(\tau \ll \tau_{\mathrm{diff}}) \approx \frac{\gamma}{\langle N\rangle_t}\,, \tag{10.4}$$

where τ_{diff} is the diffusion time of the fluorochromes through the observation volume. A measure of the zero lag time correlation function allows us to count directly the number of diffusing and fluorescent molecules. The proportionality factor γ between the fluorescence and the number fluctuations (Berland and Shen 2003) is determined by the actual shape of the observation volume. As an example, for a 3D diffusion, the geometrical factors for single-photon confocal detection and for TPE excitation are $\gamma_{\text{OPE}} = 0.35$ and $\gamma_{\text{TPE}} = 0.076$, respectively.

10.2.2 ACF in a Generic Optical Field

The information on the dynamics of the system (diffusion, drift, photodynamics, chemical kinetics, etc.) can be gained from the time evolution of the ACFs. The most simple case, the 3D diffusive motion, corresponds to a hyperbolic decay of the ACFs, which is unusual for the physical theory of random motion that predicts (Doi and Edwards 1986) exponential decays of the density fluctuations correlation functions. The reason for the hyperbolic decay in FCS ACFs lies in the fact that the FCS decay arises from the sum of an infinite number of exponential decays due to the wide **q**-vector bunch which spans the numerical aperture of the objective used to focus the laser beam on a tiny volume. The small volume needed to observe few particles at a time can be obtained only by raising the optics numerical aperture and therefore increasing the number of decay modes that are summed in the FCS ACFs. For a single component solution at the average concentration $\langle C \rangle$, this can be formally described, as done first by Aragon and Pecora (1976), as:

$$g(\tau) \approx \frac{\int d\mathbf{q} \left|\hat{W}(\mathbf{q})\right|^2 \Re(\mathbf{q}, \tau)}{\left|\hat{W}(\mathbf{0})\right|^2 \langle C \rangle^2}. \quad (10.5)$$

$$\Re(\mathbf{q}, \tau) = \int \langle \delta C(\mathbf{r}, t + \tau) \delta C(0, t) \rangle_t \exp[-i\mathbf{q} \cdot \mathbf{r}]. \quad (10.6)$$

In (10.5), $\hat{W}(\mathbf{k}) \propto \int d\mathbf{k} \exp\left[-i\mathbf{k} \cdot \mathbf{r}\right] W(\mathbf{r})$ is the Fourier transform of the effective beam profile, $W(\mathbf{r})$, which may be approximated in the OPE confocal case to a 3D Gaussian, and in the TPE case to a Gaussian–Lorentzian shape. The function $\Re(\mathbf{q}, t)$ is the Fourier transform of the ACF of the dye concentration fluctuation, $P(\mathbf{r}, \tau) = \langle \delta C(\mathbf{r}, t + \tau) \delta C(0, t) \rangle_t$ that satisfies the diffusion equation:

$$\begin{cases} \dfrac{\partial}{\partial t} P(\mathbf{r}, t) = -div(\mathbf{J}) \\ \mathbf{J} = -Dgrad[P(\mathbf{r}, t)] + v_{\text{drift}} P(\mathbf{r}, t) \end{cases} \quad (10.7)$$

where v_{drift} is a drift velocity that arises from fluxes in the samples or deterministic motions of the particles, and D is the average translational diffusion coefficient.

Due to the boundary condition, $P(\mathbf{r},0) = \langle \delta C^2 \rangle \delta(\mathbf{r}) = \langle C \rangle \delta(\mathbf{r})$, the ACF of the concentration fluctuations for each Fourier vector $\mathbf{q}$ has then the following decay:

$$\Re(\mathbf{q},t) = \langle C \rangle \exp\left[-D|\mathbf{q}|^2 t - i\mathbf{q} \cdot \mathbf{v}_{\text{drift}} t\right] \tag{10.8}$$

Several interesting issues can be discussed by starting from (10.5) to (10.8). First of all, we notice that the limit to zero lag time of the correlation function is:

$$g(0) \approx \frac{\int d\mathbf{q} \left|\hat{W}(\mathbf{q})\right|^2}{\left|\hat{W}(\mathbf{0})\right|^2 \langle C \rangle} = \frac{\int d\mathbf{q} \left|\hat{W}(\mathbf{q})\right|^2}{\left|\hat{W}(\mathbf{0})\right|} \frac{1}{\langle N \rangle} = \frac{\gamma}{\langle N \rangle} = \frac{\gamma'}{\langle C \rangle} \tag{10.9}$$

from which it is apparent the geometrical meaning of the proportionality factor γ.

The simplest dynamical process that can be analyzed through (10.5) is that of 3D free diffusion of a single species. By assuming a 3D Gaussian laser profile, a good approximation for the confocal detection optics, $W(x,y,z) \propto \exp\left[-\left(x^2/w_x^2\right) - \left(y^2/w_y^2\right) - \left(z^2/w_z^2\right)\right]$ and therefore $W(q_x,q_y,q_z) \propto \exp\left[-\left(q_x^2 w_x^2/4\right) - \left(q_y^2 w_y^2/4\right) - \left(q_z^2 w_z^2/4\right)\right]$ we obtain:

$$g(\tau) \approx \left(\frac{1}{\sqrt{2\pi}}\right)^3 \frac{1}{w_x w_y w_z} \frac{1}{\langle C \rangle} \times \frac{\exp\left[-\left(\frac{\tau}{\tau_{\text{drift},x}}\right)^2 \frac{1}{\left(1+\tau/\tau_{\text{D},x}\right)} - \left(\frac{\tau}{\tau_{\text{drift},y}}\right)^2 \frac{1}{\left(1+\tau/\tau_{\text{D},y}\right)} - \left(\frac{\tau}{\tau_{\text{drift},z}}\right)^2 \frac{1}{\left(1+\tau/\tau_{\text{D},z}\right)}\right]}{\sqrt{1+\tau/\tau_{\text{D},x}}\sqrt{1+\tau/\tau_{\text{D},y}}\sqrt{1+\tau/\tau_{\text{D},z}}}, \tag{10.10}$$

where we have defined a drift time, $\tau_{\text{drift}} = w_x/v_{\text{drift},x} \cong w_y/v_{\text{drift},y} \cong w_z/v_{\text{drift},z}$, and a diffusion relaxation time, $\tau_{D,x} = w_x^2/(4D)$. Therefore, when both diffusive and drift motions are present, the decay of the ACF is a composition of an exp $(-t^2)$ and of a hyperbolic decay. In the case of a simple drift motion in the focal plane with circularly symmetric beam, (10.10) simplifies to:

$$\begin{cases} g(\tau) \approx g(0) \dfrac{\exp\left[-\left(\frac{\tau}{\tau_{\text{drift},x}}\right)^2 \frac{1}{\left(1+\tau/\tau_{\text{D},x}\right)}\right]}{\left(1+\tau/\tau_{\text{D},x}\right)\sqrt{1+\tau/\tau_{\text{D},z}}} = g(0) G_{\text{drift}}(\tau) G_{\text{diff}}(\tau) \\ G_{\text{drift}}(\tau) = \exp\left[-\left(\dfrac{\tau}{\tau_{\text{drift},x}}\right)^2 \dfrac{1}{\left(1+\tau/\tau_{\text{D},x}\right)}\right] \\ G_{\text{diff}}(\tau) = \dfrac{1}{\left(1+\tau/\tau_{\text{D},x}\right)\sqrt{1+\tau/\tau_{\text{D},z}}} \end{cases} . \tag{10.11}$$

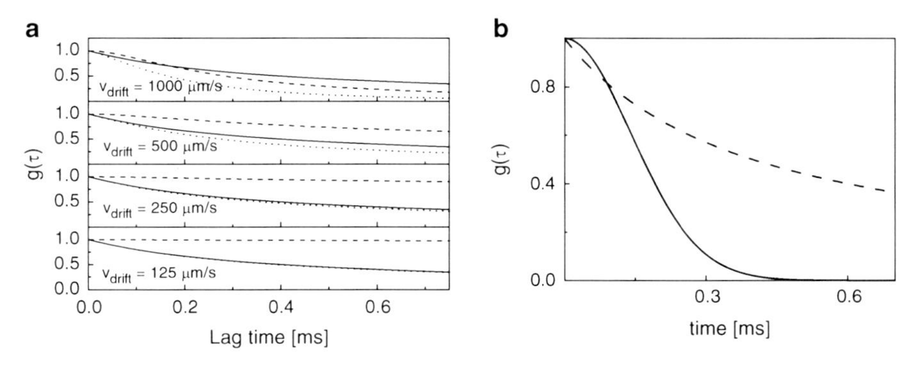

Fig. 10.1 (**a**) Simulations of the ACF of the fluorescence signal collected from fluorophores with a diffusion time $\tau_D = 400$ μs, and drift speeds $v_{drift} = 1{,}000, 500, 250, 125$ μm/s from *top* to *bottom*. The *solid*, *dashed*, and *dotted lines* are the simulation of the pure diffusive, pure drift, and total correlation functions according to (10.11). For a beam waist radius $w_0 = 0.4$ μm, with a diffusion coefficient $D = 100$ μm^2/s, the drift times are $\tau_{drift} = 0.4, 0.8, 1.6$, and 3.2 ms from *top* to *bottom*. (**b**) Simulation of the ACF decay for a pure diffusive motion (*dashed line*) with $\tau_D = 400$ μs and for an almost pure drift motion ($\tau_{drift} = 200$ μs; $\tau_D = 40$ ms, *solid line*)

As seen in Fig. 10.1a when the ratio $\tau_{drift}/\tau_D = 4D/(wv_{drift}) \gg 1$, i.e., when the drift velocity is much smaller than the ratio $4D/w$, the diffusion decay largely determines the whole ACF and the shape of the decay can easily be discriminated by a nonlinear least square fitting procedure (Fig. 10.1b). As an additional example for a microsphere 400 nm in size, and a beam radius $w \cong 0.4$ μm, the diffusion time is $\tau_D \cong 40$ ms. For a drift velocity $\cong 200$ μm/s, the effect of the drift component is evident in the ACF decay ($\tau_{drift} \cong 2$ ms, $\tau_{drift}/\tau_D = 0.05$), and it becomes overwhelming for $v_{drift} \cong 2{,}000$ μm/s ($\tau_{drift} \cong 0.2$ ms, $\tau_{drift}/\tau_D = 0.005$, Fig. 10.1b). On the other hand, for $v_{drift} \cong 20$ μm/s, the drift contribution to the ACF would be negligible ($\tau_{drift} \cong 20$ ms, $\tau_{drift}/\tau_D = 0.5$).

10.2.3 Generalized Excitation Modes

Equation (10.5) indicates that the correlation function is basically determined by two major contributions, related to the shape of the excitation beam and to the dynamics that the molecules are undergoing. For example, when scanning FCS excitation mode (Petrasek and Schwille 2008; ries and Schwille 2006; Skinner et al. 2005; Xiao et al. 2005) or double foci FCS (Xia et al. 1995; Brinkmeier et al. 1999; Dittrich and Schwille 2002) mode are adopted, we need to modify the Fourier transform (FT) of the beam profile, $\hat{W}(\mathbf{q})$.

10.2.3.1 Dual Beam Excitation: ACF

For the double-beam excitation, the beam profile FT changes to:

$$\begin{aligned}\hat{W}_{\mathrm{DB}}(\mathbf{q},\mathbf{R}) &\propto \int d\mathbf{q}\exp[-i\mathbf{q}\cdot\mathbf{r}][W(\mathbf{r})+W(\mathbf{r}+\mathbf{R})]\\ &= \hat{W}(\mathbf{q})[1+\exp(i\mathbf{q}\cdot\mathbf{R})],\end{aligned} \tag{10.12}$$

where $\mathbf{R}$ is the vector that joins the center of the two beam waists. If only one diffusing species is present, subject also to a drift with velocity v_{drift}, then by substituting (10.12) in (10.5) we find[1] that:

$$g(\tau) \approx 2\frac{\int d\mathbf{q}\Re(q,\tau)\left|\hat{W}(\mathbf{q})\right|^2[1+\cos(\mathbf{q}\cdot\mathbf{R})]}{\langle C\rangle}\gamma'. \tag{10.13}$$

By substituting (10.8) into (10.5) and integrating over $\mathbf{q}$, one can readily obtain the decay of the ACF as:

$$\begin{cases} g(\tau,R) \approx g(0)G_{\mathrm{diff}}(\tau)G_{\mathrm{DB}}(\tau) \\ G_{\mathrm{DB}}(\tau) = \left[1+\exp\left(-\dfrac{[(R-v_{\mathrm{drift}}\tau)/w]^2}{(1+\tau/\tau_{\mathrm{D},x})}\right)\right] \end{cases}. \tag{10.14}$$

In the previous relation, we have omitted a term proportional to $\exp([(R+v_{\mathrm{drift}}\tau)/w]^2)$, which is rapidly decreasing with the lag time and we have written the diffusive part of the ACF for a confocal setup. The trend of the ACF reported in (10.14) is sketched in Fig. 10.2. The presence of a second component due to the drift motion appears here as a peak at lag times $\tau \cong R/v_{\mathrm{drift}}$ (Fig. 10.2b), as can be expected from the excitation configuration.

10.2.3.2 Dual Beam Excitation: CCF

The previous case corresponds to the situation in which two beams are impinging on the solution and we are collecting the fluorescence signal from both the excitation volumes. In most cases, it is more convenient to collect the signals emitted by the two excitation volumes with two distinct detectors and compute their cross-correlation function (CCF). In this case, the cross-correlation function of the two signals becomes:

$$g(\tau) \approx \frac{\int d\mathbf{q}\left|\hat{W}(\mathbf{q})\right|^2\exp[-i\mathbf{q}\cdot\mathbf{R}]\Re(\mathbf{q},\tau)}{\left|\hat{W}(\mathbf{0})\right|^2\langle C\rangle^2}. \tag{10.15}$$

[1]We have used the relation: $\left|\hat{W}_{DB}(\mathbf{q},\mathbf{R})\right|^2 = 2\left|\hat{W}(\mathbf{q})\right|^2(1+\cos[\mathbf{q}\cdot\mathbf{R}])$

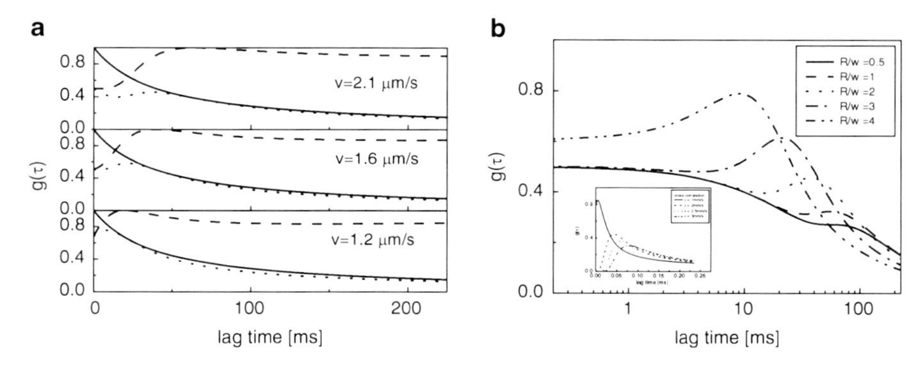

Fig. 10.2 Simulations for the dual beam case: ACFs (10.14). (**a**) Diffusive ACF (*solid line*), pure dual beam contribution (*dashed line*), and total ACF (*dotted line*) for $R/w = 2$ and $\tau_D = 40$ ms. The value of the drift velocity is indicated in the panels. (**b**) Total ACFs for variable ratios $R/w = 0.5, 1, 2, 3, 4$ at $\tau_D = 40$ ms and $v_{drift} = 2.1$ μm/s. The *inset* shows the corresponding cross-correlation acquisition mode results

If the sample is subject to a drift motion in the focal plane between the two excitation volumes, and the observation volume is axially symmetric around the optical axis, the cross-correlation function is then:

$$\begin{cases} g(\tau, R) \approx g(0) G_{xDB}(\tau) G_{\text{diff}}(\tau) \\ G_{xDB}(\tau) = \exp\left[-\dfrac{[(R - v_{\text{drift}}\tau)/w]^2}{(1 + \tau/\tau_{D,x})} \right] \end{cases} . \quad (10.16)$$

As can be seen from the decay shown in Fig. 10.2b (inset), the effect of the drift motion between the two excitation volumes is here even more pronounced than in the case of (10.14).

10.2.3.3 Scanning FCS

In this case, the excitation beam is moved on the sample according to a generic functional form $\rho(t)$. The same treatment that has led (Aragon and Pecora 1976) to (10.5) leads to the generalized case:

$$g(\tau) \approx \frac{\int d\mathbf{q} \left| \hat{W}(\mathbf{q}) \right|^2 \exp[-i\mathbf{q} \cdot (\boldsymbol{\rho}(t + \tau) - \boldsymbol{\rho}(t))] \Re(\mathbf{q}, \tau)}{\left| \hat{W}(\mathbf{0}) \right|^2 \langle C \rangle^2} . \quad (10.17)$$

Since the FT of the beam profile and the diffusion function depend on the square of the wave vector q, the correlation function can also be simplified to:

$$g_{\text{scan}}(\tau) \approx g(0)G_{\text{diff}}(\tau)G_{\text{scan}}(t,\tau)$$
$$= g(0)G_{\text{diff}}(\tau)\exp\left[-\frac{|\boldsymbol{\rho}(t+\tau)-\boldsymbol{\rho}(t)|^2}{w^2\left(1+\tau/\tau_{\text{D},x}\right)}\right]. \quad (10.18)$$

The two simplest implementations are the circular scanning FCS and the stochastic scanning FCS. In the first case, already applied in the literature (Petrasek and Schwille 2008; Ries and Schwille 2006; Skinner et al. 2005; Xia et al. 1995), the scanning function is given by $\boldsymbol{\rho}(t) = \mathbf{A}\cos(\Omega t)$ and the correlation function becomes:

$$g_{\text{scan}}(t,\tau) \approx g(0)G_{\text{diff}}(\tau)G_{\text{cScan}}(\tau)$$
$$= g(0)G_{\text{diff}}(\tau)\exp\left[-4\frac{A^2\sin^2(\Omega\tau/2)}{w^2\left(1+\tau/\tau_{\text{D},x}\right)}\right]. \quad (10.19)$$

In Eq. (10.19) the ACF displays a periodic behavior, with period $T = 2/\Omega$, superimposed to the diffusive one.

The case of a stochastic motion of the beam may appear a pure academic exercise. In fact, when investigating a fluorophore solution, stochastic motion of the laser beam should not change the shape of the correlation function of the fluorescence fluctuations. However, this excitation mode may be extremely valuable when investigating samples in which the diffusive motions are frozen, as it happens when studying fluorescent molecules immobilized in gel or glassy matrices. In this situation, bleaching is largely effective if one keeps the laser beam position fixed: by moving the excitation laser beam, instead, the molecules would have enough relaxing time between two consecutive irradiation time stretches and the laser induced photobleaching should be therefore reduced.

The scanning function can be expressed in this case only in terms of the average of the square of the beam displacement on the sample, and in the ideal case it should be described by a function of the type:

$$\left\langle|\rho(t+\tau)-\rho(t)|^2\right\rangle_t = \left\langle\left|\rho_{x,y}(t+\tau)-\rho_{x,y}(t)\right|^2\right\rangle_t = 4\alpha^2\tau\,. \quad (10.20)$$

At the same time, since the diffusive motions are frozen, the diffusive function is a constant, $\Re(\mathbf{q},\tau) = 1$, and the ACF is therefore similar in shape to the one typically detected from freely diffusing molecules in solution. Equation (10.17) becomes:

$$g(\tau) \approx \frac{\int d\mathbf{q}\left|\hat{W}(\mathbf{q})\right|^2\langle\exp(-i\mathbf{q}\cdot[\boldsymbol{\rho}(t+\tau)-\boldsymbol{\rho}(t)])\rangle_t}{\left|\hat{W}(\mathbf{0})\right|^2\langle C\rangle^2}$$
$$= \frac{\int d\mathbf{q}_{xy}\left|\hat{W}(\mathbf{q}_{xy})\right|^2\exp\left[-q_{xy}^2\left\langle\left|\boldsymbol{\rho}_{xy}(t+\tau)-\boldsymbol{\rho}_{xy}(t)\right|^2\right\rangle_t\right]}{\left|\hat{W}(\mathbf{0})\right|^2\langle C\rangle^2}. \quad (10.21)$$

The shape of the correlation function is then the usual hyperbolic one [see (10.11) with $\tau_{\text{drift}} = 0$], where the diffusion coefficient is replaced by α^2:

$$g(\tau) \approx \left(\frac{1}{\sqrt{2\pi}}\right)^3 \frac{1}{w_x w_y w_z} \frac{1}{\langle C \rangle} \frac{1}{\left[1 + (4\alpha^2\tau)/w_x w_y\right]} = g(0) G_\alpha(\tau) \,. \tag{10.22}$$

Apart from the simplified shape of the correlation function, the advantage of this excitation mode would consist mainly in the possibility to measure the whole ACF on immobilized fluorophores while limiting the bleaching effects. This is even more apparent when fluorescence fluctuations due to internal photodynamics or conformational dynamics of the fluorophores occur in the lag time spanned by the correlation function decay ($\cong w/\alpha$), as it is the case, for example, of the triplet dynamics, conformational (*cis–trans* or rotamers) dynamics of complex fluorophores, such as the GFP one, or chemical kinetics. The internal dynamics decay, $M(\tau)$, appears as a multiexponential decay superimposed on the diffusive-like trend as:

$$g(\tau) \approx \left(\frac{1}{\sqrt{2\pi}}\right)^3 \frac{1}{w_x w_y w_z} \frac{1}{\langle C \rangle} \frac{M(\tau)}{\left[1 + (4\alpha^2\tau)/w_x w_y\right]} = g(0) M(\tau) G_\alpha(\tau) \,. \tag{10.23}$$

In (10.23), the fluorescence fluctuations due to changes in the quantum yield induced by chemical equilibria are accounted for by the function $M(\tau)$. Let us then analyze briefly, as a prototype of the internal dynamics decay, the effect of chemical kinetics on the fluorescence fluctuations.

10.2.3.4 Chemical Kinetics

A simple chemical dynamics between a bright and a dark form of the same molecule,

$$B \underset{k_\text{b}}{\overset{k_\text{f}}{\leftrightarrow}} D$$

implies the presence of an additional exponential relaxation, superimposed on the diffusion term, that is determined by the solution of the following system:

$$\begin{cases} \dfrac{\partial}{\partial t} \delta C_i(\mathbf{r}, t) = -\text{div}(\mathbf{J}_i) + \sum\limits_k T_{ik} \delta C_k(\mathbf{r}, t) \\ \mathbf{J}_k = -D grad[\delta C_k(\mathbf{r}, t)] \end{cases}, \tag{10.24}$$

where $\boldsymbol{\delta} C(r, t) = \begin{bmatrix} \delta C_\text{B}(r, t) \\ \delta C_\text{D}(r, t) \end{bmatrix}$ and $i = B, D$. The matrix $\mathbf{T}$ is given by:

$$\mathbf{T} = \begin{bmatrix} -k_\text{f} & k_\text{b} \\ k_\text{f} & -k_\text{b} \end{bmatrix} \tag{10.25}$$

By performing the FT on the position coordinate, we come to a modified and generalized form of (10.8):

$$\Re(q,\tau) = \langle \delta C(q,t+\tau)\delta C^*(q,t)\rangle_t = \langle C\rangle \exp\left[-Dq^2t\right]\left[1 - T + T\exp(-\lambda t)\right]$$
$$\lambda = k_\mathrm{b} + k_\mathrm{f}\,. \tag{10.26}$$

By substituting this decay into (10.5), we obtain the ACF decay in the presence of a chemical kinetics between a dark and a bright form ($\tau_T = 1/\lambda$):

$$M(\tau) = \frac{\left[1 - T + T\exp(-t/\tau_T)\right]}{1 - T}\,. \tag{10.27}$$

10.3 Experimental Examples

10.3.1 In Vitro Experiments: Photodynamics of Fluorescent Proteins Trapped in Agarose Gels

The computation of the fluorescence correlation function on single molecules is usually hindered by the limited number of photons that can be collected before molecular bleaching occurs. Under single-photon excitation, the typical number of detected photons $N_\mathrm{ph} \cong$ 10,000–50,000, is enough to compute with less than 1% of uncertainty the excited state lifetime. However, these are not enough to compute an ACF with sufficient accuracy for a molecule with a typical brightness ≅5,000–10,000 photons/s/molecule. When we measure an ACF on highly diluted samples, the function is actually computed on a set of different dyes diffusing through the observation volume. If the dye or the fluorescent protein is instead trapped within a solid matrix, the photons that can be collected from it are at most those emitted by the single molecule before being bleached. Bleaching is a very complex phenomenon, not completely understood in its origin (Eggeling et al. 1998). It is mainly related to chemical reactions at the triplet states, to the oxygen content of the sample, and to the excitation mode (Eggeling et al. 2005). Often, the bleached molecule can recover its fluorescence after sometime spent under no excitation light (Chirico et al. 2003) or under irradiation with shorter wavelength light (Dickson et al. 1997). This suggests that, if the sample embedding the dyes were moved with respect to the excitation beam, we should be able to observe few molecules at a time while they are moved through the excitation volume. In this way, each molecule should not spend a long time under continuous excitation to be bleached, or, if the bleaching occurs, the molecule should be able to recover its fluorescence before the next excitation.

We have then devised an experiment in which fluorescein-labeled BSA (BSA-Fl, Invitrogen, A23015, solution concentration 100 nM) molecules were spin coated onto chemically etched glass slides, moved by means of a *xy* piezo-actuator (P-541. ZSL, Physik Instrumente, Karlsruhe, D) on the focal plane of the microscope objective while acquiring fluorescence from the sample. No direct control of the pH could be made in this kind of sample. Two-photon excitation has been used to prime fluorescence by exciting BSA-Fl at 800 nm with an average power of 5 mW (excitation volume $V_{exc} \cong 0.8\ \mu m^3$). The movement of the sample was determined by two independent voltage signals fed to the two analog input of the piezo-scanner electronics. At first, a buffer of (x, y) positions was created and then fed to the piezo-actuator analog inputs with a time delay, τ_{delay}, between consecutive updates. This delay time acts in addition to the minimum time delay determined by the PC speed and data transfer. The algorithm that proved to minimize bleaching in the sample for long exposure time was a "rosetta"-like circular motion composed of two out-of-phase circular paths on radii R and $r < R$:

$$\begin{aligned} V_x(t) &= V_{0x}(t) + R(t)\cos(\Omega t) + r(t)\cos(\omega t)\,, \\ V_y(t) &= V_{0y}(t) + R(t)\sin(\Omega t) + r(t)\sin(\omega t)\,, \end{aligned} \tag{10.28}$$

$$\begin{aligned} R(t) &= R_0 + \xi(0, \sigma_R)\,, \\ r(t) &= r_0 + \xi(0, \sigma_r)\,, \\ V_{0,k}(t) &= V_{0,0,k} + \xi(0, \sigma_{Vk})\,, \end{aligned} \tag{10.29}$$

where $\xi(0, \sigma)$ is a Gaussian random functions with $\langle \xi \rangle = 0$ and $\langle \xi^2 \rangle = \sigma^2$. The built-in WhiteNoise() function (LabWindows/CVI 8, National Instruments) has been used for this purpose.

Typical examples of the path that the piezo-actuator would in principle perform when subject to such a bias are given in Fig. 10.3a. A substantially uniform coverage of the plane around the center is obtained for a ratio $r/R > 0.5$. However, even in this case, the distribution of the radial distance of the beam from the center displays a (wide) maximum (Fig. 10.3a4). By raising the excitation intensity during scanning, we observe the appearance of bright spots, probably due to photo-activated products, and the simultaneous overall bleaching of the sample. Fluorescence images of the sample acquired after a long random scanning at high excitation intensities are shown in Fig. 10.3b. These brighter spots determine the appearance of oscillatory behavior in the ACFs, as shown in the computed ACF reported in Fig. 10.4a, which have then the same origin of the oscillatory component detected when performing circular scanning (Berland et al. 1995).

The shape of the ACF measured while performing random scanning is not well described by a diffusive model (10.23) but rather by a drift-like motion (10.11) when $\tau_D = 0$ (Fig. 10.4a). This motion is characterized by a $\exp[-(t/\tau_{drift})^2]$ decay, where τ_{drift} has a marked dependence on the delay time of the scanning algorithm (Fig. 10.4b, 0.1 ms $\leq \tau_{delay} \leq$ 100 ms). The actual value of the delay time must

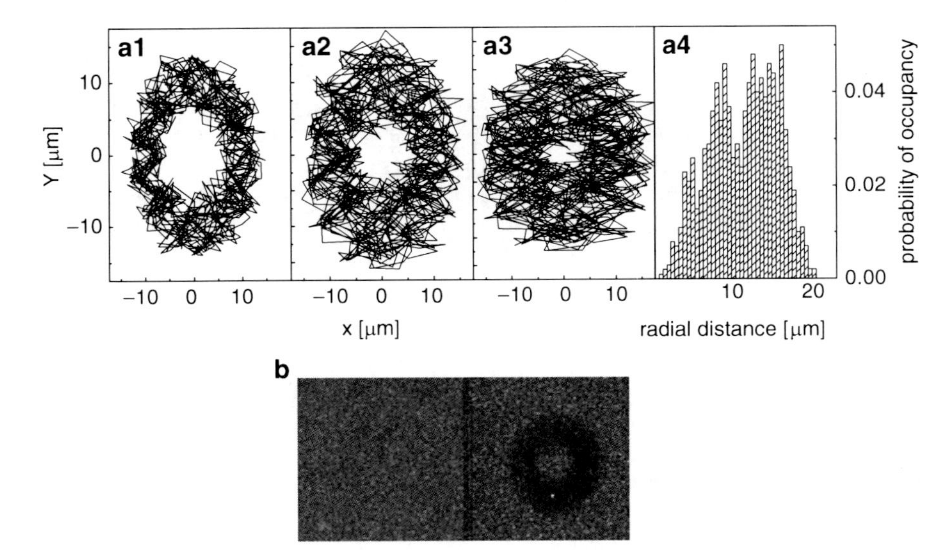

Fig. 10.3 (**a**) Simulation of the xy movement of the piezo-actuator according to (10.28) (see text) with $R = 10$ μm, $\Omega = 6.28$ Hz. (**a1**) $\omega = 125.6$ Hz , $r = 3$ μm. (**a2**) $\omega = 125.6$ Hz , $r = 4.5$ μm. (**a3**) $\omega = 1.256$ kHz, $r = 6.0$ μm. The Gaussian variable variances are $\sigma_{Vx} = \sigma_{Vy} = 1$ μm, $\sigma_R = \sigma_r = 0.001$ μm. (**a4**) The distribution of the occupancy probability versus the radial distance from the center of the scanning for the data reported in the (**a3**). (**b**) Images acquired before (*left*) and after (*right*) a long scanning at high excitation power

take into account, as said before, the time needed for the update of the analog outputs of the board (PCI 6711, National Instruments, Austin, TX) (Malengo et al. 2004) used to drive the piezo-actuator, estimated $\cong 1.0 \pm 0.1$ ms. The piezo-actuator reaches the desired position, L, exponentially according to $x(t) = L[1 - \exp(-t/\tau_P)]$. If the bias is updated every $t = \tau_{delay}$ seconds, the actual distance spanned by the actuator will be $x(\tau_{delay}) = L\left[1 - \exp\left(-\tau_{delay}/\tau_P\right)\right]$, and the corresponding average speed will be:

$$\bar{v}(\tau_{delay}) = \frac{x(\tau_{delay})}{t} = \frac{\langle L \rangle}{\tau_{delay}}\left[1 - \exp\left(-\tau_{delay}/\tau_P\right)\right]. \tag{10.30}$$

The best fit values of the drift velocity obtained by fitting the ACFs reported in Fig. 10.4a are indeed well described by the functional form in (10.30), as shown in Fig. 10.4b and the best fit values are $\langle L \rangle \cong 2 \pm 0.8$ μm and $\tau_P = 1 \pm 0.9$ ms. The meaning of the point brackets over the distance spanned by the piezo-actuator is that this distance is the average length of each of the many steps performed on the overall "rosetta" shape trajectory. The result shown in Fig. 10.4 is a clear indication that the actual motion of the excitation volume within the sample is composed by a series of stretches of rectilinear uniform motions. The actual exponential motion is sensed for the largest delay time, for which the corresponding measured ACF is actually poorly described by a drift motion correlation function (Fig. 10.4a).

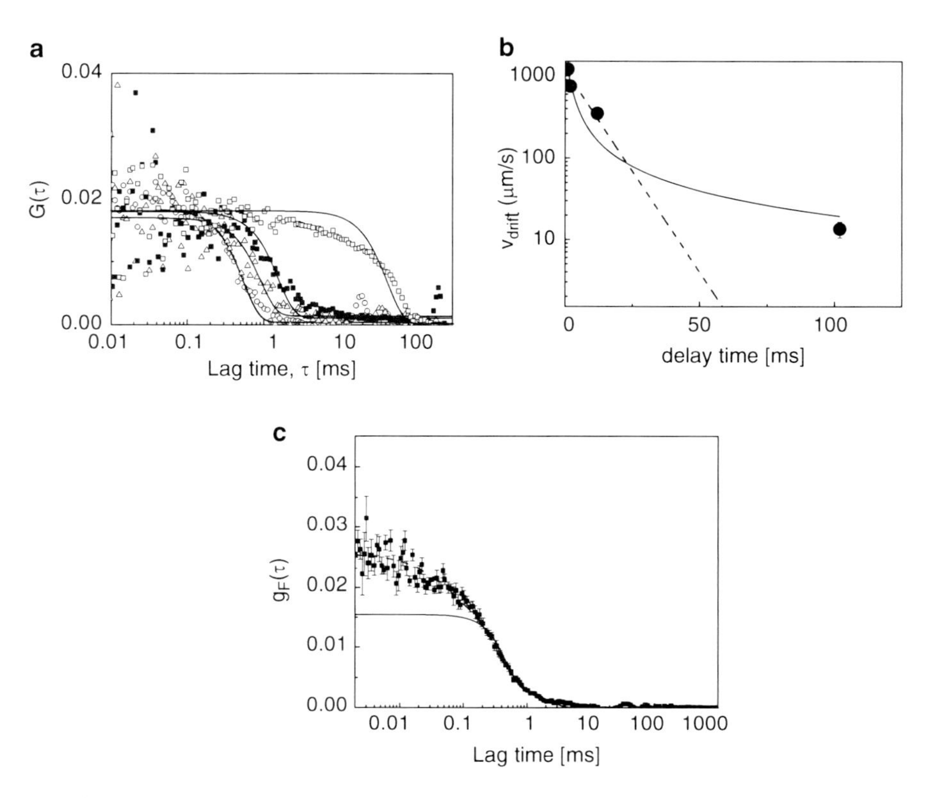

Fig. 10.4 (**a**) ACF computed while randomly scanning a spin-coated fluorescein-labeled BSA sample (100 nM) with movements described in Fig. 10.3a3. The *solid lines* are best fit of the data to the drift ACF function [(10.11) with $\tau_D = 0$]. The best fit values of the drift velocity are 1,300 ± 100 μm/s ($\tau_{delay} = 1.1$ ms, *open circles*), 720 ± 100 μm/s ($\tau_{delay} = 2$ ms, *open triangles*), 410 ± 60 μm/s ($\tau_{delay} = 11$ ms, *filled squares*), 16 ± 2 μm/s ($\tau_{delay} = 101$ ms, *open squares*). (**b**) Best fit values of the drift velocity measured versus the delay time. *Solid* and *dashed lines* are the best fit of the data to (10.29) and to a simple exponential decay. The best fit parameters are $\langle L \rangle \cong 2 \pm 0.8$ μm and $\tau_P = 1 \pm 0.9$ ms. (**c**) ACFs of fluorescein-labeled BSA samples (20 nM) in agarose gel (5% w/w) computed during random scanning. The *dashed* and *solid lines* are best fit to the data of a drift motion ACF plus a photodynamic component, $g(\tau) \approx g(0)(1 + (T/1 - T)\exp[-(\tau/\tau_T)])\exp\left[-(\tau/\tau_{drift})^2\right]$, and the pure drift motion component of the best fit ACF, respectively. The best fit parameters are $\tau_{drift} = 480$ μs, $g(0) = 0.015$, $\tau_T = 40$ μs, $T = 0.4$

The random scanning would have given a real diffusion-like shape in the ACF (Eq. 10.22) only in the case $\langle L \rangle \ll w_x$, a condition that is not satisfied here. This latter condition however, though appealing from a theoretical point of view, would not inhibit the sample bleaching.

The random scanning FCS tested here allows to explore the internal photodynamics of fluorophores trapped in gel matrices. In fact, we have measured the ACFs of samples of fluorescein-labeled BSA molecules trapped in 5% agarose gel at pH ≅ 5.5 [(BSA-Fl) = 20 nM]. At this pH, the internal photodynamics of the fluorescein dye is relevant and we expect to detect an exponential decay with relaxation times ≅10–30 μs on top of the scanning-induced drift motion as actually found in the experimental data shown in Fig. 10.4c.

10.3.2 *In Vivo Experiments: Nanoparticles Targeting of Cells, Tracking and Fluctuations*

Nanoparticles (NPs) are promising tools for medicine and biology, in terms of both basic research and direct diagnostic and therapeutic applications (De Jong and Borm 2008). The cell uptake process is characterized by several steps with a complex dynamics related to the interaction of the NPs with the cell plasmatic and nuclear membrane and the citosol structures. Beside the direct visualization of the NPs motion within the cell, an FCS characterization of the motion can provide relevant dynamic information, as shown here.

In the example discussed here, we have used solid lipid nanoparticles (SLN) produced by NANOVECTOR, in the terms of the CE Contract STREP N° LSHB-CT-2006-037639-BONSAI (bio-imaging with smart functional nanoparticles). The main component of the matrix of these particles is tripalmitin and SLN are conjugated with 3-(2-benzothazolyl)-7-(diethylamino)coumarin to allow their visualization by means of a fluorescence microscope ($\lambda_{exc} = 454$ nm under one-photon excitation and $\lambda_{exc} \cong 810$ nm under two-photon excitation). To follow the intracellular SLN distribution, experiments were done on a monolayer of human lung epithelial cells (A30 cells) (Botto et al. 2008), grown on Petri dishes in DMEM medium supplied with 10% fetal bovin serum (FBS), 1% of L-glutamine, and 1% of penicillin/streptomycin. During their growth, cells are incubated in a controlled environment at 37°C with 5% CO_2. Once checked the biocompatibility of the SLN, we performed the experiments incubating the cells with a nanoparticles concentration of 0.01 mg/ml. Since our goal was to study the intracellular distribution and the effect of the temperature on this phenomenon, the incubation was done at physiological temperature (32°C and 37°C) or at 4°C. Although the size of the NPs cannot be resolved by optical microscopy (180–230 nm; from unpublished dynamic light scattering experiments), right after the cell loading, a number of round bright bodies of size ≅0.6 μm appear within the cell (Fig. 10.5a). These particles may be the result of stable NPs aggregation, possibly driven by the interaction with some intracellular structures, as indicated by fluorescence recovery after photobleaching experiments (Fig. 10.5b) from which we can see that the bright spots recover almost completely in their fluorescence signal and, most importantly, their shape. The motion of these particles appears, at first instance, to be irregular. The application of FCS methods to the image analysis shows, however, that this is not a completely stochastic motion, but rather a collection of short interrupted drift motions.

Typically we have acquired image stacks with 700–800 images and image sampling time of 0.85 s. In order to perform a FCS analysis, we have selected on

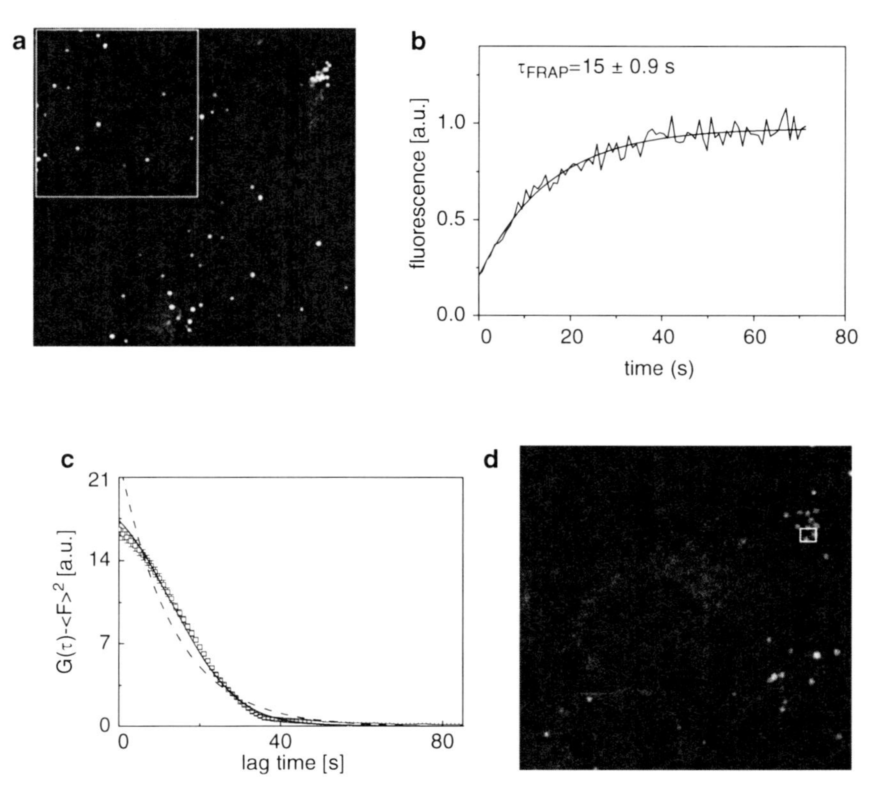

Fig. 10.5 Analysis of the motion of solid lipid nanoparticles (coumarin loaded) within A10 cells. (**a**) An image taken after 30 min from the addition of the NPs to the cell culture. Three cells are visible together with a number of well-distinct bright spots. (**b**) The trend of the fluorescence recovery after photobleaching of the spot indicated by an *arrow* in (**a**). The *solid line* is the best fit exponential growth recovery, with characteristic time $\tau_{FRAP} = 15 \pm 1$ s. (**c**) The average of 12 ACFs computed on a 600-s image stack (full experiment duration 500 s). The fluorescence signal was computed as the average over the observation volume marked in (**d**) (*white box*). The *solid* and *dashed lines* are the best fit of the drift and diffusive models to the data

the image stack a circular area whose radius was ≅2.5 times the average radius of the bright organelles (Fig. 10.5c) and computed the fluorescence signal versus time. The autocorrelation function of this signal can be well fit to a pure drift correlation function [(10.11); $\tau_D = 0$] with drift time $\tau_{drift} = 21 \pm 0.5$ s. Since the diameter of the regions on which the fluorescence is computed versus time is $w_x \cong 2$ μm, the average drift speed with which the particles appear to cross the observation area is $v_{drift} \cong 1.0 \pm 0.03$ μm/s. The drift motion character revealed here must not, however, be considered in contrast with the apparent stochastic behavior of the trajectories of the fluorescent intracellular bodies followed on the time stacks. The actual meaning of the drift speed estimated above is that over sizes of the order of two-three body diameters their motion is ballistic up to a time in which an interaction with some intracellular structures occurs that changes abruptly the body motion. Over much larger distances, the motion of the bright bodies appears to be stochastic. However, due to the large number of interactions suffered by the fluorescent bodies, their motion may appear, as frequently reported in the literature, better described by an anomalous diffusion model in which the average $\left\langle [r(t) - r(0)]^2 \right\rangle$ increases sublinearly with time.

10.4 Conclusions

The content of information of the fluorescence fluctuations arising from molecules in solutions and within cells is huge, and it appears that it has not been fully explored completely yet. In this report, we have outlined a theoretical treatment that is amenable to the analysis of a wide class of possible experimental configurations and physico-chemical conditions. The examples that have been discussed are related to two critical issues in current spectroscopic research applied to life science: single molecule and intracellular dynamics. We hope that the general treatment given here would be valuable for the analysis of future experimental realization of FCS.

Acknowledgments We acknowledge the Fondazione Cariplo (fund 2005-1079 to G.C.) and the MIUR Prin fund (2006027587 to G.C.).

References

Aragon SR, Pecora R (1976) Fluorescence correlation spectroscopy as a probe of molecular-dynamics. J Chem Phys 64:1791–1803

Berland K, Shen GQ (2003) Excitation saturation in two-photon fluorescence correlation spectroscopy. Appl Opt 42:5566–5576

Berland KM, So PT, Gratton E (1995) Two-photon fluorescence correlation spectroscopy – method and application to the intracellular environment. Biophys J 68:694–701

Berne BJ, Pecora R (2000) Dynamic light scattering: with applications to chemistry, biology, and physics. Dover, Mineola

Bismuto E, Gratton E, Lamb DC (2001) Dynamics of ANS binding to tuna apomyoglobin measured with fluorescence correlation spectroscopy. Biophys J 81:3510–3521

Botto L, Beretta E, Bulbarelli A, Rivolta I, Lettiero B, Miserocchi G, Palestini P (2008) Hypoxia-Induced modifications in plasma membranes and lipid microdomains in A 549 cells and primari human alveolar cells. J Cell Biochem 105(2):503–513

Brinkmeier M, Dorre K, Stephan J, Eigen M (1999) Two beam cross correlation: a method to characterize transport phenomena in micrometer-sized structures. Anal Chem 71:609–616

Chen Y, Muller JD, Ruan QQ, Gratton E (2002) Molecular brightness characterization of EGFP in vivo by fluorescence fluctuation spectroscopy. Biophys J 82:133–144

Chirico G, Gardella M (1999) Photon cross-correlation spectroscopy to 10-ns resolution. Appl Opt 38:2059–2067

Chirico G, Cannone F, Baldini G, Diaspro A (2003) Two-photon thermal bleaching of single fluorescent molecules. Biophys J 84:588–598

De Jong WH, Borm PJ (2008) Drug delivery and nanoparticles: applications and hazards. Int J Nanomedicine 3(2):133–149

Dertinger T, Pacheco V, von der Hocht I, Hartmann R, Gregor I, Enderlein J (2007) Two-focus fluorescence correlation spectroscopy: a new tool for accurate and absolute diffusion measurements. Chemphyschem 8:433–443

Dickson RM, Cubitt AM, Tsien RY, Moerner WE (1997) On/off blinking and switching behaviour of singlemolecules of green fluorescent protein. Nature 388:355–358

Dittrich PS, Schwille P (2002) Spatial two-photon fluorescence cross-correlation Spectroscopy for controlling molecular transport in microfluidic structures. Anal Chem 74:4472–4479

Doi M, Edwards SF (1986) The theory of polymer dynamics. Oxford University Press, New York

Eggeling C, Widengren J, Rigler R, Seidel CAM (1998) Photobleaching of fluorescent dyes under conditions used for single-molecule detection: evidence of two-step photolysis. Anal Chem 70:2651–2659

Eggeling C, Volkmer A, Seidel CAM (2005) Molecular photobleaching kinetics of Rhodamine 6G under the conditions of one- and two-photon induced confocal fluorescence microscopy. Chemphyschem 6:791–804

Enderlein J, Gregor I, Patra D, Fitter J (2004) Art and artefacts of fluorescence correlation spectroscopy. Curr Pharm Biotechnol 5:155–161

Gosch M, Rigler R (2005) Fluorescence correlation spectroscopy of molecular motions and kinetics. Adv Drug Deliv Rev 57:169–190

Haustein E, Schwille P (2003) Ultrasensitive investigations of biological systems by fluorescence correlation spectroscopy. Methods 29:153–166

Hobel M, Ricka J (1994) Dead-time and afterpulsing correction in multiphoton timing with nonideal detectors. Rev Sci Instrum 65:2326–2336

Kohl T, Haustein E, Schwille P (2005) Determining protease activity in vivo by fluorescence cross-correlation analysis. Biophys J 89:2770–2782

Magde D, Webb WW, Elson E (1972) Thermodynamic fluctuations in a reacting system – measurement by fluorescence correlation spectroscopy. Phys Rev Lett 29:705–708

Malengo G, Milani R, Cannone F, Krol S, Diaspro A, Chirico G (2004) High sensitivity optical microscope for single molecule spectroscopy studies. Rev Sci Instrum 75(8):2746–2751

Muller JD, Chen Y, Gratton E (2003) Fluorescence correlation spectroscopy. Biophotonics, PT B 361:69–92

Petrasek Z, Schwille P (2008) Precise measurement of diffusion coefficients using scanning fluorescence correlation spectroscopy. Biophys J 94:1437–1448

Ries J, Schwille P (2006) Studying slow membrane dynamics with continuous wave scanning fluorescence correlation spectroscopy. Biophys J 91:1915–1924

Skinner JP, Chen Y, Muller JD (2005) Position-sensitive scanning fluorescence correlation spectroscopy. Biophys J 89:1288–1301

Weidemann T, Wachsmuth M, Knoch TA, Muller G, Waldeck W, Langowski J (2003) Counting nucleosomes in living cells with a combination of fluorescence correlation spectroscopy and confocal imaging. J Mol Biol 334:229–240

Xia KQ, Xin YB, Tong P (1995) Dual-beam incoherent cross-correlation spectroscopy. J Opt Soc Am A 12:1571–1578

Xiao Y, Buschmann V, Weston KD (2005) Scanning fluorescence correlation spectroscopy: a tool for probing microsecond dynamics of surface-bound fluorescent species. Anal Chem 77:36–46

Chapter 11
Interference X-ray Diffraction from Single Muscle Cells Reveals the Molecular Basis of Muscle Braking

L. Fusi, E. Brunello, M. Reconditi, R. Elangovan, M. Linari, Y.-B. Sun, T. Narayanan, P. Panine, G. Piazzesi, M. Irving, and V. Lombardi

11.1 Introduction

Skeletal muscle fibres are elongated cells generating force and shortening when electrically stimulated. A large proportion of the cytoplasm of these cells is devoted to contractile material, mainly represented by the contractile proteins actin and myosin II polymerized in overlapping arrays of filaments in each sarcomere, the structural unit of the muscle (Fig. 11.1a). Actin molecules form a double helical strand with a 37 nm period, which constitutes the thin filament (Fig. 11.1b). Myosin molecules form a three stranded structure, the thick filament (Fig. 11.1b), with a helical periodicity of 43 nm. The two globular motor domains of each myosin molecule project from the thick filament in crowns of three pairs every 14.5 nm. Each thick filament is composed of two symmetrical halves due to the anti-parallel arrangement of the molecules.

The motor domain contains the catalytic site and the actin binding site. Force and shortening in muscle are generated by cyclic ATP-driven interactions of the myosin motors with the overlapping thin filaments (Huxley 1957; Huxley 1969; Kushmerick and Davies 1969; Lymn and Taylor 1971; Linari and Woledge 1995). During the interaction, the myosin motors undergo a conformational change (the working stroke) that pulls the thin filaments towards the centre of the sarcomere. Crystallographic models indicate that the working stroke consists of a 70° rotation of the light chain domain (LCD) of the motor (the lever arm) pivoted on the catalytic domain (CD) firmly attached to actin. This rotation corresponds to an axial shift of ~11 nm, between the tip of the lever arm (the head–rod junction) attached to the

L. Fusi (✉), E. Brunello, M. Reconditi, R. Elangovan, M. Linari, G. Piazzesi and V. Lombardi
Università degli Studi di Firenze, Florence, Italy
e-mail: luca.fusi@unifi.it

Y.-B. Sun and M. Irving
King's College London, London, UK

T. Narayanan and P. Panine
ESRF, Grenoble, France

A. Diaspro (ed.), *Optical Fluorescence Microscopy*,
DOI 10.1007/978-3-642-15175-0_11,

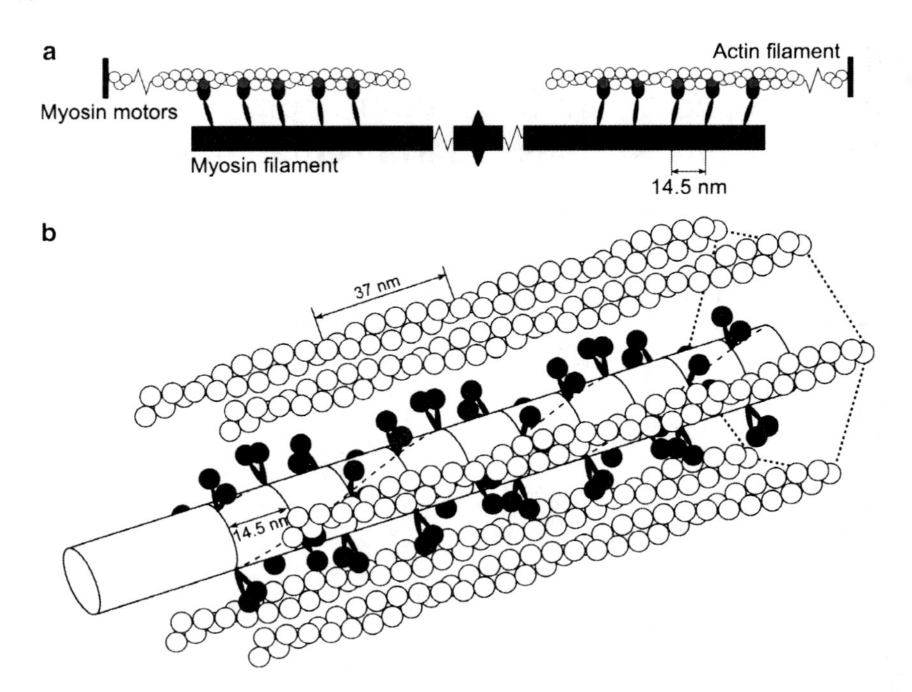

Fig. 11.1 (**a**) Organisation in the muscle sarcomere of myosin and actin filaments and myosin heads (emerging from the myosin filament with 14.5 nm axial periodicity). The actin monomers to which the myosins are attached are grey. Adapted from Reconditi et al. 2004. (**b**) Three-dimensional arrangement of the filaments. Each myosin filament, with the myosin motors emerging in crown of three pairs (*black*) every 14.5 nm, is surrounded by six neighbouring actin filaments. The monomers in the actin filament form a double helix. The axial repeat of actin monomers in each strand is 5.5 nm

myosin filament and the CD attached to actin. The anti-parallel arrangement of the myosin molecules in each thick filament and the opposite orientation of the actin filaments in the two halves of the sarcomere combine to produce steady sarcomere force and filament sliding. Thus, the array of motors in each half thick filament works as a collective motor, converting metabolic energy into mechanical work.

When an external load larger than the array force is applied to the active muscle, the sarcomere opposes the lengthening, as the motor array acts as a brake to resist the load with reduced metabolic cost (Katz 1939; Infante et al. 1964). The braking action of muscle is a matter of everyday experience, for example, when the extensor muscles of the legs have to decelerate the body at the end of a jump. Previous experiments have shown that the steady lengthening of active muscle induces a sharp increase in force up to twice the isometric value accompanied by an increase in the stiffness, suggesting the recruitment of an additional elastic structure (Linari et al. 2000), that could consist of either new myosin motors or elastic structures in parallel to the motor array.

We used time-resolved X-ray diffraction from synchrotron light to investigate the structural basis of the mechanism of muscle resistance to stretch. This structural

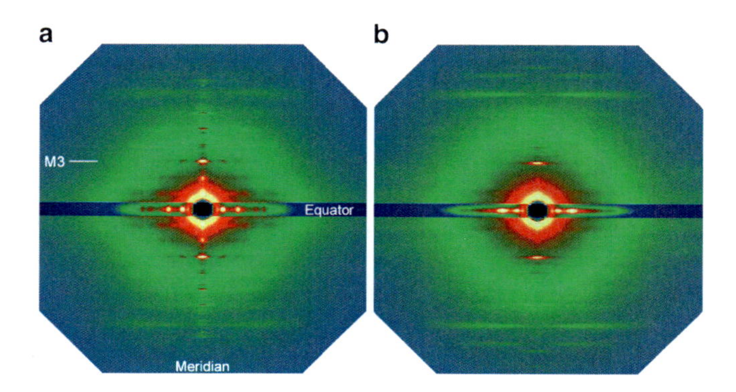

Fig. 11.2 X-ray diffraction patterns from a single muscle fibre (sarcomere length 2.1 μm, temperature 4°C) at rest (**a**) and during isometric contraction (**b**), collected on a CCD detector (total exposure time 1 s). The reflections along the meridional axis (parallel to the fibre axis) are due to quasi-helical symmetry of myosin motors emerging from the myosin filament; the reflections along the equatorial axis (orthogonal to the fibre axis) are due to the double hexagonal lattice formed by the actin and myosin filaments

technique is uniquely powerful in single fibres from skeletal muscle, thanks to the quasi-crystalline arrangement of the motor proteins in the three-dimensional lattice. The main feature of the diffraction pattern of a single fibre from frog muscle in the resting state (Fig. 11.2a) is represented by a series of regularly spaced reflections on the meridional axis (parallel to the fibre axis), originating from the 43 nm quasi-helical arrangement of the myosin motors along the thick filament (Fig. 11.1b). The M3 reflection that arises from the 14.5 nm axial repeat of the myosin heads remains strong even during isometric contraction, when the other reflections become very weak (Fig. 11.2b). The intensity of the M3 reflection (I_{M3}) is sensitive to the changes in mass density projections of the myosin motors associated with axial movement of the motors (Irving et al. 2000), but I_{M3} changes can be interpreted only by using structural models or mechanical constraints, because the X-ray diffraction signals lack phase information. With the high collimation of the beam at ID2 (ESRF, Grenoble, France), it is possible to collect the fine structure of the M3 reflection, due to the interference between the two mirror arrays of motors in the myosin filament (Huxley and Brown 1967). During isometric contraction, the M3 fine structure shows two closely spaced peaks with the ratio of the high angle peak to that of the low angle peak (R_{M3}) circa 0.8 (Fig. 11.3c black line). Changes in R_{M3} record with sub-nanometer precision, the extent and direction of the axial movements of the motors (Piazzesi et al. 2002; Reconditi et al. 2004).

11.2 Experimental Protocol and Results

We measured the changes in intensity and fine structure of the M3 reflections following stretches of 2–6 nm per half-sarcomere superimposed on isometric contraction of fibres isolated from frog muscle, a preparation that allows hundreds of contractions to be elicited with the preservation of sarcomere homogeneity. In this way X-ray frames with satisfactory signal-to-noise ratio can be collected with sub-millisecond time resolution by signal averaging from thousands of repeats of the mechanical manoeuvre. Single intact fibres were dissected from frog skeletal muscle (tibialis anterior, *Rana temporaria*) and then mounted in a trough containing Ringer's solution between the lever arms of a force transducer, with resonant frequency 50 kHz (Huxley and Lombardi 1980), and a loudspeaker motor. The sarcomere length was adjusted to 2.1 μm and temperature to 4°C. During the experiments the half-sarcomere length changes were measured in a selected sarcomere population along the fibre with the striation follower (Huxley et al. 1981). When a fibre electrically stimulated with voltage pulses of alternate polarity (20 Hz frequency) is suddenly stretched, the force increases during the stretch and then partially recovers in the next few milliseconds (Fig. 11.3a). The peak of force reached at the end of the step (T_1) represents the undamped elasticity of the half-sarcomere, due to the compliance of myofilaments and of the motors that cross-link them (Ford et al. 1981). The recovery of force within the first 3–4 ms (T_2) is due to the reversal of the myosin motor stroke that drives muscle shortening (Huxley and Simmons 1971). Once the filament compliance ($C_f = 0.013$ nm kPa^{-1}; from Piazzesi et al. 2007) is taken into account, it is possible to calculate the axial distortion of the motors attached to actin (Δz; Fig. 11.3a, b) following the stretch ($\Delta z = \Delta L - \Delta T\ C_f$, where ΔL is the stretch amplitude and ΔT the force change). For example, the force increase at T_1, produced by a step stretch of 3.8 nm hs^{-1}, strains the myofilaments by 2.8 nm and induces an elastic distortion of the motors

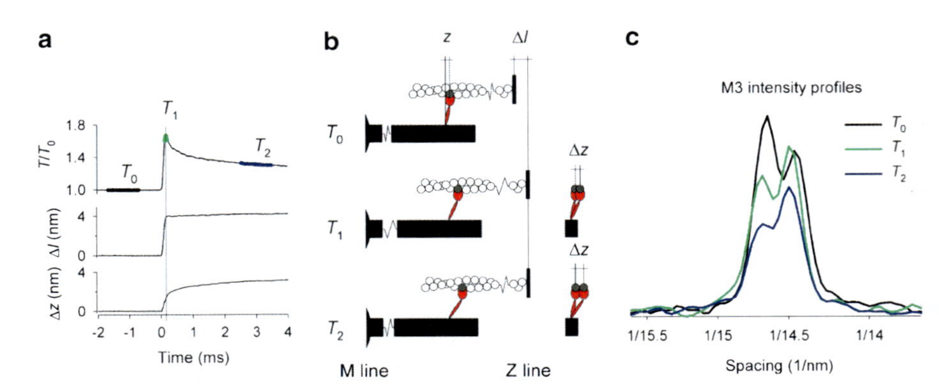

Fig. 11.3 (**a**) Time course of force (T) relative to isometric force (T_0), length change of the half-sarcomere (Δl), and axial motion of myosin motors (Δz) associated with tilting of the lever arm of the attached myosin motor. Times of X-ray exposures at T_0 (*black*), T_1 (*green*) and T_2 (*blue*) indicated on the force trace. (**b**) Myosin motor (red) with its CD attached to a monomer (*grey*) in the actin filament (*white*) and LCD attached to the myosin filament backbone (*black*). Axial motion following a stretch is accompanied by tilting of the LCD, and quantified by the axial separation z between the two ends of the LCD. The right-hand panels show the motor conformations at T_1 and T_2 superimposed on that at T_0. (**c**) Intensity profiles of the M3 reflection along the meridional axis. *Colours* as in (**a**). Replotted from Brunello et al. (2007)

by 1 nm. During the force recovery, the filaments shorten because the strain reduces by 1.6 nm, inducing an additional distortion of the motors by 1.6 nm.

After mechanical measurements with the striation follower, the trough was sealed and the fibre was mounted vertically at the beamline, to have the fibre axis parallel to the smaller dimension of the X-ray beam. To minimise the X-ray path in the solution, two moveable mica windows were put as close as possible to the fibre. X-ray data were recorded on a CCD detector, placed at 10 m from the fibre, during the isometric period preceding the step (1 ms exposure), at T_1 (0.1 ms exposure) and T_2 (1 ms exposure) (Fig. 11.3a). Two-dimensional intensity distributions in the region of the M3 were integrated across the meridional axis to give the intensity distribution along the meridian (Fig. 11.3c). The total intensity of the M3 reflection (I_{M3}) decreases after the step stretch and decreases further during the force recovery, while R_{M3} increases during both the stretch and the force recovery (Fig. 11.4b, c). These responses can be explained by tilting of the actin-attached motors away from the centre of the sarcomere that is caused by the stretch and continues during the force recovery (Fig. 11.3b). However, the structural model which was able to fit the changes in I_{M3} and R_{M3} following shortening steps (Piazzesi et al. 2002; Reconditi et al. 2004) cannot reproduce the I_{M3} and R_{M3} changes following a stretch for all the stretch amplitudes used (2–6 nm per half-sarcomere): the calculated I_{M3} reductions are larger and calculated R_{M3} increases are smaller than those observed. These discrepancies arise from the assumption that the number of actin-attached motors remains constant during phase 1 and 2 of the response to stretch, and can be removed by assuming that the stretch induces rapid attachment, to the next actin monomer in M-ward direction, of the second motor domain of some of the myosin molecules that have the first motor domain already attached during isometric contraction (Fig. 11.4a–c). The only free parameter of the model simulation is the number of partner motor domains recruited at T_1 and T_2 following the stretch.

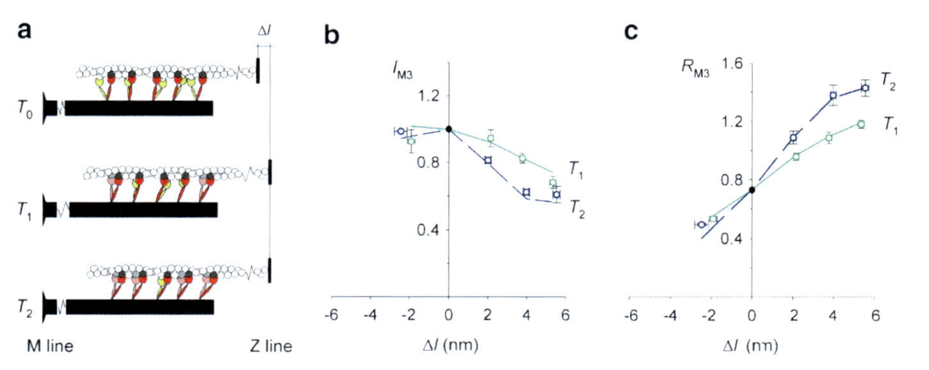

Fig. 11.4 (**a**) Structural model of motors and filaments including stretch-induced attachment of partner motors. *Red*, motors attached to actin monomers (*dark grey*) at T_0; *yellow*, detached partner motors; *pink*, partner motors attached to next actin monomer on the M-ward side (*light grey*). (**b**) I_{M3} and (**c**) R_{M3}: data from stretches of 2–6 nm amplitude, lines from model with stretch-induced attachment described in the text. *Green symbols and line*, T_1; *blue symbols and line*, T_2. Replotted from Brunello et al. (2007)

The results of the simulation indicate that the number of newly attached motors increases with the size of the stretch and increases from phase 1 to phase 2 of the response. For a 4 nm stretch, the number of additional motors is 20% of the isometric number at T_1 and becomes 40% at T_2, and, for a 5.5 nm stretch, the number of additional motors increases to 40% and 100%, respectively. The reason for the further increase in the number of motors attached during quick force recovery is the simultaneous increase in motor distortion (Fig. 11.3a, b). In these experiments, Δz for the 4 nm stretch increases from 1 nm at T_1 to 2.8 nm at T_2; for the 5.5 nm stretch Δz increases from 1.4 nm at T_1 to 3.8 nm at T_2.

11.3 Conclusions

X-ray interference between the two bipolar arrays of myosin motors in each sarcomere of a muscle cell provides a powerful tool for investigating the structural basis of the response of muscle to stretch. The results indicate that the high resistance of active muscle to a sudden increase in load is due to the rapid attachment to the actin of the partner myosin motor, promoted by the distortion of the motor already attached in the isometric contraction. The underlying mechanism implies that the first motor works as a strain sensor in a feedback system that limits the stress undergone by individual motors.

References

Brunello E, Reconditi M, Elangovan R, Linari M, Sun YB, Narayanan T, Panine P, Piazzesi G, Irving M, Lombardi V (2007) Skeletal muscle resists stretch by rapid binding of the second motor domain of myosin to actin. Proc Natl Acad Sci U S A 104:20114–20119

Ford LE, Huxley AF, Simmons RM (1981) The relation between stiffness and filament overlap in stimulated frog muscle fibres. J Physiol 311:219–249

Huxley AF (1957) Muscle structure and theories of contraction. Prog Biophys Biophys Chem 7:255–318

Huxley HE (1969) The mechanism of muscular contraction. Science 164:1356–1365

Huxley HE, Brown W (1967) The low-angle X-ray diagram of vertebrate striated muscle and its behaviour during contraction and rigor. J Mol Biol 30:383–434

Huxley AF, Lombardi V (1980) A sensitive force transducer with resonant frequency 50 kHz. J Physiol 305:15–16

Huxley AF, Simmons RM (1971) Proposed mechanism of force generation in striated muscle. Nature 233:533–538

Huxley AF, Lombardi V, Peachey LD (1981) A system for fast recording of longitudinal displacement of a striated muscle fibre. J Physiol 317:12P–13P

Infante AA, Klaupiks D, Davies RE (1964) Adenosine triphosphate: changes in muscles doing negative work. Science 144:1577–1578

Irving M, Piazzesi G, Lucii L, Sun YB, Harford JJ, Dobbie IM, Ferenczi MA, Reconditi M, Lombardi V (2000) Conformation of the myosin motor during force generation in skeletal muscle. Nat Struct Biol 7:482–485

Katz B (1939) The relation between force and speed in muscular contraction. J Physiol 96:45

Kushmerick MJ, Davies RE (1969) The chemical energetics of muscle contraction. II. The chemistry, efficiency and power of maximally working sartorius muscles. Appendix. Free energy and enthalpy of atp hydrolysis in the sarcoplasm. Proc R Soc Lond B Biol Sci 174:315–353

Linari M, Woledge RC (1995) Comparison of energy output during ramp and staircase shortening in frog muscle fibres. J Physiol 487(Pt 3):699–710

Linari M, Lucii L, Reconditi M, Casoni ME, Amenitsch H, Bernstorff S, Piazzesi G, Lombardi V (2000) A combined mechanical and X-ray diffraction study of stretch potentiation in single frog muscle fibres. J Physiol 526(Pt 3):589–596

Lymn RW, Taylor EW (1971) Mechanism of adenosine triphosphate hydrolysis by actomyosin. Biochemistry 10:4617–4624

Piazzesi G, Reconditi M, Linari M, Lucii L, Sun YB, Narayanan T, Boesecke P, Lombardi V, Irving M (2002) Mechanism of force generation by myosin heads in skeletal muscle. Nature 415:659–662

Piazzesi G, Reconditi M, Linari M, Lucii L, Bianco P, Brunello E, Decostre V, Stewart A, Gore DB, Irving TC, Irving M, Lombardi V (2007) Skeletal muscle performance determined by modulation of number of myosin motors rather than motor force or stroke size. Cell 131:784–795

Reconditi M, Linari M, Lucii L, Stewart A, Sun YB, Boesecke P, Narayanan T, Fischetti RF, Irving T, Piazzesi G, Irving M, Lombardi V (2004) The myosin motor in muscle generates a smaller and slower working stroke at higher load. Nature 428:578–581

Chapter 12
Low Concentration Protein Detection Using Novel SERS Devices

Gobind Das, Francesco Gentile, Maria Laura Coluccio, G. Cojoc, Federico Mecarini, Francesco De Angelis, Patrizio Candeloro, Carlo Liberale, and Enzo Di Fabrizio

12.1 Introduction

Surface enhanced Raman scattering (SERS) permits the detection of adsorbed molecules on noble metal surfaces, such as Au, Ag, and Cu, at subpicomolar concentration. It is a phenomenon that results in strongly increased Raman signals when molecules are in close proximity to nanometer-sized metallic structures

G. Das
Lab. BIONEM, Dipartimento di Medicina Sperimentale e Clinica, Università "Magna Graecia" di Catanzaro, Catanzaro, Italy,
and
CalMED s.r.l., Catanzaro C.da Mula Loc. Germaneto, Campus Universitario Edificio Area Medicina, 88100 Catanzaro, Italy,
and
Istituto Italiano di Tecnologia, via Morego 30 Genova, Italy

F. Gentile, M.L. Coluccio, F. Mecarini, F. De Angelis, and C. Liberale
Lab. BIONEM, Dipartimento di Medicina Sperimentale e Clinica, Università "Magna Graecia" di Catanzaro, Catanzaro, Italy,
and
CalMED s.r.l., Catanzaro C.da Mula Loc. Germaneto, Campus Universitario Edificio Area Medicina, 88100 Catanzaro, Italy,

G. Cojoc, and P. Candeloro
Lab. BIONEM, Dipartimento di Medicina Sperimentale e Clinica, Università "Magna Graecia" di Catanzaro, Catanzaro, Italy,

E. Di Fabrizio (✉)
Lab. BIONEM, Dipartimento di Medicina Sperimentale e Clinica, Università "Magna Graecia" di Catanzaro, Catanzaro, Italy,
and
CalMED s.r.l., Catanzaro C.da Mula Loc. Germaneto, Campus Universitario Edificio Area Medicina, 88100 Catanzaro, Italy,
and
INFM–TASC–S.S., 163.5 Area Science Park, Basovizza, Trieste, Italy,
and
Istituto Italiano di Tecnologia, via Morego 30 Genova, Italy
e-mail: Enzo.Difabrizio@iit.it

A. Diaspro (ed.), *Optical Fluorescence Microscopy*,
DOI 10.1007/978-3-642-15175-0_12,

(Haynes et al. 2005). SERS intensity depends on localized surface plasmon resonance (LSPR) and because the size, shape, and interparticle spacing of the material as well as the dielectric environment influence the LSPR, these factors should be chosen carefully, to ensure the reproducibility of the SERS substrate. The distinct advantage regarding the application of SERS measurement is because of the increasing importance of biological studies since this technique is well suited for the analysis carried out in aqueous environment.

Substrates exhibiting SERS characteristics can be of wide variety, i.e., metal island films (Constantino et al. 2001; Weitz et al. 1983), colloids (Hildebrandt and Stockburger 1984; Kneipp et al. 1997; Nie and Emory 1997), and recently reproducible nanostructured substrates (Perney et al. 2006; Zhang et al. 2006). In the past, some efforts have been made to fabricate the SERS substrates, based on the periodic nanostructures using e-beam lithography (EBL) and nanosphere lithography techniques (Kahl et al. 1998; Haes et al. 2005).

In this work, we report two different novel nanostructure plasmonic devices; (1) plasmonic gold agglomerated device fabricated by using electro-plating and electron beam lithography techniques and its application as a novel SERS device ("Device1"), (2) site-selective metal (Ag, Ag/Au bimetal, Au) deposition using electroless technique on the prior selected location ("Device2").

Kahl et al. (1998) presented his SERS device based on Ag metal showing the SERS effect on thiobenzene (an organic compound) but because of the inherent properties of Ag, the durability of such device as a SERS substrate is limited due to the fact that the Ag surface oxides or absorbs sulfur species in air. Recently, few works on SERS substrate based on Au-nanoparticles (Li et al. 2007) have been carried out for Rhodamine 6G (R6G) and is verified for 10^{-6} M.

Here, we report a novel device for generation of plasmon polaritons (PPs) made of array of gold nanoaggregates on gold–chromium base plated Si wafer ("Device1"). The presence of gold nano-aggregates and high-resolution e-beam patterning down to 10 nm of interspatial distance between two nano-aggregates are the major fabrication novelty of this present work. SERS effect using our noble device based on gold nano-aggregates is verified on myoglobin protein and then SERS measurements have been carried out for various proteins [BSA, lysozyme, ribonuclease-B (RNase-B), ferritin] by changing the temperature (–65°C to 90°C).

In addition, SERS device ("Device2") is fabricated by utilizing EBL coupled with electroless metal (Ag, Au and Ag/Au bimetal) deposition techniques. Active SERS substrate is realized on site selective silver/gold nanoparticles utilizing electroless technique, assembled on nano-patterned Si substrate with different shape and size. The nano-pattern based substrate, obtained by electron beam lithography, has the function of controlling the diffusion of the metal particles on the surface. On the other hand, the deposition of metallic (silver or gold) nanospheres, permits plasmonic enhancement. This is the first time that Coluccio et al. (2009) is able to produce a SERS substrate without dendritic structures and well-reproducible substrates exploiting EBL and site selective electroless deposition technique. R6G with concentration down to 10^{-20} M was observed using our SERS substrate. Later on, the work is extended to deposit Ag/Au bimetal in order to overcome its (Ag nanoparticles) lack of durability property and enhancement in Raman signal of R6G as well.

12.2 Experimental

12.2.1 Device Fabrication

12.2.1.1 Periodic Gold Nanoarray SERS Device ("Device1")

SERS device based on periodic gold nanoarray is fabricated onto bi-layer of Au (20 nm)–Cr (10 nm) base-plated silicon wafer following a high resolution fabrication techniques, using electro-plating and EBL. Several patterns of nano-holes were designed, using a "Crestec CABL–9000C" EBL system, on a layer of electronic-resist spinned onto the wafers. Disk-like structures were obtained from a nominal pattern comprising arrays of concentric squares due to proximity effects. The sample was then immersed for 35 s in ZEP developer (ZED–N50®) to selectively remove the exposed resist. Si wafer was then processed using reactive ion etching (RIE) in oxygen for 40 s to remove any residual resist from the disks. The successive gold growth was accomplished in an electrolytic system using a solution of gold–pottasium cynide, keeping density current of 10 mA/cm^2 for 8 s. Due to resist confinement, the gold nanograins could grow solely inside the holes. The grains assemblies have a 100 nm diameter and an 80 nm thickness, with an interspatial gap of 10–30 nm. Several SEM micrographs of the grain assemblies were taken to assess uniformity and reproducibility (Fig. 12.1a–d). The nanograin size and the mutual distance between one nanoaggregate to the other are 80 nm diameter and 10–30 nm gap, respectively.

12.2.1.2 Site Selective Electroless SERS Device ("Device2")

Silver nitrate and R6G are purchased from Sigma. All chemicals, unless mentioned otherwise, were of analytical grade and were used as received. Si wafer (100) was cleaned with acetone and ethanol to remove possible contaminant and then etched with a 4% wt HF solution. The wafers were rinsed with deionized water and dried with N_2. A reproducible pattern of an array of nano-structures is written using ultra high-resolution electron-beam lithography (CRESTEC CABL–9000C, 50 KeV acceleration voltage) onto a clean silicon wafer (100) spin-coated with 50 nm thick ZEP layer. Substrates of different shapes are prepared with a maximum dimension size in the range of 50–1000 nm.

Silver nano-spheres are deposited on the substrate by means of the electroless method, in which the patterned silicon wafer is dipped in a 0.15 M HF (hydrofluoric acid) solution containing 1 mM silver nitrate for different times (10–60 s). After the

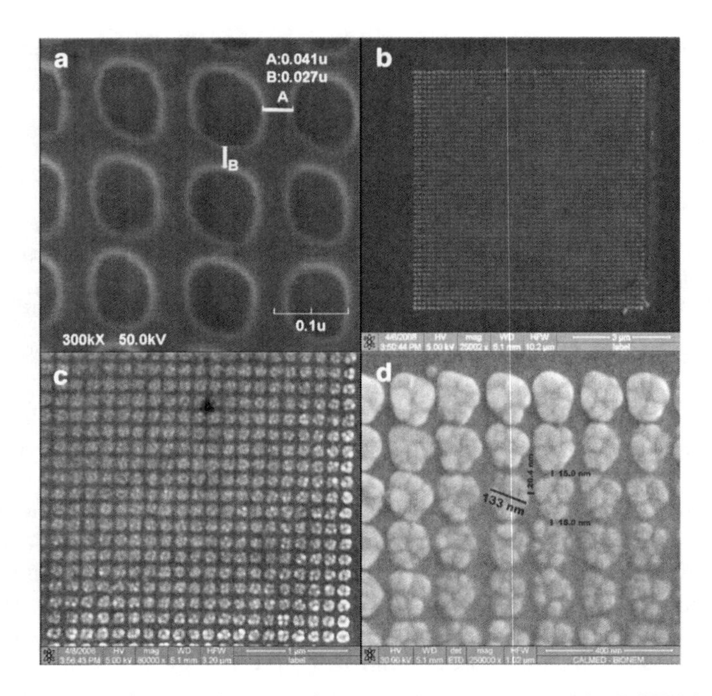

Fig. 12.1 (**a**) Array of empty disk ready for successive gold growth; (**b–d**) SEM-images with different magnifications

etching process, the silicon wafer is rinsed with water and dried under nitrogen flux. The samples with 60 s deposition time are used for Raman analysis as these samples show better Raman response. The device is shown in Fig. 12.2. The zoomed area of Fig. 12.2a, b are also shown.

Substrates with bimetallic electroless deposition (silver over gold and vice versa) (Yang et al. 2008; Bulovas et al. 2007) and substrates with only gold deposition are also prepared to compare their SERS activity with Ag deposited samples. Au is electroless deposited on nano-patterned silicon from a solution of an Au salt 1 mM and 0.15 M HF, for 3 min. The bimetallic deposition is obtained with the same solution utilized in two steps: in the case of the Ag/Au deposited substrate, a first step is a silver electroless deposition for 20 s and the last step is an Au electroless deposition for 30 s; on the contrary for the Au/Ag sample. R6G is deposited by soaking the substrate in an aqueous solution at known concentration ranging from 10^{-4} to 10^{-20} M for 30 min and then the sample is rinsed with water and dried with N_2.

12.2.2 Sample Preparation

All the proteins (BSA, lysozyme, RNase-B, and ferritin) areprepared with 7 μM concentration in high purity water. These solutions are analyzed using DCDR

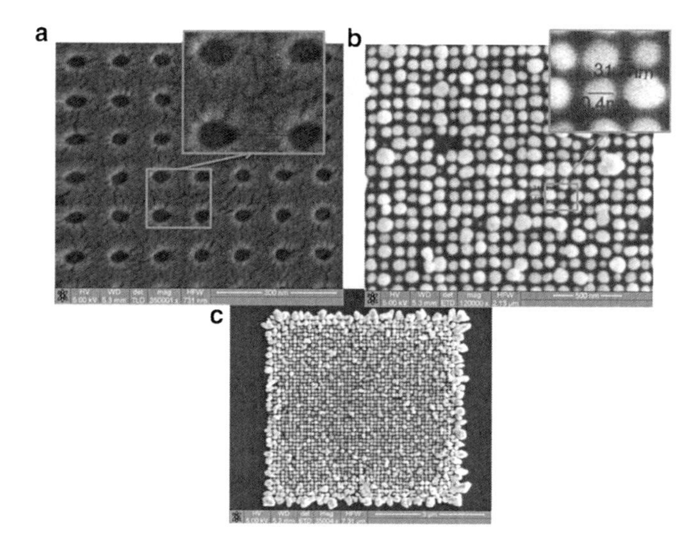

Fig. 12.2 SEM images of (**a**) lithography on a silver substrate and (**b**, **c**) silver deposition on a Si substrate with different magnifications

method in which the solution of interest is microdeposited (2 μl) on SERS substrate. The solution is left for around 10 min in a closed microcryostat chamber (to avoid any impurity from the surroundings) to evaporate the superficial water. Protein Raman spectra are taken from the visible ring of protein (outer edge of the deposition region). The DCDR samples are in a well-hydrated state after deposition and evaporation (Zhang et al. 2003). The measurements are performed at fixed temperatures in the range between −65°C and 90°C. DCDR technique, in detail, is reported elsewhere (Zhang et al. 2003).

12.2.3 Characterization Technique

Microprobe Raman spectra were excited by visible laser with 830 nm ("Device1", Fig. 12.1) and 514 nm ("Device2," Fig. 12.2) laser line in backscattering geometry through 50× long focal range objective (NA–0.50, "Device1") and 50× (NA–0.75, "Device2"). The laser power was fixed to 0.018 mW (514 nm) and 13 mW (830 nm). The proteins, in case of "Device1," were always verified through optical image before and after the measurements to ensure that there was no protein damage through laser radiation. Block diagram of microRaman setup and SERS spectrum of myoglobin are shown in Fig. 12.3. Figure 12.3a shows the block diagram of microRaman set-up. A microcryostate chamber, which is used to vary the protein temperature, is also shown . Figure 12.3b shows the Raman spectrum of myoglobin protein deposited over the Device1. In the inset of Fig. 12.3b, Raman spectrum of virgin SERS device ("Device1") is illustrated.

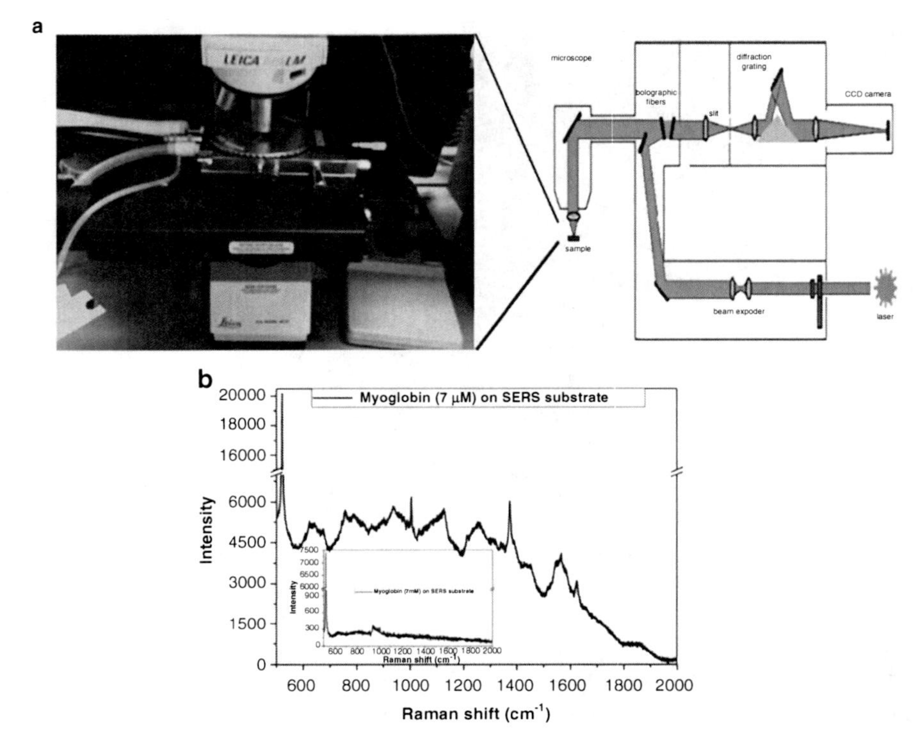

Fig. 12.3 (**a**) The block diagram of microRaman set-up. The hot/cold microsample chamber is shown in zoomed area; (**b**) SERS spectrum of myoglobin protein with concentration of 7 μM. In the inset, Raman spectrum of SERS background is shown

12.2.4 Data Analysis

All the SERS protein spectra are firstly baseline-corrected using four to five order polynomials and, thereafter, are normalized to protein marker band centered at 1,450 cm^{-1}, attributed to C–H_x bending vibration (Thomas et al. 1983). Difference Raman spectra were derived by subtracting the Raman spectra taken at different temperature to the Raman spectrum at room temperature (RT) and generalized 2D-correlation analysis was carried out for all the proteins at fixed interval of temperature in two regions of Raman spectrum: 600–1,060 cm^{-1} and 1,600–2,000 cm^{-1} by selecting the RT Raman spectrum as a reference. Synchronous 2D-correlation spectra are derived employing standard numerical procedures as described in Noda et al. (2000), as shown in Fig. 12.4 for one particular protein in the range between 600 and 1,060 cm^{-1} by changing only the coefficient of determination. It can be observed clearly the simplified plot of 2D-correlation spectra by changing the coefficient of determination from +0.3 (Fig. 12.4a) to +1 (Fig. 12.4b), which retains all the necessary information needed.

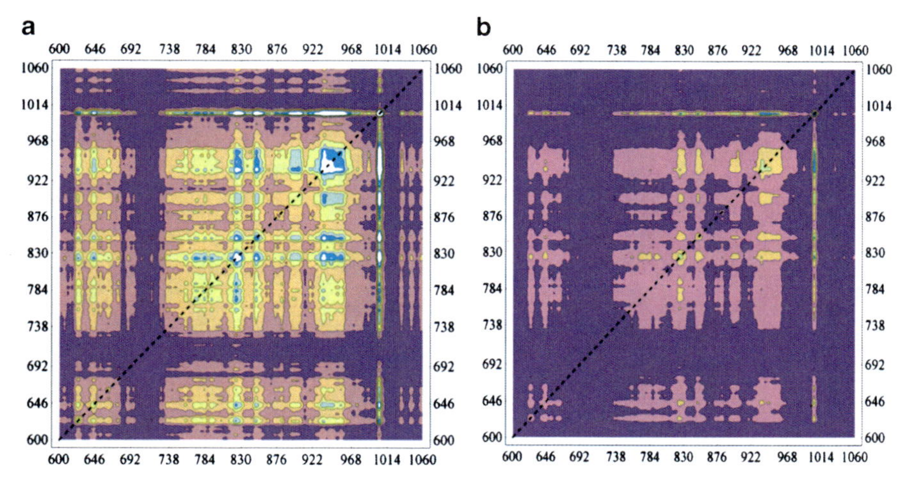

Fig. 12.4 2D-correlation spectra with the short range of coefficient of determination. 2D-correlation spectra with coefficient of determination +0.3 (**a**) and with coefficient of determination 1.0 (**b**)

12.3 Results and Discussions

12.3.1 Proteins on "Device1"

Raman spectroscopy provides the structural and chemical information of substances. Because of very low scattering cross-section of water, this technique is very much suitable for biological substances. Myoglobin is one of the basic proteins to be tested for SERS substrate due to the presence of one planar Fe protoporphyrin prosthetic heme group implanted in polypeptide chains. The presence of ring π-bonding vibrational band lies in the range of 1,000–1,700 cm^{-1}. The SERS spectrum for myoglobin with 7 μM concentration in region 500–2,000 cm^{-1} is shown in Fig. 12.3b. The protein Raman spectrum shows well-known Mb marker vibrational bands of protein centered at around 1,125, 1,373, and 1,560 cm^{-1}, which are attributed to the C–N stretching (Spiro 1985), an oxidation marker band of heme group (Sato et al. 2001), and C–C vibrational band, respectively. In addition, the bands at around 760, 1,013, 1,365, and 1,555 cm^{-1} in the same spectrum are related to the tryptophan (Trp), whereas the bands around 1,005 and 1,033 cm^{-1} are due to the phenylalanine (Phe) residues. The sharp band centered at 520 cm^{-1} in Raman spectrum is due to the Si substrate. The amount of protein under observation is calculated as discussed by Zhang et al. (2003) by assuming the size of myoglobin to be around 20 nm^2. It is found that the minimum amount of myoglobin to be investigated under laser spot of 1 μm diameter is estimated to be around 10–240 attomole, depending on the percentage of molecules present within the ring.

12.3.1.1 Bovine Serum Albumin

BSA (66,500 Da, 583 amino acids) has two Trp residues, which are located in grossly different microenvironments, and so even the interaction from Trp to water alone is inherently complex. Among six tyrosin (Tyr) residues, three are buried in the native molecules, whereas the other three are located on the protein surface. BSA is the principal carrier of fatty acids that are insoluble in circulating plasma and performs many other functions, such as, sequestering oxygen free radicals and inactivating various toxic lipophilic metabolites.

Figure 12.5a shows the SERS spectra for BSA in the region of 600–2,000 cm^{-1}, performed at different temperatures. The Raman spectrum at RT is dominated by bands due to the amide I (around 1,650 cm^{-1}, mainly by ν–C–O vibration), $C–H_3$ and $C–H_2$ deformation vibration from side chains of different amino acids (in the range of 1,440–1,480 cm^{-1}), the symmetry stretching of $–COO^-$ group (around 1,420 cm^{-1}) (Gilmanshin et al. 1996), amide III band from in-plane δ–NH group and ν–C–N (1,230–1,300 cm^{-1}), the contribution from different amino acids; Phe (~620, 1,005, and 1,033 cm^{-1}) (Li et al. 1990), Trp (1011 and 1560 cm^{-1}) (Miura et al. 1991), Tyr (645, 825, 855, 1,210, and 1,620 cm^{-1}) (Miura and Thomas

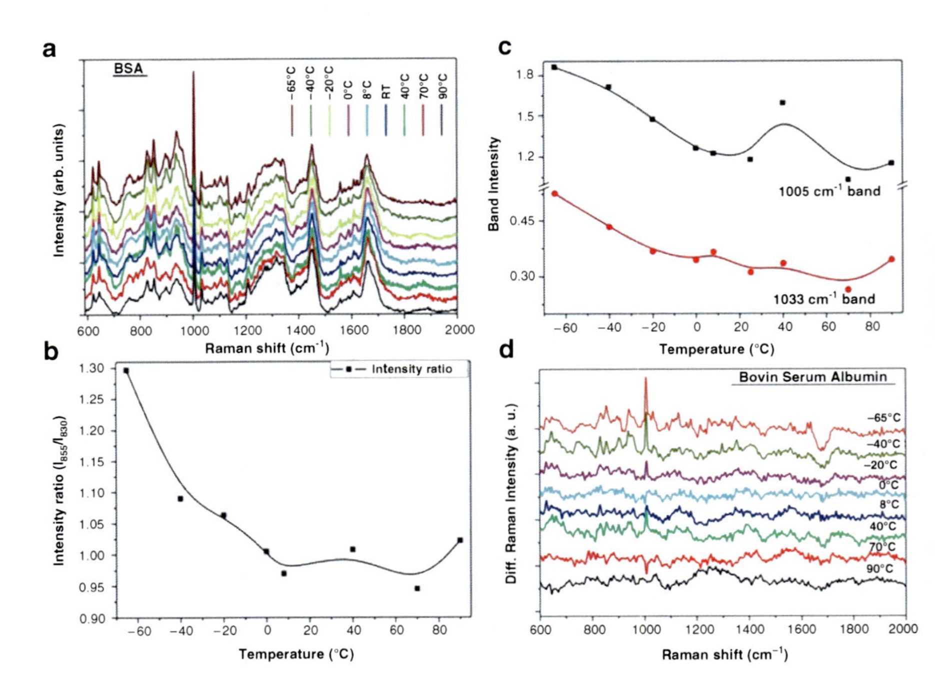

Fig. 12.5 (**a**) SERS spectra of BSA by varying the temperature in the range between −65°C and 90°C; (**b**) intensity ratio (I_{855}/I_{830}) vs. temperature; (**c**) conformational change of Phe bands centered at 1,005 and 1,033 cm^{-1} by varying the temperature; and (**d**) difference Raman spectra of BSA protein with respect to the BSA spectrum at RT

1995), cysteine (655, 672, and 720 cm^{-1}) (Li et al. 1990; Tuma et al. 1995). SERS spectra for BSA at a glance do not show any remarkable change with temperature except for two extreme temperatures (−65°C and 90°C).

The intensity ratio of two vibrational bands at ~855 and ~830 cm^{-1} (I_{855}/I_{830}), attributed to the Fermi resonance for Tyr residue, is very sensitive to the environment of H-bonding at the phenolic –OH group (Miura et al. 1991; Siamwiza et al. 1975). When the phenolic oxygen is an acceptor atom in a strong H-bond for the Tyr residues in proteins, the intensity ratio (I_{855}/I_{830}) stands at around 2.5, whereas when the phenolic oxygen is a proton donor in a strong H bond, the value lies around 0.7. In addition, for the surface Tyr in a protein, the intensity ratio stands around 1.25. The intensity ratio (I_{855}/I_{830}) for BSA protein vs. temperature is shown in Fig. 12.5b. From the intensity ratio curve, the protein always shows that the phenolic OH of Tyr residues serve both as acceptor and donor of H-bond (Miura and Thomas 1995). However, going toward higher temperatures, the phenolic –OH shows to be a stronger donor than at low temperatures. Figure 12.5c shows the Raman intensity for Phe residue bands centered at 1,005 and 1,033 cm^{-1}. Figure 12.5c, d show that there is an overall decrease in intensity of Phe band with the increase of temperature, indicating the gradual change of Phe's surrounding with temperature.

The change in secondary structure of BSA with temperature can be observed in Fig. 12.5d in difference Raman spectra. BSA has 68% of α-helix, 18% of β-sheet, and 14% of random coil secondary structures (Keiderling 1993). The secondary structure changes drastically on cooling down below –20°C. Two bands, one hump centered at 1,675 cm^{-1} and one sharp peak centered at 1,635 cm^{-1}, are observed by cooling down the protein. However, the increase in temperature up to 70°C shows very little changes regarding unfolding of BSA protein. Murayama and Tomida (2004) suggested that BSA starts showing the change in secondary structure on treating it around 60°C. Here, we observed that the BSA at 40°C (Fig. 12.5d) shows a positive broad band in difference Raman spectra in the region of 1,630–1,680 cm^{-1}. Lower frequency band (1,630 cm^{-1}) could be related to the intermolecular α-sheet, whereas the higher frequency band (1,675 cm^{-1}) is related to the intramolecular β-sheet in the protein. With further increase in temperature ($T \geq 70°C$), the spectra start showing this difference clearly, supporting the recent observation of protein unfolding using FT-IR absorption spectroscopy at about 60°C or 70°C (as in our case) depending on the protein concentration (Celej et al. 2003; Guo et al. 2006). In addition, the Phe band at about 1,005 cm^{-1}, (Fig. 12.5d), shows interesting behavior with temperature. It reveals that the intensity of Phe band rises with the decrease in temperature which could be due to the possibility that the Phe residues show better exposure to the solvent.

Synchronous 2D-correlation spectra with coefficient of determination equals to 1, based on the spectral changes with the temperature, are shown in Fig. 12.6a, b for BSA protein in the range of 600–1,060 cm^{-1} and 1,600–2,000 cm^{-1}, respectively. The 2D-correlation autopeaks are located at around 625, 645 (both C–S stretching), Tyr doublet at 830 and 855, 940 (C–C stretching), 1,005 and 1,033 (Phe ring breathing) cm^{-1} (Fig. 12.6a) and in the range between 1,650 and 1,700 cm^{-1}

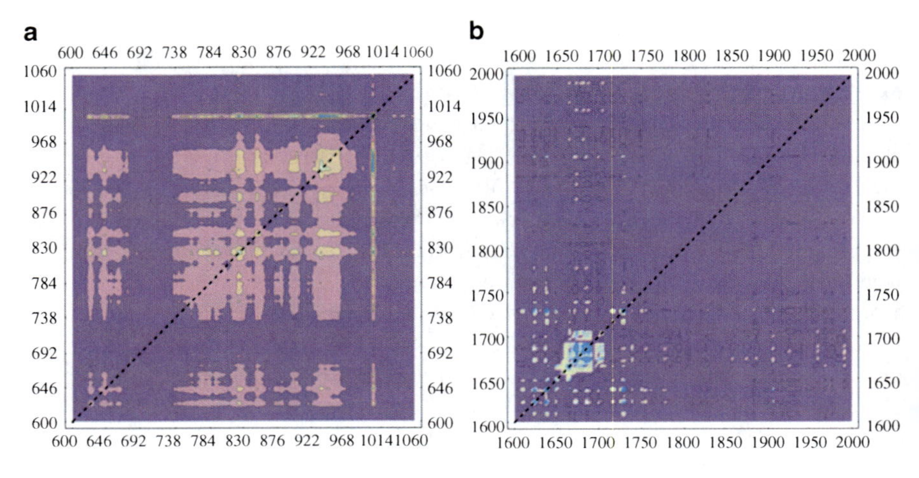

Fig. 12.6 Synchronous 2D-correlation spectra for BSA protein in the range between (**a**) 600–1,060 cm^{-1}, and (**b**) in the range of 1,600–2,000 cm^{-1}, respectively

(secondary structure) (Fig. 12.6b), representing the maximum conformation alteration with temperature.

Autopeaks describe the conformational changes that occur over the temperature interval (T_{min}–T_{max}). Any anomalous behavior that the protein may exhibit at a given temperature is disregarded or distributed over a wider range of values. The convenience of bidimensional Raman spectroscopy over conventional analysis resides in the unambiguous interpretation of data. Positive remarkable cross-peaks centered at (625, 645), (830, 855), (830, 1,005), (855, 1,005), and (1,005, 1,033) in Fig. 12.6a clearly indicate that the simultaneous changes in the range (600–1,060) are consistent; within the domain of interest, all spectral intensities increase (decrease) together with T. The information that one may infer from Fig. 12.6a perfectly agrees with the vibrational band intensity variation as in Fig. 12.5b, in which both the bands (1,005 and 1,033 cm^{-1}) undergo the steady decrease in temperature. This shows that when the temperature is decreased, it causes the access of water to surface residues of protein (Nara et al. 2001) as expected by water–protein interaction as a function of temperature. The original information, which 2D-correlation spectra carry about, is also that the 830 and 855 cm^{-1} vibrational bands decrease with temperature. As it is well known that these two bands are attributed to the Tyr residues, both vibrational bands behave in the same way. Such coincidental changes suggest the possible existence of a common origin for variations, i.e., the variation in temperature causes the similar variation of conformational change for Tyr and Phe residues. There are major intensity variations (illustrated by white spots) with respect to RT SERS spectrum in the region of 1,650–1,700 cm^{-1} (Fig. 12.6b). These findings well agree with the results derived through conventional difference Raman (Fig. 12.5d).

12.3.1.2 Lysozyme

Lysozyme (14,400 Da) serves as bactericidal agents (i.e., used to degrade the bacterial cell wall). Native lysozyme secondary structure consists of 30% α-helix, 6% β-sheet, and 64% random coil (Artymiuk et al. 1982). SERS spectra for lysozyme vs. Temperature are shown in Fig. 12.7a ranging between 600 and 2,000 cm^{-1}. The amide I band with major contribution of C–O stretching vibration is clearly observed ranging between 1,630 and 1,700 cm^{-1}, whereas amide II (N–H bending coupled with C–N stretching) and amide III (C–N stretching mixed with N–H bending vibration) are observed in the range between 1,530–1,580 and 1,220–1,300 cm^{-1}, respectively. Raman spectra show high signal noise ratio with respect to the lysozyme Raman spectrum, carried out on the gold-coated Teflon substrate (concentration –100 μM, integration time –100 s) (Zhang et al. 2003) even though the lysozyme, in our case, has much less concentration (7 μM) and less integration time (50 s). The increase in temperature after 70°C causes the increase in broadness of amide I vibrational band. This broad band contribution, coming from different secondary structures, is because of conversion of helix to coil structure of protein.

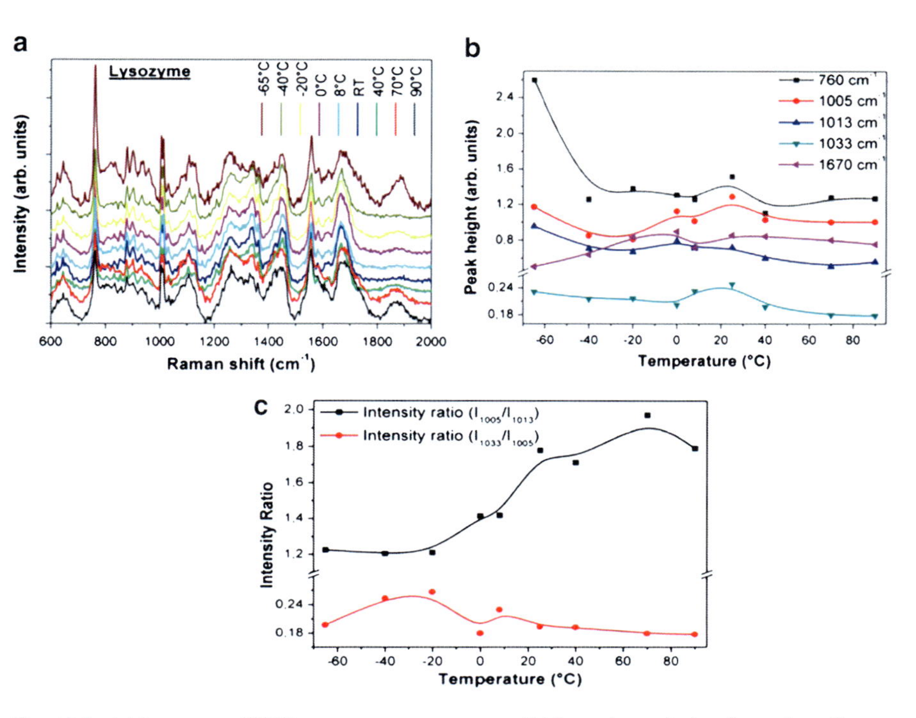

Fig. 12.7 (**a**) Lysozyme SERS spectra vs. temperature; (**b**) intensity variation for various Raman vibration bands (760, 1,005, 1,013, 1,033 and 1,670 cm^{-1}); (**c**) intensity ratio (I_{1005}/I_{1013} and I_{1033}/I_{1005}) vs. temperature

The bands around 625 and 645 cm^{-1} are related to the C–S stretching vibration of cystein residues, showing quite stable in the temperature range between –40°C and 40°C. The intense peak centered at 760 cm^{-1} is clearly observed for lysozyme protein. This band, sensitive to the hydrophobic environment of Trp residues, is attributed to the indole ring breathing vibration of Trp residues (W18).

Interesting behavior in Trp residues (W17) vs. temperature is observed in the range of 872–885 cm^{-1} with regard to their H-bonding strength (peak at 883 cm^{-1} represents the higher H-bonding strength of Trp, whereas the peak at 871 cm^{-1} signifies the nonH bonding Trp) (Tuma et al. 1995). The increase in temperature up to 40°C causes the appearance of one more band centered at around 880 cm^{-1} in addition to the 877 cm^{-1} which could be the combination of different Trp residues present within the protein. On further increase in temperature, the Raman band sharpens and comes up with one narrow band centered at 878 cm^{-1} for sample treated at 90°C. Hence, it seems that the increase in temperature causes different Trp residues exposed to the solvent. However, the decrease in temperature from RT to –40°C does not show any remarkable change although at –65°C, the Raman band shows an increase in width of this band. The broad and asymmetric band shape arises from different H-binding strength with various Trp residues (Trp-18, Trp-62, Trp-63, Trp-108, Trp-111 and Trp-123) (Jung et al. 1980).

Figure 12.7b shows the peak height variation with temperature for various bands, 760, 1,005, 1,013, and 1,033 cm^{-1} attributed to different vibrations associated with Trp, Phe, Trp, and Phe residues, respectively. The peak height of broad band centered at 1,670 cm^{-1}, attributed to secondary structure of protein, is also shown. The increase in intensity of indole ring breathing mode of Trp residues centered at 760 cm^{-1} signifies the decrease in hydrophobicity of the Trp indole ring environment. It is observed that: (1) the intensity of this band drastically decreases for temperature going from −65°C to −40°C (increase in hydrophobicity); (2) the intensity remains almost constant in the temperature range of –40°C to 90°C. The band intensity of 1,005, 1,013, and 1,033 cm^{-1}, attributed to different residues, does not show remarkable changes, which signifies no noticeable conformational changes in Phe, and Trp residues for this protein with temperature. Guo et al. (2006) performed various measurements on different concentrations of lysozyme (10 and 400 mg/ml) using various techniques and revealed that lysozyme of a lower concentration shows the protein stability up to 80°C, whereas lysozyme of a higher concentration shows a conformational change in protein secondary structure profile even at 70°C. Similar trend is also found by Nara et al. (2001) up to RT and is attributed to the modification in Phe or their surroundings. Figure 12.7c shows the intensity ratio for I_{1005}/I_{1013} (Phe to Trp ring) and I_{1033}/I_{1005} (both attributed to the Phe ring). The intensity ratio (I_{1005}/I_{1013}) shows no variation up to around −20°C and it starts increasing with the increase in temperature. The intensity ratio (I_{1033}/I_{1005}) reveals that it increases with the temperature up to −20°C and then starts decreasing with increase in temperature.

Figure 12.8a, b show the 2D-correlation spectra for lysozyme in the range of 600–1,060 cm^{-1} and 1,600–2,000 cm^{-1}. Autopeaks for lysozyme are observed at about 645 (cystein), 760 (Trp), 1,005 (Phe), and 1,013 cm^{-1} (Trp) and around

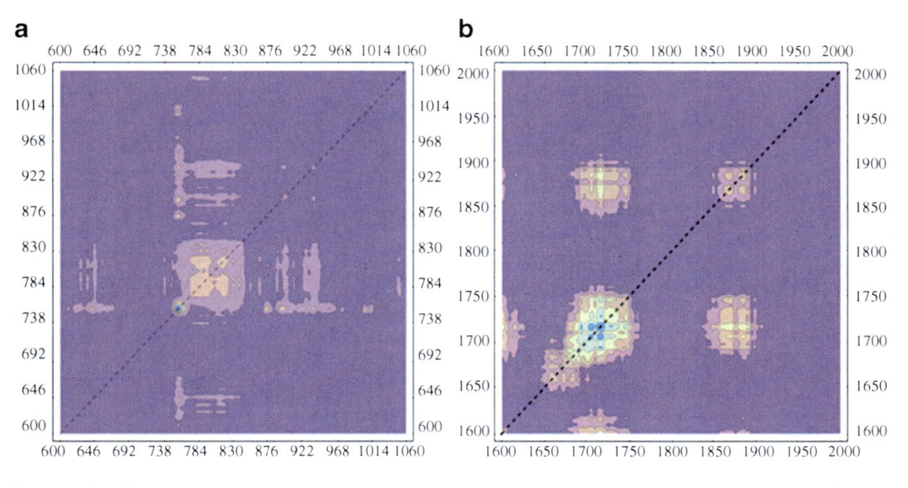

Fig. 12.8 2D-correlation spectra for lysozyme protein in the range between 600–1,060 and 1,600–2,000 cm^{-1}, respectively

1,700 cm^{-1} (increase in width of secondary structure region) (Fig. 12.8a, b), showing major variations at these frequencies. The alteration of 760 cm^{-1} band intensity shows the appreciable variation of hydrophobicity of lysozyme protein with the temperature. The lysozyme has three Phe residues which are located at the surface of protein (Imoto et al. 1972). The slight variation in Phe band intensity could be due to the protein–water interaction with temperature. Though, Fig. 12.7b does not show the conformational changes in Phe (1,005 and 1,033 cm^{-1}) and Try (1,013 cm^{-1}) breathing modes, 2D-correlation spectra show clear variation with the temperature. Remarkable positive cross peak at (760, 1,013) is observed due to the fact that both the bands are related to Trp residue. In addition, the positive peak at (760, 1,005) is also observed, showing that the Phe amino acids follows the same trend as of Trp residue.

12.3.1.3 Ribonuclease-B

RNAse-B (13,700 Da with 124 amino acids) is an enzyme secreted by the pancreas into the small intestine, where it catalyzes the hydrolysis of certain bonds in the ribonucleic acids present in the ingested food. Its secondary structure contains around 40% α-helix, 23% β-sheet and the rest as a random coil structure. Fig. 12.9a shows the Raman spectra for RNase-B vs. temperature in the range from 600 to 2,000 cm^{-1}. It illustrates various vibrational bands attributed to different amino acids, specially from Phe residues at around 1005, 1033, and 1210 cm^{-1}, Tyr residues at around 830 and 855 cm^{-1}, different scissor and wagging vibration related to the C–H$_x$ (x = 1–3) at about 1,450 and 1,320 cm^{-1}, respectively.

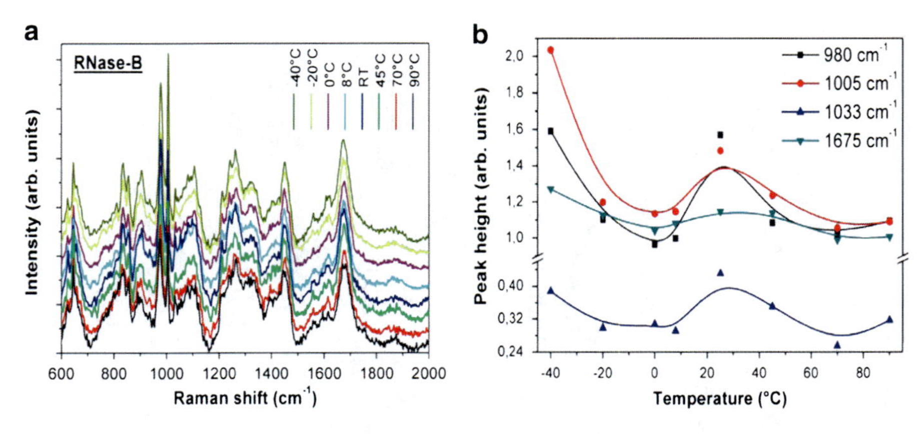

Fig. 12.9 (**a**) Raman spectra for RNase-B protein in the range of 400–1,800 cm^{-1} for different temperature, (**b**) Raman peak height of different bands centered at 980, 1,005, 1,033, and 1,675 cm^{-1} vs. temperature for RNase-B

In Fig. 12.9a, C–S vibrational bands centered at 625 and 645 cm^{-1} decrease with temperature which could be because of breaking the C–S band (these bands also represent the stability of protein). One peak centered at around 980 cm^{-1} is observed only for RNase-B among the other proteins described above. This band is attributed to the symmetric stretching vibration of sulfate ion which acts as an inhibitor of enzymatic action. It was found in the past (Wyckoff et al. 1970) that the reactive sites for the sulfate ion of RNase B are histidine residues. It is observed in our analysis that the 980 cm^{-1} band intensity decreases with respect to the intensity at RT. Daly et al. (1972) discusses that the loss of symmetry of sulfate ion by protonation causes the disappearance of band at around 980 cm^{-1}. Therefore, we suggest that the increase in temperature, in our case, causes the binding of free sulfate ion to the histidine residues. Figure 12.9b shows the peak height of different Raman bands centered at around 980, 1,005, 1,033, and 1,675 cm^{-1}. The peak height of 1,005 cm^{-1} (Phe) band is continuously varying with temperature but 1,033 cm^{-1} (Phe) band does not show remarkable variation which could be related to the change of Phe residues surrounding. Interestingly, our analysis reveals that the temperature effect on Phe band at 1,005 cm^{-1} follows the same trend of sulfate ions, as shown in Fig. 12.9b. Hence, it seems that with the increase in temperature, as the sulfate ion starts to bind with the histidine ions, the Phe residues in the vicinity start to have less access to the solvent, causing the decrease in intensity. The change in secondary structure for this protein with temperature is not clearly observed from amide I band (1,675 cm^{-1}).

Two-dimensional correlation spectra for RNase-B are illustrated in Fig. 12.10a, b in the range between 600–1,060 cm^{-1} and 1,600–2,000 cm^{-1}, respectively. Autopeaks in 2D-correlation spectra for RNase-B are observed at 980 (sulfate ion), 1,005 (Phe), 1,033 (Phe) and at around 1,670 cm^{-1} (secondary structures). However, the major changes in 2D-correlation spectra, due to the temperature

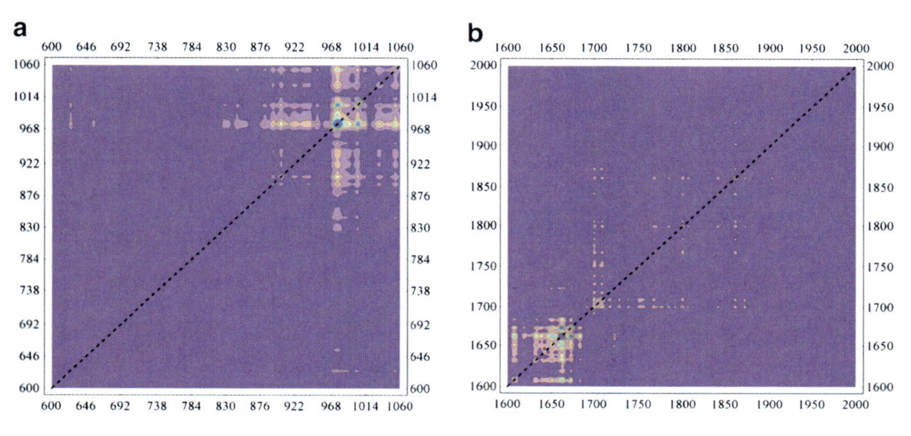

Fig. 12.10 2D-correlation spectra for RNase-B protein in the range of 600–1,060 and 1600–2000 cm^{-1}, respectively

variation, are observed in the range of 970–1,015 cm^{-1}. These noticeable variations in the intensity of these bands with the temperature, as shown in Fig. 12.9b, are clearly followed by the 2D-correlation analysis. Remarkable cross-correlation vibrational bands (980, 1,005), (980, 1,033) and (1,005, 1,033) can be clearly observed from Fig. 12.10a, which represent the vibrational bands 980, 1,005, and 1,033 cm^{-1} following the same trend with change in temperature. In other way, the binding of sulfate ion to histidine residue and the access of water to protein with temperature follow the same trend with temperature. Though, the secondary structure region does not show the clear variation in the intensity (Fig. 12.9b), the 2D-correlation spectra show a major variation in this range. The conformational changes in the secondary structure of protein by varying the temperature are clearly visible in Fig. 12.10b by showing the autopeaks in the range of 1,640–1,680 cm^{-1}.

12.3.1.4 Ferritin

Ferritin is an iron storage protein found in many living organisms, including bacteria, insects, plants etc. It is the only protein that can bind the metal ions in solution and convert them into a solid-phase mineral. The protein consists of an iron core surrounded by a shell of 24 protein subunits. The Raman and difference Raman spectra for ferritin protein vs. temperature are shown in Fig. 12.11a, b in the range between 400 and 1,800 cm^{-1}. Raman spectrum of ferritin at RT shows many vibrational bands attributed to different amino acids as stated above for other proteins. In addition, a faint band is observed at around 1,372 cm^{-1} which could be related to C–N vibration from porphyrin group, as discussed by Precigoux et al. (1994) using his XRD measurement about the possibility of porphyrin presence in ferritin protein and a very clear secondary structure band in the range of 1,620–1,700 cm^{-1}. By varying the temperature, there are the major changes in

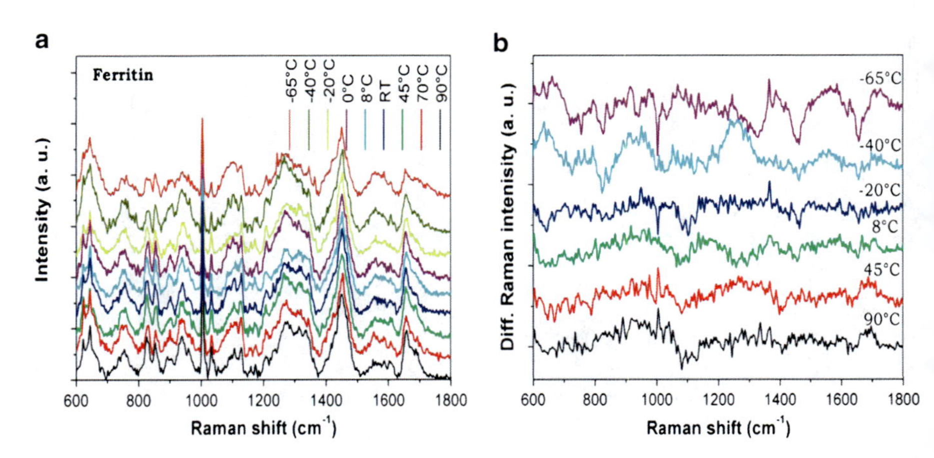

Fig. 12.11 (**a**) Raman and (**b**) difference Raman spectra for Ferritin protein in the range of 400–1800 cm^{-1} for different temperature

S–S vibrational band (450–550 cm^{-1}), C–S stretching vibrations (600–670 cm^{-1}), Trp band (830–870 cm^{-1}), aromatic breathing of Phe (1005 cm^{-1}), and secondary structure (1,610–1,700 cm^{-1}). From Fig. 12.11b, it can be observed that the relative intensity of 623 cm^{-1} (Phe) and 645 cm^{-1} (Tyr) vibrational band with temperature in the range between –20°C and 90°C does not show any remarkable change but on further decrease in temperature ($T < -20$°C) causes the decrease in Phe band at around 623 cm^{-1}. Stefanini et al. (1996) carried out the UV circular dichroism measurements on horse spleen apoferritin protein and found its thermal stability up to 93°C. The difference in this denaturation temperature could be because of the difference in concentration as it was suggested by Guo et al. (2006) that the denaturation temperature for proteins varies with the concentration. It is found that the intensity of Phe (1,005 cm^{-1}) band shows that the there is a continuous increase in its intensity with the temperature but the Phe (1,035 cm^{-1}) does not show remarkable variation, except for two extreme temperatures –65°C and 90°C, so we suggest that instead of any variation in Phe amino acid, there is the change in Phe surroundings. The increase in N–H and C–N vibrational band region between 1,200 and 1,380 cm^{-1}, where we observed the increase in broadening with increase, could be related to weakening H-bond as suggested by Grdadolnik and Marechal (2001). The phosphate contribution in the protein Raman spectrum is not distinguishable in the region 980 and 1,045 cm^{-1} as these regions are covered by various amino acids. Though it is suggested by Grdadolnik and Marechal (2001) about thermal stability higher than 90°C, we observed the conformational change in protein secondary structure at temperature –65°C and 90°C. For the temperature of –65°C, there is a sharp decrease in the α-helix structure (Fig. 12.11a), whereas for 90°C (Fig. 12.11b), we observe the positive band in the frequency range of 1,660–1,700 cm^{-1}, suggesting an increase in random coil and β-sheet at the expense of α-helical structure.

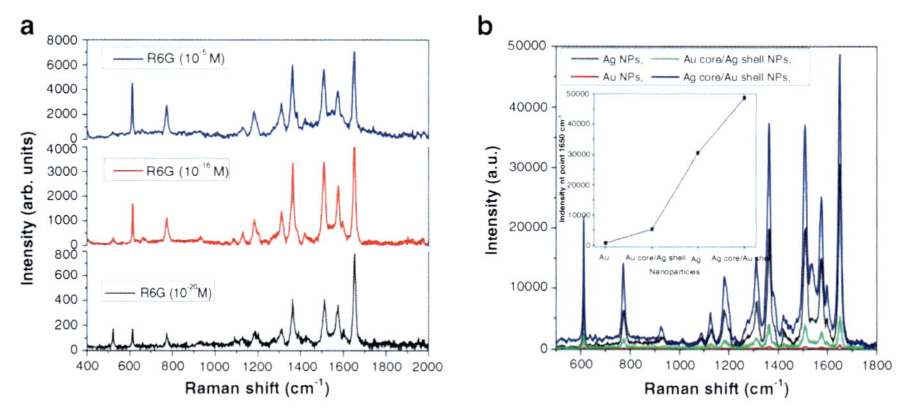

Fig. 12.12 (**a**) SERS spectra acquired from 10^{-5}, 10^{-16}, and 10^{-20} M R6G, absorbed on the silver structures, (**b**) SERS spectra acquired from 10^{-12} M R6G, absorbed on the silver, gold, Ag/Au, and Au/Ag structures

12.3.2 *Rhodamine 6G (R6G) on "Device2"*

Raman measurements are carried out for R6G with different concentrations from 10^{-2} to 10^{-20} M in the spectral range of 400–2,000 cm^{-1}. R6G is deposited by soaking the Ag substrate into the R6G aqueous solution for 30 min, then rinsing with water and drying with N_2. In Fig. 12.12a, Raman spectra of R6G molecules, obtained using a substrate of a 60 s silver deposition, are shown for fixed concentrations (10^{-5}, 10^{-16} and 10^{-20} M). There are various bands for R6G compound at around 1,650, 1,509 and 1,361 cm^{-1}, attributed to the xanthenes ring stretching of C–C vibrations, 1,575 cm^{-1} attributed to C–O stretching; while the band at 1,183 cm^{-1} is related to the C–H bending and N–H bending vibration of xanthenes ring. It can be clearly observed in Fig. 12.12a, the Raman intensity decreases with the decreasing concentration of R6G, as expected. Though, the concentration is very low, various bands are clearly visible even for R6G molecule with concentration of 10^{-20} M, keeping accumulation time 60 s and laser power 0.018 mW.

As it is well-known that:

(a) Ag-based SERS substrate shows very high Raman scattering enhancement, while the silver metal exhibits the surface oxidation in air due to its inherent properties. The device durability as a SERS substrate is limited;
(b) Au-based SERS substrates show better durability but exhibit lower SERS enhancement. In order to estimate SERS behavior with different metal nanoparticles (Ag, Au, and bimetals), Raman measurements have been carried out on Ag, Au, Au-core/Ag-shell, Ag-core/Au-shell metallic SERS substrates.

SERS response of the silver deposited substrates for R6G monolayer is compared with that of gold, gold/silver, and silver/gold deposited substrates. Raman spectra of 10^{-12} M R6G, deposited on various kinds of above described substrates are shown in Fig. 12.12b. Raman spectra show similar vibrational bands for individual and mixed metallic nanospheres based substrates. It is clearly evident from Fig. 12.12 that Raman signal intensity is increasing, respectively, for Au, Au-core/ Ag-shell, Ag and Ag-core/Au-shell samples. As it is shown in the inset of Fig. 12.12b, the Raman intensity of reference vibrational band centered at 1,649 cm^{-1} shows around 800 counts for gold metal and around 30,000 counts for Ag nanometals, whereas it shows about 48,700 counts for Ag-core/Au-shell mixed nanometallic structures. In this last substrate, it seems that gold has a protection function against the Ag oxidation and sulfur passivation, allowing so a better SERS activity.

12.4 Conclusions

We have fabricated two kinds of well-controllable and reproducible SERS devices using e-beam and electro-plating techniques ("Device1"), and e-beam and electroless techniques ("Device2"). Various proteins have been analyzed by varying the temperature ranging between −65°C and 90°C using "Device1" with concentration down to attomole. 2D-correlation and difference Raman analysis were performed in order to understand the conformational changes with the temperatures. "Device2" based on Ag and Au metallic nanoparticles are grown by means of electroless technique. We were able to investigate the R6G with concentration down to 10^{-20} M. By modifying the device with bimetallic layer instead of single metal layer we were able to enhance the Raman signal and the device durability, as well. This work would further enhance the research towards single molecule device with these innovative technological combinations to fabricate a better SERS device.

References

Artymiuk PJ, Blake CCF, Rice DW, Wilson KS (1982) The structures of the monoclinic and orthorhombic forms of hen egg–white lysozyme at 6 Å resolution. Acta Crystallogr B Struct Crystallogr Cryst Chem 38:778–783

Bulovas A, Talaikyte Z, Niaura G, Kazemekaite M, Marcinkeviciene L, Bachmatova I, Meskys R, Razumas V (2007) Double layered Ag/Au electrode for SERS spectroscopy: preparation and application for adsorption studies of chromophoric compounds. CHEMIJA 10(4):9–15

Celej MS, Montich GG, Fidelio GD (2003) Protein stability induced by ligand binding correlates with changes. Protein Sci 12:1496–1506

Coluccio ML, Das G, Mecarini F, Gentile F, Pujia A, Bava L, Tallerico R, Candeloro P, Liberale C, De Angelis F, Di Fabrizio E (2009) Silver–based surface enhanced Raman scattering (SERS) substrate fabrication using nanolithography and site selective electroless deposition. Microelectron Eng. doi:10.10161/j.mee.2008.12.061

Constantino CJL, Lemma T, Antunes PA, Aroca R (2001) Single–molecule detection using surface–enhanced resonance Raman scattering and Langmuir–Blodgett monolayers. Anal Chem 73(15):3674–3678

Daly FP, Brown CW, Kester DR (1972) Sodium and magnesium sulfate ion pairing: evidence from Raman spectroscopy. Biochemistry 76:3664–3668

Gilmanshin R, Beek VB, Callender R (1996) Study of the ribonuclease S–peptide/S–protein complex by means of Raman difference spectroscopy. J Phys Chem 100(41):16754–16760

Grdadolnik J, Marechal Y (2001) Bovin serum albumin observed by infrared spectroscopy. 1. Methodology, structural investigation, and water uptake. Biopolymers 62:40–53

Guo J, Harn N, Robbins A, Dougherty R, Middaught CR (2006) Stability of helix–rich proteins at high concentrations. Biochemistry 45(28):8686–8696

Haes AJ, Zhao J, Zou S, Own CS, Marks LD, Schatz GC, Duyne RPV (2005) Solution–phase, triangular Ag nanotriangles fabricated by nanosphere lithography. J Phys Chem B 109(22): 11158–11162

Haynes CL, McFarland AD, VanDuyne RP (2005) Surface–enhanced Raman spectroscopy. Anal Chem 77(17):338A–346A

Hildebrandt P, Stockburger M (1984) Surface–enhanced resonance Raman spectroscopy of Rhodamine 6G adsorbed on colloidal silver. J Phys Chem 88(24):5935–5944

Imoto T, Johnson LN, North ACT, Phillips DC, Rupley JA (1972) In the enzymes, 7, 3rd edn. Academic Press, New York

Jung A, Sippel AE, Grez M, Schfjtz G (1980) Exons encode functional and structural units of chicken lysozyme. Proc Natl Acad Sci USA 77(10):5759–5763

Kahl M, Voges E, Kostrewa S, Viets C, Hill W (1998) Periodically structured metallic substrates for SERS. Sens Actuators B Chem 51:285–291

Keiderling TA (1993) In: Cark RJH, Hester RE (eds), Biomolecular spectroscopy, Part B. John Wiley and Sons, New York, pp 267–312

Kneipp K, Wang Y, Kneipp H, Perelman LT, Itzkan I, Dasari RR, Feld MS (1997) Single molecule detection using surface–enhanced Raman scattering (SERS). Phys Rev Lett 78(9):1667–1670

Li T, Chen Z, Johnson JE, Thomas GJ (1990) Structural studies of bean pod mottle virus, capsid, and RNA in crystal and solution states by laser Raman spectroscopy. Biochemistry 29(21):5018–5026

Li Y, Zhou J, Zhang K, Sun C (2007) Gold nanoparticle multilayer films based on surfactant films as a template: preparation, characterization, and application. J Chem Phys 126(9):094706

Miura, T., Thomas Jr., G.J., (1995). In: Biswas, B.B., Roy, S. (Eds.), Proteins: Structure, Fuction and Engineering, 24. Plenum Press, New York, pp. 55–97.

Miura T, Takeuchi H, Harada I (1991) Raman spectroscopic characterization of tryptophan side chains in lysozyme bound to inhibitors: role of the hydrophobic box in the enzymatic function. Biochemistry 30(24):6074–6080

Murayama K, Tomida M (2004) Heat–induced secondary structure and conformation change of bovine serum albumin investigated by Fourier transform infrared spectroscopy. Biochemistry 43(36):11526–11532

Nara M, Sakamoto A, Yamamichi J, Tasumi M (2001) Temperature dependence of near–IR excited Raman spectra of crystalline hen egg–white lysozyme. Biopolymers 62:168–172

Nie S, Emory SR (1997) Probing single molecules and single nanoparticles by surface–enhanced Raman scattering. Science 275:1102–1106

Noda I, Dowrey AE, Marcott C, Story GM (2000) Generalized Two Dimensional Correlation Spectroscopy. Applied Spectroscopy 54(7):236A–248A

Perney N, Baumberg J, Zoorob M, Charlton M, Mahnkopf S, Netti M (2006) Tuning localized plasmons in nanostructured substrates for surface enhanced Raman scattering. Opt Express 14(2):847–857

Precigoux G, Yariv J, Gallois B, Dautant A, Courseille C, d'Estaintot BL (1994) A crystallographic study of heme binding to ferritin. Acta Crystallogr D Biol Crystallogr 50:739–743

Sato H, Chiba H, Tashiro H, Ozaki Y (2001) Excitation wavelength–dependent changes in Raman spectra of whole blood and hemoglobin: comparison of the spectra with 514.5–, 720–, and 1064–nm excitation. J Biomed Opt 6(3):366–370

Siamwiza MN, Lord RC, Chen MC (1975) Interpretation of the doublet at 850 and 830 cm^{-1} in the Raman spectra of tyrosyl residues in proteins and certain model compounds. Biochemistry 14(22):4870–4976
Spiro TG (1985) Resonance Raman spectroscopy as a probe of heme protein structure and dynamics. Adv Protein Chem 37:111–159
Stefanini S, Cavallo S, Wang CQ, Tataseo P, Vecchini P, Giartosio A, Chiancone E (1996) Thermal stability of horse spleen apoferritin and human recombinant H apoferritin. Arch Biochem Biophys 325:58–64
Thomas GJ, Prescott B, Day LA (1983) Structure similarity, difference and variability in the filamentous viruses fd, If1, IKe, Pf1 and Xf: investigation by laser Raman spectroscopy. J Mol Biol 165:321–356
Tuma R, Russell M, Rosendahl M, Thomas GJ Jr (1995) Solution conformation of the extracellular domain of the human tumor necrosis factor receptor probed by Raman and UV-resonance Raman spectroscopy: structural effects of an engineered PEG linker. Biochemistry 34(46): 15150–15156
Weitz DA, Garoff S, Gersten JI, Nitzan A (1983) The enhancement of Raman scattering, resonance Raman scattering and fluorescence from molecules adsorbed on a rough silver surface. J Chem Phys 79:5324
Wyckoff HW, Tsernoglou D, Hanson AW, Knox JR, Lee B, Richanrds MF (1970) The three-dimensional structure of ribonuclease-S. Interpretation of an electron density map at a nominal resolution of 2 a. J Biol Chem 245:305–328
Yang Y, Shi J, Hawamura G, Nogami M (2008) Preparation of Au–Ag, Ag–Au core–shell bimetallic nanoparticles for surface–enhanced Raman scattering. Scr Mater 58:862–865
Zhang D, Xie Y, Mrozek MF, Ortiz C, Davisson VJ, Ben-Amotz D (2003) Raman detection of proteomic analytes. Anal Chem 75(21):5703–5709
Zhang X, Yonzon CR, Duyne RPV (2006) Nanosphere lithography fabricated plasmonic materials and their applications. J Mater Res 21:1083–1092

Chapter 13
Near Infrared Three-Dimensional Nonlinear Optical Monitoring of Stem Cell Differentiation

Uday K. Tirlapur and Clarence Yapp

13.1 Introduction

Mouse embryonic stem cells (mESCs) have been used as a model to understand differentiation into the cartilaginous, neuronal and pancreatic lineages with the hope that this basic knowledge gained can be applied to human embryonic stem cells (hESCs) for subsequent applications in cell-based therapies and regenerative medicine. Normally, these mESCs are simply cultured while attached to the surface of Petri dishes or tissue culture plates (2D culture). Only recently concerns have been raised whether these cells could be cultured while suspended in liquid or in a 3D scaffolds, which mimics the actual environment of the body more closely. Although there are other studies that have performed similar research, there are differences in the use of materials and technology, which might show a more complete analysis of the cell growth over time.

13.1.1 Stem Cells

Stem cells, in general, are precursors of all other cell types that can be found in the body. In tissue engineering, they are being extensively used for repair as they are able to divide numerous times while retaining their ability to differentiate into any mature cell type (under certain experimental conditions) when needed. This property is known as a cell's potency. Like hESC, those from mice are also able to differentiate into various cells such as cartilaginous (Hwang et al. 2006a), neuronal (Fraichard et al. 1995) and pancreatic (Nakanishi et al. 2007).

There are basically two types of stem cells, namely ESCs and adult stem cells. ESCs are derived from the blastocyst's inner cell mass during the first weeks of

U.K. Tirlapur (✉) and C. Yapp
Stem Cell Epigenetics and Biomolecular Imaging Group, The Botnar Research Centre, Institute of Musculoskeletal Sciences, Nuffield Orthopaedic Centre, University of Oxford, Oxford OX3 7LD, UK
e-mail: uday.tirlapur@ndorms.ox.ac.uk, uday.tirlapur@gmail.com

A. Diaspro (ed.), *Optical Fluorescence Microscopy*,
DOI 10.1007/978-3-642-15175-0_13,

pregnancy (Buttery et al. 2001; Nieden et al. 2003). According to Hwang et al. (2006a), these cells are pluripotent and are able to differentiate into any of the three germ layers found in the body. These include the ectoderm, endoderm and mesoderm. Skin cells are an example of ectoderm tissue, whereas endoderm includes cells that make up the digestive tract. The mesoderm includes connective tissue such as bone, cartilage and tendon. ESCs also have the advantage that they proliferate more rapidly than adult stem cells, thus, improving the time for repair (Kramer et al. 2006).

When stem cells differentiate into cartilage cells (chondrocytes), several genes are expressed and these have been identified. They are aggrecan, Sox9, collagen IIA, collagen IIB and collagen X (Hwang et al. 2006a). If the chondrocytes are allowed to mature further, they undergo hypertrophy and become bone-cells. In this case, the osteocalcin and osteopontin genes (Nieden et al. 2003) are expressed.

For effective pancreatic differentiation, St-Onge et al. (1999) identifies several important genes in the mouse. By removing these genes, the mice experienced poor pancreatic development. For example, p48 is necessary in the early development of the pancreas and in its absence, the endocrine cells developed in the spleen. Without Nkx2.2, the size of the islets of Langerhans was reduced and affected the amount of hormones being released from the endocrine cells, especially insulin. Without Isl1 and pdx1, there was an absence of endocrine cells and an entire pancreas, respectively. Miyazaki et al. (2004) also report that pdx-1 expression is crucial for pancreatic development and that mice which are missing this gene have a low number of endocrine cells.

Pax 4 and 6 both affect islet formation. However, both genes seem to have opposite effects on the development of glucagon producing cells; thus, an optimal balance needs to be achieved.

Neuronal differentiation and morphology can be observed by measuring gene expression of microtubule-associated protein 2 (MAP2) and glial fibrillar acidic protein (GFAP). For example, the differentiation of neuronal stem cells into astrocytes causes a change in GFAP expression (Tang et al. 2002). According to Riederer and Matus (1985), MAP2 is only expressed in dendrites. It is believed that MAP2 supports the maturation of dendrites (Huber and Matus 1984). Also, MAP2 is essential in the assembly of tubulin into microtubules (Huber and Matus 1984).

Overall, by characterizing the expression of cartilage-, neuron- and pancreatic-specific genes at different time points, it is possible to determine the state of differentiation that they are in. Changes in culturing parameters (such as growth factors) may shift the onset of differentiation earlier or later and may shorten the time needed for this process to occur.

13.1.2 Differentiation into Chondrocytes

With the right signals, ESCs can be induced to differentiate into chondrocytes. Various methods have been used such as plating a single cell suspension in a Petri dish and allowing the cells to aggregate together in clumps called embryoid bodies

(EBs), as was performed by Martin and Evans (1975), Buttery et al. (2001), Bourne et al. (2004) and Hwang et al. (2006a, b). However, these formations were too fragile to handle when transferring between culture plates. Furthermore, the size and number of cells in each embryoid body would not be the same, making comparisons difficult. Hwang et al. (2006a) report that each embryoid body could contain anywhere between 1,436 and 4,524 cells. Another method is through “hanging drops,” which is used extensively by Kramer et al. (2000, 2003, 2005a, and 2005b) and Nieden et al. (2003, 2005). Hanging drop method involves plating drops of a single cell suspension on the inside cover of a bacteriological dish. Again, the cells in each drop aggregate together to form a more robust embryoid body of predetermined size.

The use of certain chemicals is another factor that is used to drive stem cells towards the chondrocyte lineage (Nieden et al. 2005; Kramer et al. 2000). Transforming growth factor-beta (TGF-B_1) and bone morphogenic protein-2 and -4 (BMP-2 and BMP-4) are possible induction factors although BMP-2 is the more popular reagent of choice. According to Wang et al. (1988), it is a purified form of bone mineral extract and has been shown to induce chondrogenesis. Both Kramer et al. (2000) and Nieden et al. (2005) compared the onset of the formation of cartilage using different concentrations of 2 and 10 ng/ml. It was found that as the concentration was increased, there was an earlier and larger extent of chondrocyte differentiation.

13.1.3 Differentiation into Neurons

Bain et al. (1995) proposed culturing embryoid bodies in ES cell culture conditions without retinoic acid for 4 days. This was followed by supplementing the medium with retinoic acid for another 4 days. Other successful growth factor conditions for embryoid bodies include: basic fibroblast growth factor, epidermal growth factor and L-ascorbic acid or platelet-derived growth factor (Lau et al. 2006; Brustle et al. 1999); hepatocyte growth factor (Kato et al. 2004); sonic hedgehog, BMP-2, Wnt3A and retinoic acid (Murashov et al. 2005); and fibroblast growth factor-2 alone (Hancock et al. 2000). Neuronal differentiation has also been successful without the use of embryoid bodies with growth factors (Ying et al. 2003; Kitazawa and Shimizu 2005; Kitajima et al. 2005). A less common technique is by transfecting a mono-culture of ES cells with a noggin expression plasmid resulting in neuronal differentiation within 24 h (Gratsch and O’Shea 2002).

13.1.4 Differentiation into Pancreatic Cells

Raikwar and Zavazava (2009) have compared various methods of directing ESCs to differentiate towards the pancreatic lineage. Two established techniques for generation of insulin producing cells (IPC) are embryoid body formation and definitive endodermal (DE) approach using different growth factors. With the DE method, ESCs are treated with activin A followed by a combination of retinoic acid and FGF. The current challenge faced by many scientists is finding a method for generating a homogeneous set of IPCs. Insulin production is also only about one-tenth as reported by Raikwar and Zavazava (2009) compared to actual pancreatic islets.

Previously, many studies on ESC differentiation to the pancreatic lineage were performed in 2D. One study by Wang and Ye (2009) showed that when mESCs underwent differentiation in a 3D collagen structure, the resulting clusters appeared closer to adult islet cells found in the pancreas. These clusters were reported as being spherical without necrosis in the central regions. Upon performing an immunohistochemistry on the clusters, Wang and Ye (2009) reported that clusters grown in 2D and 3D express pancreatic-specific markers in different places. In 2D, there were no islets that were formed and the localization of insulin is scattered. In 3D, however, insulin was expressed in the centre of the cluster, which is a closer representation of actual adult islet clusters. Somatostatin and glucagons were found in the periphery of the clusters. Unlike in the 2D case, pdx1 was upregulated in the 3D cultures. It was also reported that the 3D cultures produced more insulin when exposed to glucose solution. However, this test is not definitive since the detected insulin could have actually originated from the various supplements found in the differentiation medium (ITSFn, N2 and B27), which could potentially mask the insulin produced by the cells themselves by 1,000-folds (Raikwar and Zavazava 2009). These growth factors with the addition of bFGF were also used in a study by Lumelsky et al. (2001) where embryoid bodies were generated by suspension rather than via the hanging drop method. Both Shi et al. (2005) and Nakanishi et al. (2007) have also relied on subjecting embryoid bodies to different growth factors such as a combination of retinoic acid and activin for different durations.

13.1.5 Nonlinear Optical/Second Harmonic Generation Imaging

The fibre-forming collagen is known to have a crystalline triple-helical structure that is non-centrosymmetric and possesses a strong second-order nonlinear susceptibility, a nonlinear optical property characterized by second harmonic generation (SHG) of incident light (Zoumi et al. 2002; Zipfel et al. 2003).

It can be hypothesized that SHG can as well be adopted to monitor formation of such fibrillar collagen type II and X during various stages of chondrogenesis from mESC expressing eGFP. To test this hypothesis, a near infrared femtosecond pulsed laser at various NIR wavelengths (800–900 nm) was adopted for simultaneous imaging of backward-SHG (b-SHG) and forward-SHG (f-SHG) from chondrogenic areas/nodules within the EBs, as well as identification of two-photon excited cellular eGFP fluorescence associated with the cell clusters.

Boerboom et al. (2007) describes an alternative and new method that uses fluorescent probes, which when excited by the laser, emit light in return. The advantage of using SHG is that photobleaching does not occur. Thus, the intensity of the fluorescence is better maintained with SHG and allows for easier imaging without the need for expensive probes.

13.2 Materials and Methods

13.2.1 Cell Culture

The mESC expressing the green fluorescent protein (GFP) was cultured on a monolayer of feeder cells at 37°C and with 5% CO_2.

The cells were grown in ES cell culture medium which consisted of 80 ml of DMEM (Dulbecco's modified Eagle's medium, Sigma-Aldrich, UK, stock supplemented with 2.2 g/l of $NaHCO_3$), 15 ml of batch-tested FCS (foetal calf serum), 1 ml of 200 mM L-glutamine stock (Sigma-Aldrich, UK), 1 ml of beta-mercaptoethanol stock (7 μl of beta-mercaptoethanol in 10 ml PBS), 1 ml of 100× nonessential amino acid stock (Sigma-Aldrich, UK) and 1 ml of antibiotic stock (5×10^5 units of penicillin and 500 mg of streptomycin in 100 ml of PBS). The medium was changed every second day. Cells were passaged every 3 days.

13.2.2 Hanging Drop Cultures and Induction of Chondrogenic/ Pancreatic Differentiation

ESCs were induced to differentiate into different lineages using the hanging drop culture method (as shown in Fig. 13.1). Fifteen nanograms per millilitre of BMP-2 and a combination of 1 μM of activin and 25 ng/ml of retinoic acid (RA) were used to direct mESCs into becoming chondrocytes and pancreatic cells, respectively. These growth factors were separately added to a single cell suspension of 6,000 cells in 20 μl of medium. This was then plated onto the inside of a bacteriological dish cover and placed back onto the dish. A small amount of ES medium was also added to the bottom of the dish to prevent drying. The hanging drops were left for 2 days.

After that, an embryoid body, which was an aggregation of cells, could be seen at the bottom of each hanging drop. Each embryoid body was transferred to a new bacteriological dish with ES medium containing the aforementioned growth factors at the same concentrations for an additional 3 days. At the end of the 5 days, each embryoid body was transferred onto a glass coverslip in culture well plates containing growth factor free medium.

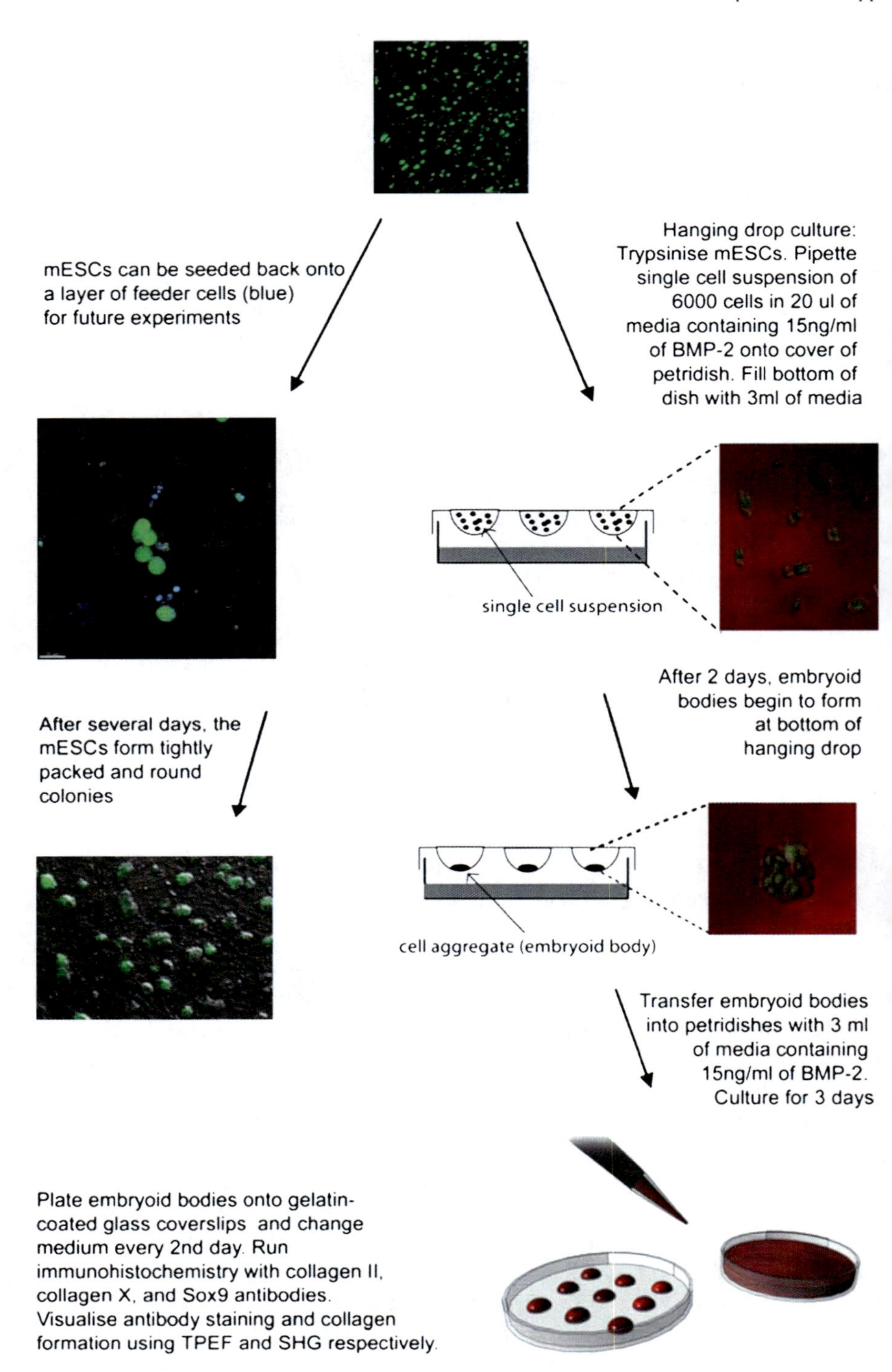

Fig. 13.1 Semi-schematic diagram showing procedure for induction of chondrogenic/pancreatic differentiation from mESCs using hanging drop technique

13.2.3 Induction of Neuronal Differentiation

For neuronal induction, 1×10^6 mESCs were seeded onto 60-mm bacteriological Petri dishes containing ES medium supplemented with 1 μM of retinoic acid (Sigma-Aldrich, UK). The mESCs aggregated together into embryoid bodies after 2 days and were subsequently transferred onto glass coverslips in culture plates. The embryoid bodies were cultured in regular ES medium without retinoic acid.

For imaging 2D attached EBs, they were placed on coverslips in six-well plates. Between two and three EBs were placed on each coverslip, which had been coated with gelatin. Floating 3D cultured EBs, on the other hand, were taken from their Petri dishes and placed inside microscope cavity slides so as not to crush them. Cultures embedded in a 3D scaffold of 6 mg/ml of rat-tail-derived collagen I (Becton Dickinson, UK) were similarly scooped onto these slides.

13.2.4 Immunohistochemical Localisations

The protocol for immunohistochemistry was adapted from Hegert et al. (2002). Briefly, samples were fixed with methanol and acetone (70:30) for 5 min at room temperature. After rinsing with PBS three times, cultures were blocked with goat serum diluted to 10% for 30 min at 37°C in an incubator. The primary antibody was then added with the following dilution factors for 2 h at 37°C. Samples were then incubated for 1.5 h in secondary antibody. The primary and secondary antibodies were used at the dilutions shown in Table 13.1.

The collagen II and collagen X primary antibodies were obtained from the Developmental Studies Hybridoma Bank from the University of Iowa. The glucagon, insulin and amylase antibodies were from Sigma, UK. The GFAP, MAP2 and Sox 9 antibodies were obtained from Abcam, UK.

When ready, the coverslips were mounted and glued onto cavity slides to preserve the integrity of the sample. To prevent the sample from drying, a small amount of medium was also added.

Table 13.1 Antibodies and dilutions used for Immunohistochemistry

Marker proteins	Primary antibodies	Dilution factor	Secondary antibodies	Dilution factor
Collagen II	CIIC1-s	1:10	IgG2a	1:500
Collagen X	X-AC9-s	1:10	IgG1	1:500
Sox-9	ab3697	1:500	IgG	1:500
Glucagon	G2654-.2ML	1:500	IgG1	1:500
Insulin	I2018-.2ML	1:500	IgG1	1:500
Amylase	A8273-1VL	1:500	IgG	1:500
GFAP	ab10062	1:2,500	IgG2b	1:500
MAP2	ab11267	1:300–1:500	IgG1	1:500

13.2.5 Imaging and 3D Monitoring

13.2.5.1 High-Resolution Two/Multiphoton Imaging

A tunable (700–1,000 nm) near infrared (NIR) Ti:sapphire laser (Mira-Coherent, Ely, UK) was used to obtain a femtosecond pulsed (150 fs) laser beam that was directly coupled to a modified (Mulholland 2006) multiphoton imaging system as described recently (Tirlapur et al. 2006). Briefly, it consisted of a BioRad Radiance 2100 multiphoton dedicated (MPD) laser-scanning and control system (BioRad Microscience Ltd., now Carl Zeiss GmbH, Jena, Germany) coupled to a Nikon E600 FN upright microscope (Nikon UK Ltd, Surrey, UK). Sub-femtolitre excitation beam focusing was achieved using either a 20× (N.A. 0.55) or a high-numerical-aperture (N.A. 1.3) 60× water immersion objective (Nikon), with a working distance of about 1 mm. Stepper motor control of the objective lens focus enabled scanning along the optical *z*-axis with a minimum step size between 0.45 and 1.5 μm.

In the MPD imaging system, a single 670-nm ultraviolet optimized long-pass dichroic mirror (Chroma Technology Corp, Vermont, USA) placed in the excitation path within the infinity focus of the microscope head directed fluorescence emission signal in the UV to visible wavelength range towards non-descanned bi-alkaline and multi-alkaline photomultiplier tubes (PMTs). Serial optical sections of EBs were obtained using NIR excitation at either 800 or 900 nm through a depth of 200–500 μm. Furthermore, the 3D organization of fibrillar collagen was ascertained by femtosecond NIR laser evoked b-SHG imaging (Zoumi et al. 2002) from less-ordered immature collagen (Williams et al. 2005).

Appropriate mean NIR laser powers were electronically regulated via the pockels cell and were recorded using a Fieldmaster FM power meter (Coherent, Ely, UK) at the back aperture of the objective lens. For all specimens, two/multiphoton excitation of eGFP within the cells and generation of SHG signal from structural collagen required less than 6 mW of mean laser power.

A transmission detection system consisting of a 1.4 NA oil condenser lens, fibre-coupled multi-alkaline PMT and dual filter wheels provided infrared transmission and f-SHG imaging of mature collagen (Williams et al. 2005) from the EBs.

13.2.5.2 Image Processing and Analysis

All serial optical sections from individual datasets were initially processed with BioRad Lasersharp software. Alternatively they were loaded into Imaris 4.2.0 (Bitplane Ag, Zurich, Switzerland) software suite for advanced processing and analysis. The Imaris software suite offers a range of highly advanced three-dimensional image processing capabilities including volume rendering and statistical analysis of three-dimensional surface-rendered objects. Both these techniques were used to evaluate the three-dimensional distribution of differentiating chondrocytes and organization of collagen within the EBs at different time points.

13.3 Results

13.3.1 Evidence for Two-Photon Excitation (TPE) and Second Harmonic Generation (SHG)

To ensure that the measured signal from eGFP fluorescence and the collagen SHG signal appeared only at the expected wavelengths, two different controls were performed. First, the Ti:sapphire laser was taken out of mode-locking operation (during serial optical sectioning of the EBs), and neither eGFP fluorescence nor collagen-associated SHG was observed, indicating that the signal arose from a nonlinear optical process (results not shown). This is because when in continuous wave (cw) lasing mode, the Ti:sapphire laser has insufficient peak power to produce either two-photon excited fluorescence (of eGFP) or SHG (from collagen) with any measurable efficiency (Campagnola and Loew 2003). Second, no fluorescence background was observed below 400 nm, demonstrating that the observed SHG signal was indeed free of autofluorescent components.

The wavelength for exciting the cells and collagen in the sample was also optimized to obtain the best images. As shown in Fig. 13.2, wavelengths of 800,

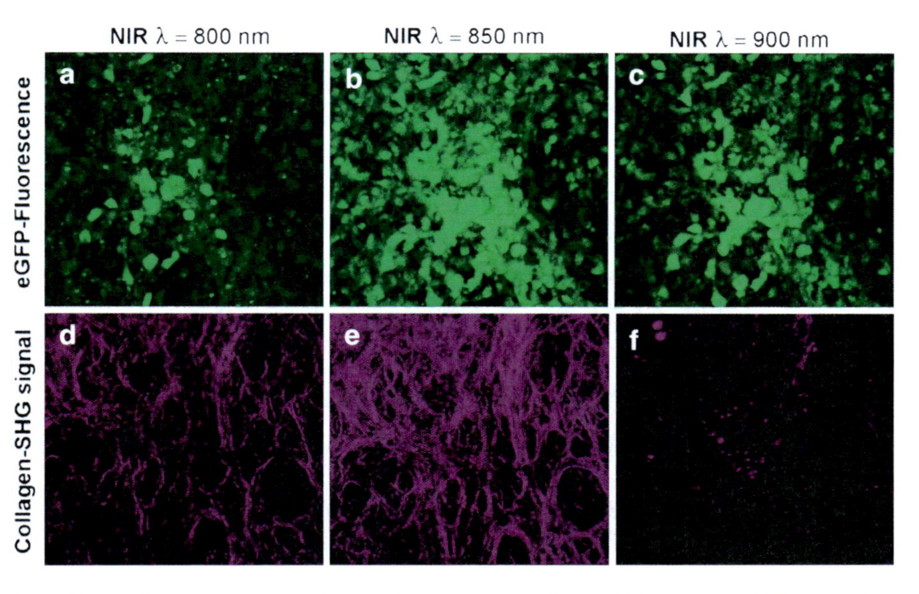

Fig. 13.2 Optimization of excitation wavelength at 800, 850 and 900 nm. (**a–c**) eGFP expression in *green* and (**d–f**) fibrillar collagen from SHG in *violet*

850 and 900 nm were tried. At 850 nm (Fig. 13.2b–e), the sample was excited optimally shown by the enhanced brightness compared to when the other wavelengths were used. At this wavelength, it was easier to discern the cells and identify single strands of collagen distinctly.

13.3.2 Cell Morphology and Organization of Fibrillar Collagen

13.3.2.1 Attached Cartilaginous Embryoid Bodies (2D Cultures)

At day 4 (Fig. 13.3a), there was a significant amount of fibrillar collagen. The thickness of the fibres was measured using the BioRad Lasersharp software. Thin fibres were measured to be about 0.80 ± 0.02 μm, while thick fibres were 2.43 ± 0.06 μm as ascertained from the microscope images.

At day 17 (Fig. 13.3b), the collagen fibrillar matrix is quite extensive showing thick fibres. The fibres have also begun to organize themselves into the typical hexagonal structures suggesting that collagen X has been produced along with the usual collagen II. Finally, the cell count has dropped as they have undergone apoptosis (programmed cell death). This is usual during bone formation and is characterized by cell fragmentation causing the collagen matrix to remain.

13.3.2.2 Attached Pancreatic Embryoid Bodies (2D Cultures)

The mESCs were differentiated towards the pancreatic lineage using a combination of retinoic acid and activin A in a hanging drop culture. After 3 days post-induction, the EBs already displayed spherical-like structures (Fig. 13.4a), which have also

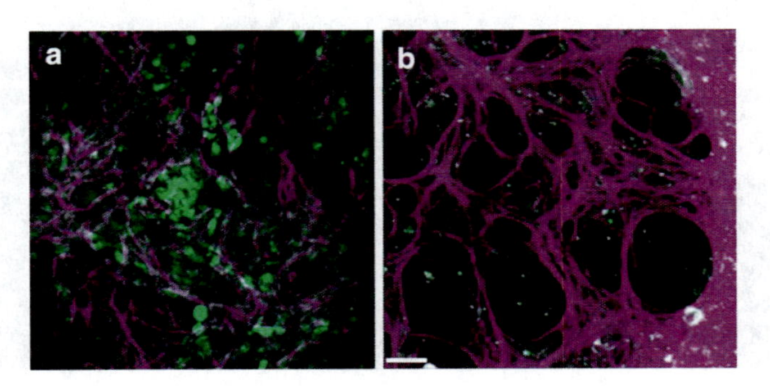

Fig. 13.3 Simultaneous imaging of fibrillar collagen (SHG, *violet*) and eGFP expression (*green*) in (**a**) 4-day-old EB and (**b**) 17-day-old EB

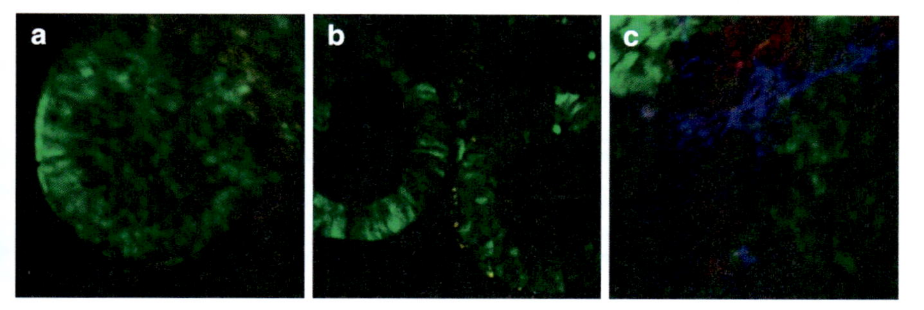

Fig. 13.4 Simultaneous imaging of (**a**) 3-day-old embryoid bodies with spherical structures consisting of differentiated pancreatic tissue. The structures are hollow as shown in (**b**). Similar to the chondrogenic nodules, formation of collagen can also be observed in (**c**) within these EBs.

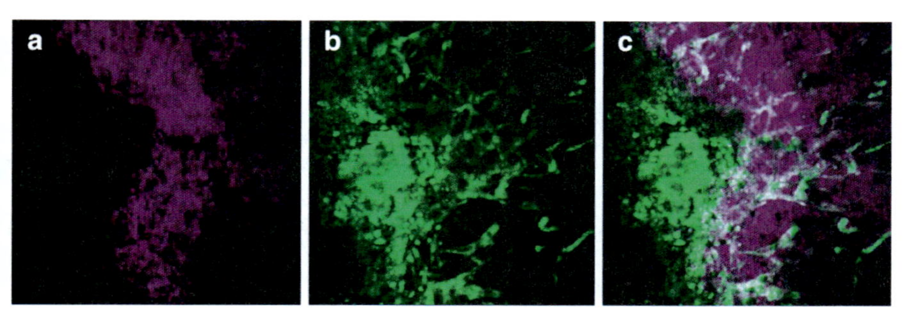

Fig. 13.5 Simultaneous imaging of 3-day-old 3D EB containing (**a**) diffuse collagen type I (SHG, *violet*), (**b**) eGFP cells (*green*) and (**c**) the merged image

been reported by Nakanishi et al. (2007). When a z-stack image of these structures were taken, we observed that they were hollow (Fig. 13.4b) with elongated cells in the periphery. Each embryoid body contained several of these hollow spherical structures. More importantly, the cells also produced structural collagen as shown in Fig. 13.4c.

13.3.2.3 Embryoid Bodies in 3D Scaffold

Figure 13.5 shows a culture grown in 3D using a scaffold made from collagen type-1. The 3D scaffold is shown in Fig. 13.5a and appears diffuse rather than the usual fibrillar structure. An embryoid body was then placed in the collagen scaffold. Already by day 3, the peripheral cells of the embryoid body (Fig. 13.5b) have begun stretching and changing their morphology through the collagen scaffold. They have also started to migrate and spread from the bottom-left to the top-right. The images of the collagen scaffold and the cells were merged and are shown in Fig. 13.5c.

13.3.3 Immunolocalization Studies

13.3.3.1 Nanog

The GFP-positive mESCs (Fig. 13.6a) were checked for their pluripotency, which was characterized by nanog antibody staining (Fig. 13.6b shown in red). It was found that there was an inhomogeneous level of nanog expression within the colony while they were grown on a feeder layer. For example, there was more nanog expression in the peripheral ells compared to those located in the centre.

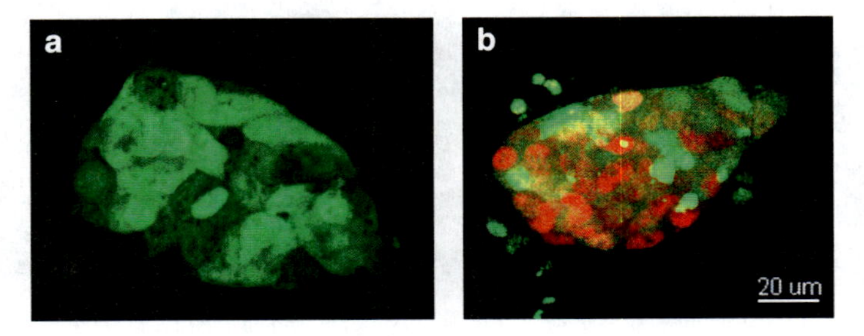

Fig. 13.6 (**a**) GFP-positive mESCs grown on gelatin-coated coverslips. (**b**) Nanog staining (shown in *red*) of undifferentiated mESCs grown on a feeder layer. Nanog expression varies within colony. Magnification ×60

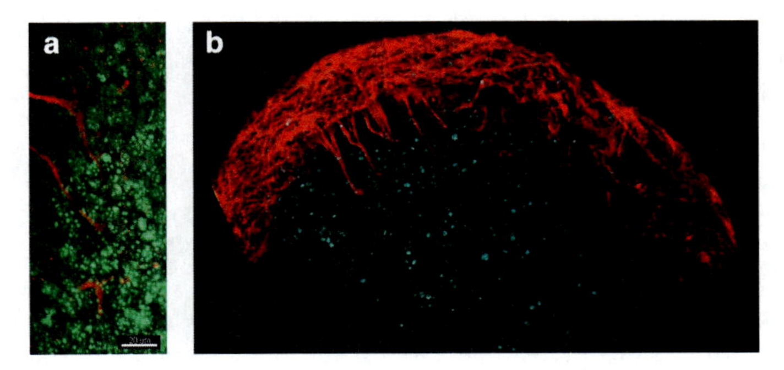

Fig. 13.7 (**a**) Simultaneous imaging of 10-day-old EB undergoing neuronal differentiation showing eGFP (*green*) and GFAP-positive expression (*red*) (×60). (**b**) An 11-day-old EB showing positive staining for MAP2 (×20). *Insert* shows EB at ×60

13.3.3.2 Attached Neuronal Embryoid Bodies (2D Cultures)

Embryoid bodies were induced to differentiate into neural cells using retinoic acid and activin. As shown in Fig. 13.7a, by day 10, the EBs began to show the characteristic morphology of neurons (presence of axons and somas) as reported by Fraichard et al. (1995), Murashov et al. (2005), and Pagani et al. (2006). Fraichard et al. (1995) observed positive staining for GFAP which is consistent with our findings (Fig. 13.7a, shown in red); however, GFAP in our experiments declined after day 12. As shown in Fig. 13.7b, the EBs developed into an extensive network of interconnecting neurons which stained positively for MAP2. The above-mentioned studies have also patch clamped these cells to characterize their electrophysiological properties. This could be a further way to compare the functional properties of ESC-differentiated neuronal cells versus ex vivo neurons.

13.3.3.3 Collagen II

Collagen II was immunolocalized in chondrogenic nodules within the embryoid bodies (Fig. 13.8a). These chondrogenic nodules consisted of dense collagen fibers that could be conveniently imaged by second harmonic generation (SHG). Reticulate collagen fibers were invariably found to surround the differentiating cells. In particular the indirectly visualized red collagen II immunofluorescence (Fig. 13.8a), was exclusively associated with the cluster of cells that also retained high GFP fluorescence (Fig. 13.8b) and peripheral association of fibrillar collagen (violet). This finding is consistent with our RT-PCR observation (unpublished) indicating that expression of GFP mRNA remains constant and is not influenced by increases in collagen II mRNA during chondrogenic differentiation of mESC.

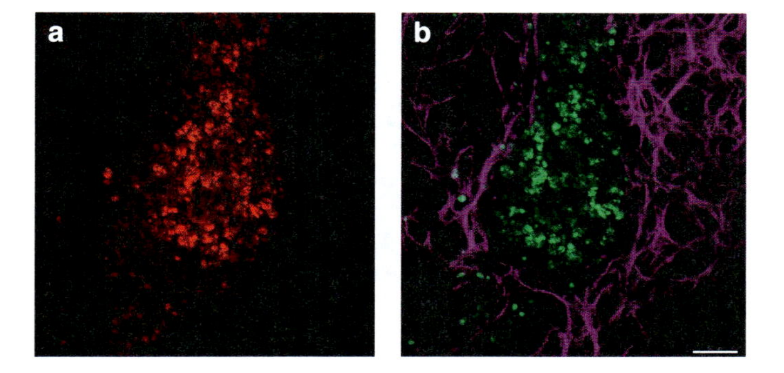

Fig. 13.8 Non-linear 3D imaging of a chondrogenic nodule within a 6-day old EB showing (**a**) immunolocalized collagen type II (red) and (**b**) the corresponding merged image of fibrillar collagen (violet) and eGFP expressing (green) cells.

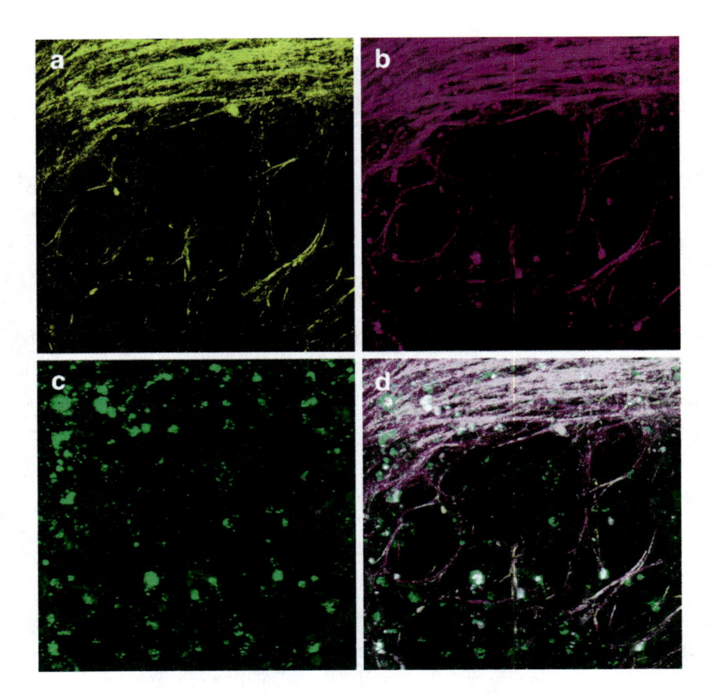

Fig. 13.9 Forward and backward SHG-signals of 12-day-old EB consisting of (**a**) new collagen (*yellow*), (**b**) old collagen (*violet*) and (**c**) eGFP expressing cells (*green*). (**d**) The merged image.

13.3.4 Forward and Backward Second Harmonic Generation Signals

Day 12 samples were imaged using the nonlinear optical imaging (NLOI) system by observing the f-SHG and b-SHG signals. As demonstrated by Williams et al. (2005), newly formed collagen is more prominent in the b-SHG signal (shown in violet in Fig. 13.9b). On the other hand, old and mature collagen can be better visualized as f-SHG signal (shown in yellow in Fig. 13.9a). This technique offers the possibility of discriminating between these two types of collagen and observing their amounts over time. Although there is some degree of co-localisation (overlap) between the two types (b-SHG and f-SHG signals) of collagen, there are subtle but distinct differences that can be seen.

13.4 Discussion

To date, the earliest time point for collagen formation during chondrogenic induction of mESC cells was reported as day 4 by Kramer et al. (2000). However, the NLOI results included herein (Fig. 13.10a–c) demonstrates that collagen fibres are

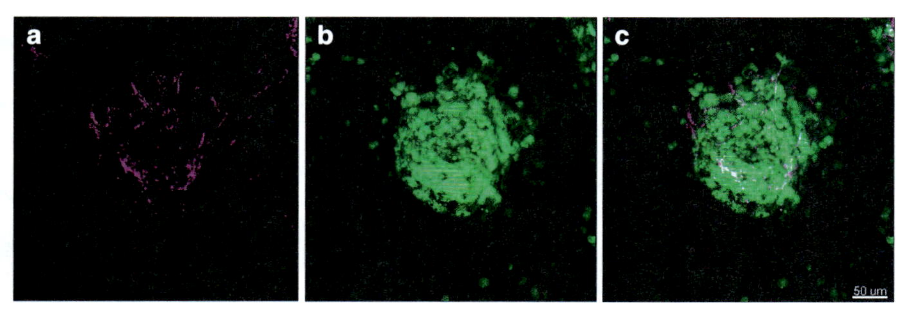

Fig. 13.10 Simultaneous imaging of (**a**) fibrillar collagen visualized by SHG from a 1-day-old EB consisting of (**b**) eGFP expressing nodules (*green*). (**c**) The merged image of SHG signal and eGFP fluorescence.

already produced in minute quantities within the EBs as early as day 1 as evidenced using SHG signal (Fig. 13.10a).

For Sox 9, Hwang et al. (2006a) and Nieden et al. (2005) also show that the expression fluctuates during the 17 and 35 days, respectively, of their experiments. Kramer et al. (2000) goes on to say that there is very little change. Both Crombrugghe et al. (2000) and Boon et al. (2004) have shown that Sox 9 is a gene required for chondrogenesis to occur.

Collagen X is present in the extracellular matrix in the form of hexagonal-shaped fibres during the later stages of differentiation. It is the gene for causing hypertrophy and bone mineralization (Elima et al. 1993) and was seen to have a sharp increase after day 6 and maintained higher than the initial value throughout the entire experiment (data not shown). Figure 13.3b also shows that there is an abundance of hexagonal-shaped fibres occurring around this time point as well. These results are also reported by Hwang et al. (2006a) who claim that this increase would be even higher if only chondrogenic medium was to be used, instead of any growth factors such as BMP-2.

To date, the effect of chondrogenesis on the expression of eGFP has not been reported. The results show that eGFP is consistently maintained throughout. This is important because during the exposure of BMP-2 medium, chondrogenic-specific genes have been upregulated which could mean that expression of other genes such as eGFP may have to be compromised. This, however, is not the case and means that this gene can be reliably used to study the morphology of cartilage-differentiating cells. Even more interesting is that the ratio between eGFP and the housekeeping gene was about 1, which suggests that their expressions are very similar.

13.4.1 Chondrogenic Nodules

The nodules shown in Fig. 13.8b have been reported by Kramer et al. (2000) as well. They mention that these cell aggregates are densely surrounded by collagen fibres, and this description matches the observations found in Fig. 13.8d. These nodules are also a characteristic stage during chondrogenesis and their frequency increases as the differentiation proceeds further.

13.5 Conclusions

In conclusion, one can successfully characterize various stages of differentiation of ESCs over several time points using NLOI that involves TPE and SHG. Importantly, the NLOI imaging techniques facilitate simultaneous visualization of the cell morphology, perform immunolocalization studies and observe the organization of fibrillar collagen.

1. SHG presents an alternative method for observing collagen II and X fibres. Although other techniques such as the use of fluoro probes and electron microscopy exist, these are time consuming and have high running costs. Furthermore, SHG-imaging is less damaging to cells and sample preparation is not necessary.
2. Also relevant to SHG, the current system has shown its ability in detecting such signals in both the f-SHG and b-SHG directions. Previously, this technique was reported elsewhere but only with collagen type I. However, results presented herein demonstrate that visualization of collagen type II is also possible and, hence, its application for in situ vital imaging. On the basis of this finding, it would be possible to discriminate between newly formed and mature collagens (also such as type II and type X). One could then ascertain exactly where in the cell this newly formed collagen begins forming and where the mature fibres are eventually polymerized to become fibrillar structures.
3. Sox 9 has been reported to act as a "master switch" for the regulation of other important genes in chondrogenesis. We have noted high level of Sox9 immunofluorescence in differentiating cells within the chondrogenic nodules. On the other hand, eGFP is not affected by the differentiation process. It has been verified that collagen II is expressed throughout all time points and is partly responsible for the extensive collagen fibrillar matrix. Through the use of SHG, it was determined that structural collagen type II is already produced after 1 day of induction. This is earlier than other reports in the past by other research groups. Collagen X is upregulated several days later and is responsible for the collagen X fibres with its characteristic hexagonal structure.
4. Apart from growing embryoid bodies attached in a 2D setting, we have also ascertained the effects of culturing them in a 3D scaffold of collagen type I. Although cell migration was seen in the attached setting, the peripheral cells of the 3D cultures showed even more spreading. This could allow for better structural support and anchorage between the EB and the scaffold.

Acknowledgements The authors are grateful to Dr. Frances Brook for her help in various ways during the completion of studies on chondrogenic differentiation of mESC. Professor Andy Carr and Professor Udo Oppermann are thanked for their support during the completion of this manuscript. UKT is presently supported by seed funding from Oxford Stem Cell Institute (OSCI), grant code: HFROVJ0.

References

Bain G, Kitchens D, Yao M, Huettner JE, Gottlieb DI (1995) Embryonic stem cells express neuronal properties in vitro. Dev Biol 168(2):342–357

Boerboom RA, Krahn KN, Megens RTA, van Zandvoort MAMJ, Merkx M, Bouten CVC (2007) High resolution imaging of collagen organization and synthesis using a versatile collagen specific probe. J Struct Biol 159(3):392–399

Boon CH, Cao T, Lee EH (2004) Directing stem cell differentiation into the chondrogenic lineage in vitro. Stem Cells 22:1152–1167

Bourne SJ, Polak JM, Hughes SPF, Buttery LDK (2004) Osteogenic differentiation of mouse embryonic stem cells: differential gene expression analysis by cDNA microarray and purification of osteoblasts by cadherin-11 magnetically activated cell sorting. Tissue Eng 10(5–6): 796–806

Brustle O, Jones KN, Learish RD, Karram K, Choudhary K, Wiestler OD, Duncan ID, McKay RD (1999) Embryonic stem cell-derived glial precursors: a source of myelinating transplants. Science 285(5428):754–756

Buttery LDK, Bourne S, Xynos JD, Wood H, Hughes FJ, Hughes SPF, Episkpou V, Polak JM (2001) Differentiation of osteoblasts and in vitro bone formation from murine embryonic stem cells. Tissue Eng 7(1):89–99

Campagnola PJ, Loew LM (2003) Second-harmonic imaging microscopy for visualizing biomolecular arrays in cells, tissues and organisms. Nat Biotechnol 21(11):1356–1360

Crombrugghe B, Lefebvre V, Behringer RR, Bi W, Murakami S, Huang W (2000) Transcriptional mechanisms of chondrocyte differentiation. Matrix Biol 19(5):389–394

Elima K, Eerola I, Rosati R, Metsäranta M, Garofalo S, Perälä M, Crombrugghe BD, Vuorio E (1993) The mouse collagen X gene: complete nucleotide sequence, exon structure and expression pattern. J Biochem 289:247–253

Fraichard A, Chassande O, Bilbaut G, Dehay C, Savatier P, Samarut J (1995) In vitro differentiation of embryonic stem cells into glial cells and functional neurons. J Cell Sci 108(10):3181–3188

Gratsch TE, O'Shea KS (2002) Noggin and chordin have distinct activities in promoting lineage commitment of mouse embryonic stem (ES) cells. Dev Biol 245(1):83–94

Hancock CR, Wetherington JP, Lambert NA, Condie BG (2000) Neuronal differentiation of cryopreserved neural progenitor cells derived from mouse embryonic stem cells. Biochem Biophys Res Commun 271(2):418–421

Hegert C, Kramer J, Hargus G, Muller J, Guan K, Wobus AM, Muller PK, Rohwedel J (2002) Differentiation plasticity of chondrocytes derived from mouse embryonic stem cells. J Cell Sci 115(23):4617–4628

Huber G, Matus A (1984) Differences in the cellular distributions of two microtubule-associated proteins, MAP1 and MAP2, in rat brain. J Neurosci 4(1):151–160

Hwang NS, Kim MS, Sampattavanich S, Baek JH, Zhang Z, Elisseeff J (2006a) Effects of three-dimensional culture and growth factors on the chondrogenic differentiation of murine embryonic stem cells. Stem Cells 24(2):284–291

Hwang NS, Varghese S, Theprungsirikul P, Canver A, Elisseeff J (2006b) Enhanced chondrogenic differentiation of murine embryonic stem cells in hydrogels with glucosamine. Biomaterials 27(36):6015–6023

Kato M, Yoshimura S, Kokuzawa J, Kitajima H, Kaku Y, Iwama T, Shinoda J, Kunisada T, Sakai N (2004) Hepatocyte growth factor promotes neuronal differentiation of stem cells derived from embryonic stem cells. NeuroReport 15(1):5–8

Kitajima H, Yoshimura S, Kokuzawa J, Kato M, Iwama T, Motohashi T, Kunisada T, Sakai N (2005) Culture method for the induction of neurospheres from mouse embryonic stem cells by coculture with PA6 stromal cells. J Neurosci Res 80(4):467–474

Kitazawa A, Shimizu N (2005) Differentiation of mouse embryonic stem cells into neurons using conditioned medium of dorsal root ganglia. J Biosci Bioeng 100(1):94–99

Kramer J, Hegert C, Guan K, Wobus AM, Muller PK, Rodwedel J (2000) Embryonic stem cell-derived chondrogenic differentiation in vitro activation by BMP-2 and BMP-4. Mech Dev 92(2):193–205

Kramer J, Hegert C, Hargus G, Rohwedel J (2003) Chondrocytes derived from mouse embryonic stem cells. Cytotechnology 41:177–187

Kramer J, Hegert C, Hargus G, Rohwedel J (2005a) Mouse ES cell lines show a variable degree of chondrogenic differentiation in vitro. Cell Biol Int 29(2):139–146

Kramer J, Klinger M, Kruse C, Faza M, Hargus G, Rohwedel J (2005b) Ultrastructural analysis of mouse embryonic stem cell-derived chondrocytes. Anat Embryol 210(3):175–185

Kramer J, Bohrnsen F, Schlenke P, Rohwedel J (2006) Stem cell-derived chondrocytes for regenerative medicine. Transplant Proc 38:762–765

Lau T, Adam S, Schloss P (2006) Rapid and efficient differentiation of dopaminergic neurons from mouse embryonic stem cells. NeuroReport 17(10):975–979

Lumelsky N, Blondel O, Laeng P, Velasco I, Ravin R, McKay R (2001) Differentiation of embryonic stem cells to insulin-secreting structures similar to pancreatic islets. Science 292:1389–1394

Martin GR, Evans MJ (1975) Differentiation of clonal lines of teratocarcinoma cells: formation of embryoid bodies in vitro. Proc Natl Acad Sci USA 72(4):1441–1445

Miyazaki S, Yamato E, Miyazaki J (2004) Regulated expression of pdx-1 promotes in vitro differentiation of insulin-producing cells from embryonic stem cells. Diabetes 53 (4):1030–1037

Mulholland WJ (2006) Targeted epidermal DNA vaccine delivery. DPhil Thesis, Department of Engineering Science, University of Oxford, Oxford

Murashov AK, Pak ES, Hendricks WA, Owensby JP, Sierpinski PL, Tatko LM, Fletcher PL (2005) Directed differentiation of embryonic stem cells into dorsal interneurons. FASEB J 19 (2):252–254

Nakanishi M, Hamazaki TS, Komazaki S, Okochi H, Asashima M (2007) Pancreatic tissue formation from murine embryonic stem cells in vitro. Differentiation 75:1–11

Nieden N, Kempka G, Ahr HJ (2003) In vitro differentiation of embryonic stem cells into mineralized osteoblasts. Differentiation 71(1):18–23

Nieden N, Kempka G, Rancourt DE, Ahr HJ (2005) Induction of chondro-, osteo- and adipogenesis in embryonic stem cells by bone morphogenetic protein-2: effect of cofactors on differentiating lineages. BMC Dev Biol 5(1):1–15

Pagani F, Lauro C, Fucile S, Catalano M, Limatola C, Eusebi F, Grassi F (2006) Functional properties of neurons derived from fetal mouse neurospheres are compatible with those of neuronal precursors in vivo. J Neurosci Res 83:1494–1501

Raikwar SP, Zavazava N (2009) Insulin producing cells derived from embryonic stem cells: are we there yet? J Cell Physiol 218(2):256–263

Riederer B, Matus A (1985) Differential expression of distinct microtubule-associated proteins during brain development. Proc Natl Acad Sci USA 82(17):6006–6009

Shi Y, Hou L, Tang F, Jiang W, Wang P, Ding M, Deng H (2005) Inducing embryonic stem cells to differentiate into pancreatic beta cells by a novel three-step approach with activin A and all-trans retinoic acid. Stem Cells 23(5):656–662

St-Onge L, Wehr R, Grusst P (1999) Pancreas development and diabetes. Curr Opin Genet Dev 9(3):295–300

Tang F, Shang K, Wang X, Gu J (2002) Differentiation of embryonic stem cell to astrocytes visualized by green fluorescent protein. Cell Mol Neurobiol 22(1):95–101

Tirlapur UK, Mulholland WJ, Bellhouse BJ, Kendall M, Cornhill JF, Cui ZF (2006) Femtosecond two-photon high-resolution 3D imaging, spatial-volume rendering and microspectral characterization of immuno-localized MHC-II and mLangerin/CD207 antigens in the mouse epidermis. Microsc Res Tech 69:767–775

Wang X, Ye K (2009) Three-dimensional differentiation of embryonic stem cells into islet-like insulin-producing clusters. Tissue Eng A 15(8):1941–1952

Wang EA, Rosen V, Cordes P, Hewick RM, Kriz MJ, Luxenberg DP, Sibley BS, Wozney JM (1988) Purification and characterization of other distinct bone-inducing factors. Proc Natl Acad Sci USA 85(24):9484–9488
Williams RM, Zipfel WR, Webb WW (2005) Interpreting second-harmonic generation images of collagen I fibrils. Biophys J 88:1377–1386
Ying QL, Stavridis M, Griffiths D, Li M, Smith A (2003) Conversion of embryonic stem cells into neuroectodermal precursors in adherent monoculture. Nat Biotechnol 21(2):183–186
Zipfel WR, Williams RM, Christie R, Nikitin AY, Hyman BT, Webb WW (2003) Live tissue intrinsic emission microscopy using multiphoton-excited native fluorescence and second harmonic generation. Proc Natl Acad Sci USA 100:7075–7080
Zoumi A, Yeh A, Tromberg BJ (2002) Imaging cells and extracellular matrix in vivo by using second-harmonic generation and two-photon excited fluorescence. Proc Natl Acad Sci USA 99:11014–11019

Chapter 14
A Correlative Microscopy: A Combination of Light and Electron Microscopy

Umberto Fascio and Anna Sartori-Rupp

14.1 Introduction

Many researchers have often considered that it would be interesting to describe tissues or cells starting from the observation of the whole sample and then investigating its subcellular components, by zooming into the same sample.

In recent years, fluorescence light microscopy (FLM) methods have allowed several advances in the clarification of the structure–function relationship existing in cells and tissues. Improved optical and laser equipments (Diaspro 2001), fluorescence labelling methods, computer technology and image analysis software have made it possible to study the location and dynamics of specifically labelled structures, from organelles to single particles. The spatial resolution of FLM techniques, which was until recently limited by optical diffraction to ~200 nm, can now reach a few tenth of nanometers thanks to the new super-resolution light microscopy-based techniques, such as photo-activated localisation microscopy (PALM; Betzig et al. 2006), stimulated emission depletion (STED; Willig et al. 2006) microscopy and structured illumination (Gustafsson 2005). However, only the labelled molecules of interest can be detected, while the information on their structure and on their structural organisation with respect to other cellular components is missing. Thus, we are led to ask the following question: what is the true size and shape of a structure seen through the FLM? This complementary structural information can be obtained by investigating the samples with transmission electron microscopy (TEM). An additional question arises: how do the structures seen with one method (FLM) relate to those observed with the other method (TEM)? Obviously, the same subcellular structure on the same sample has to be imaged with both microscopy

U. Fascio (✉)
CIMA, Interdepartmental Center for Advanced Microscopy, University of Milan, Via Celoria 26, 20133 Milano, Italy
e-mail: umberto.fascio@unimi.it

A. Sartori-Rupp
Institut Pasteur – Imagopole, 25 rue du Dr Roux, 75015 Paris, France
e-mail: anna.sartori-rupp@pasteur.fr

A. Diaspro (ed.), *Optical Fluorescence Microscopy*,
DOI 10.1007/978-3-642-15175-0_14,

techniques, meaning that the sample preparation will need to be compatible with both observation methods.

Such combination of microscopy techniques, referred to as "correlative light and electron microscopy" (CLEM), is a powerful method for correlating dynamic functional information from FLM with high-resolution structural information from TEM. The labelled structures of interest can be imaged by FLM and subsequently relocated and imaged in 2D or in 3D with electron tomography (ET) in the electron microscope. The relocation of these structures is a crucial step in a correlative microscopy approach. It is usually achieved by using sample supports provided with position markers, such as cell-locate glass slides or finder grids (Fig. 14.1), in combination with dedicated software allowing the transfer of coordinates of the labelled features of interest between the light and the electron microscope (Nickell et al. 2005, 2007). In particular, correlation of a montage of stitched micrographs recorded by LM with a similar montage recorded by EM facilitates a guided recovery of the features of interest. The accuracy of the

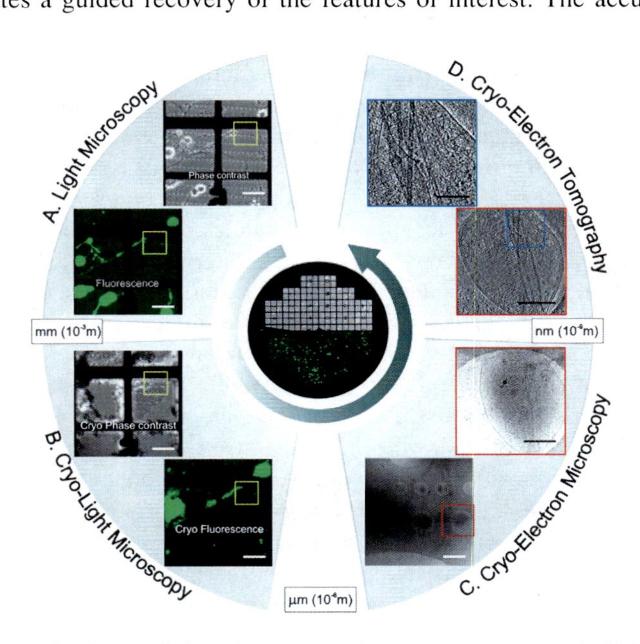

Fig. 14.1 Scheme for the correlative microscopy cycle at cryo temperatures. **A**. Light microscopy: growth of cell cultures (NG108-15 neuronal cells) on EM marker grids, live fluorescent labelling with Fura-2 calcium indicator, phase contrast and fluorescent imaging at 37°C. The yellow square points at the correlative area, a labelled cell neurite. **B**. Cryo-light microscopy: embedding of the sample in vitreous ice, cryo-phase contrast and cryofluorescent imaging (at -196°C) in the cryo-holder device and identification of the labelled neurite. **C**. Cryo-electron microscopy: transfer of vitrified grid and recovery of the coordinates of the area of interest in the EM. Low (3500x) and high (27500x) magnification micrographs of a neurite region. **D**. Cryo-electron tomography of the targeted area. The 3D tomographic reconstruction clearly reveals the cytoskeletal organization with the complex interplay between microtubules and a dense actin network. These features are barely visible in the corresponding 2D EM image (scale bars: **A**. and **B**. 50 mm; **C**. lower image 1 mm; **C**. upper image and **D**. lower image 350 nm; **D**. upper image 100 nm)

relocation depends on the type of molecular labels used, on the resolution at the FLM level and on the error on the coordinates transfer. The highest accuracy is reached with probes that are both fluorescent and electron dense, such as antibodies coupled simultaneously to a fluorescent dye and to a gold nanoparticle [e.g. fluoronanogold (FNG)] or to quantum dots (QD).

A CLEM method can be based either on ambient temperature (here denoted with classical CLEM) or on cryogenic (cryo-CLEM) approaches.

14.2 Classical CLEM Approaches

In classical CLEM approaches, the functional information obtained from high-resolution light microscopy – typically from confocal microscopy – is combined with the ultrastructural information obtained by classic TEM (Mironov et al. 2008). The sample is embedded in a support medium (e.g. resin) and cut into thin (~50–100 nm) or semi-thin (~100–500 nm) sections that are subsequently imaged with both FLM and EM. In some cases, adjacent sections are used: the fluorescence signal is collected by FLM from semi-thin section and the electron microscopy signal is collected by TEM from the adjacent thin section.

The need of imaging the same sample both with fluorescence and with TEM leads to more restrictive requirements for sample preparation than in the case of pure TEM observations.

Fluorescent and/or electron-dense labelling can be performed prior embedding in the support medium (e.g. by expressing proteins coupled to fluorescent proteins, by using vital stains or immunolabelling techniques, Suzuki et al. 2007) or post-embedding by immunolabelling the proteins of interest directly on the sections.

Fluorescent probes are incompatible with regular gluteraldehyde (GA) fixation for TEM due to fluorescent background noise generation (only low concentrations of GA ~0.05–0.5% can be used) and with osmium tetroxide post-fixation, which quenches the fluorescent signal and affects the antigenicity of many proteins. The partial or total omission of these fixation steps leads normally to poorer ultrastructural preservation and to lipid extraction.

The quality and preservation of cells or tissue sections are very important to obtain a good resolution of CLEM. High-pressure freezing (HPF) (Studer et al. 1989) followed by low-temperature resin embedding has become a standard method, but the use of thin-thawed cryosection (TCS), prepared following the Tokuyasu method or variations of this technique (Tokuyasu 1980), gives the best results. In fact, in the case of TCS, the preservation of the ultrastructural details is very high, the immunolabelling efficiency is improved, the antigenicity is preserved and the antibodies penetration into the TCS is facilitated in comparison with thin sections obtained from resin-embedded samples (Robinson et al. 2001). The use of TCS is ideally suited for the visualisation of cell membranes and for the immunolocalisation of multiple antigens with different sizes of gold particles, while it is less suited for cytoskeletal elements visualisation.

Various classic CLEM approaches have been proposed to answer different biological questions.

In an example of classical correlative microscopy, semi-thin sections mounted on coverslips were used for immunofluorescence microscopy (IFM) studies with fluorochrome-labelled secondary antibodies while the ultrathin cryosections mounted on EM grids were used for immunoelectron microscopy (IEM) with colloidal gold as the detection system. These approaches have some limitations; in fact the semi-thin sections are still thick enough to contribute to false colocalisation as the structures are stacked closely together and the adjacent ultrathin section that is examined by TEM is linked to the semi-thin section images with FLM but does not necessarily give the same structural information. Finally, different immunolabelling detection methods were used in these studies, a fluorochrome-labelled immunoprobes for IFM and colloidal gold- or ferritin-labelled immunoprobes for IEM (Takizawa and Robinson 2003).

With the aim of investigating the complexity of endocytic pathways in tobacco protoplast, Moscatelli, Onelli and Santo of the University of Milan (personal communication) have set up a CLEM method based on positively charged nanogold as a marker for the endocytic pathway and on immunolabelling of the antibody p28 to identify late endosomes. The researchers have embedded the tobacco protoplasts in LR gold resin, cut the resin blocks into thin sections (80 nm) and put the sections on nickel grids. The advantage of using thin sections – thinner than the *z*-dimension of an optical section in a confocal microscope (i.e. 200–500 nm) – consists of physically eliminating the background noise contribution coming from the out-of-focus fluorescence. Once determined the position of the fluorescently labelled structures of interest, they performed silver enhancement and observed the sections in the TEM. Even if this method requires an abundant handling of grids, it can have some significant advantages, such as for instance the possibility of observing exactly the same structure in the same section combining high-resolution confocal imaging and high-resolution TEM.

In recent years, a revolution in using fluorescence techniques to study biological systems came with the discovery of green fluorescent protein (GFP) from the jellyfish *Aequoria victoria*. The use of GFP-tagged proteins for CLEM offers exciting opportunities to probe further into the relationship between cell structure and function, and so far, the detection of GFP at ultrastructure level has proved to be very difficult. Grabenbauer and collaborators (Grabenbauer et al. 2005) have developed the GRAB method – where GRAB stands for "GFP recognition after bleaching" – to observe GFP in an electron microscope. A GRAB method uses oxygen radicals generated during the GFP photobleaching process to photo-oxidise diaminobenzidine (DAB) into an electron-dense precipitate that can be readily visualised by routine electron microscopy.

The recent development of probes that are both fluorescent and electron dense - like quantum dots (QD) and fluoronanogold (FNG) - have opened new possibilities for correlating light and electron microscopy. QD are semi-conductor nanocrystals that combine the properties of fluorescent molecules and of high-molecular weight particles detectable in the EM, with the advantage of featuring a narrow, high quantum yield and a resistance to photobleaching (Giepmans et al. 2005).

Fluoronanogold (FNG), a unique ultrasmall immunogold probe, has been developed to be used as a secondary antibody in immunocytochemical applications. It consists of a 1.4-nm gold cluster compound to which antibodies and fluorochrome are covalently conjugated. The ultrathin section mounted on EM grid and labelled with FNG can be used for correlative microscopy. The fluorescence signal is first collected and then the FNG is silver enhanced before examination by EM. Exactly the same immunolabelled structures can be imaged both with fluorescence and with EM (Robinson et al. 2001; Schwarz and Humbel 2007).

These CLEM approaches have some limitations; in particular, fluorescence photooxidation of DAB allows the labelling of recombinantly expressed proteins but not of endogenous antigens (Gaietta et al. 2002). Frequently, in the FNG method, the correlative analysis with FLM requires long observation times that cause a significant reduction in silver enhancement efficiency (Baschong and Stierhof 1998). Finally, QD's fluorescence is quenched by heavy metal staining so that correlative microscopy has to be performed on samples without a GA fixation and an osmium postfixation with a non-optimal ultrastructural preservation (Giepmans et al. 2005).

Vicidomini et al. (2008) report on a new high data output CLEM method based on the use of TCSs of rough endoplasmic reticulum and ER–Golgi intermediate compartments. The major advantages of their method are the possibility to correlate several hundreds of events at the same time and to combine the high analysis capability of fluorescence microscope with the high precision accuracy of TEM. To this aim, they have also developed a CLEM "package" (HDO-CLEM) that combines multiple labelling of both endogenous and recombinantly expressed antigens with 3D CLEM of the labelled structures of interest with a high data output rate.

Classic CLEM approaches are becoming increasingly popular but still suffer from the potential structural artefacts associated with EM preparations (Murk et al. 2003) due to the use of chemical fixation, dehydration, embedding in support media and the use of heavy ions as contrast agents. These artefacts are minimised when the biological sample is cryo-immobilised by HPF or other ultra-rapid freezing methods (such as plunge-freezing, slam-freezing, HPF, propane-jet freezing, etc.) that provide the highest structural preservation of the sample.

14.3 Cryo-CLEM Approaches

To investigate a biological sample as close as possible to its native state and image its real mass distribution in 3D, CLEM approaches at cryogenic temperatures have been recently introduced.

In the cryo-CLEM methods, the sample in its hydrated state is cryo-immobilised in vitreous (amorphous) ice by ultra-rapid freezing methods and kept frozen throughout the entire imaging process. Fluorescently labelled structures of interest are localised and imaged at cryogenic temperatures with cryo-FLM directly on the frozen hydrated sample. The frozen sample is then transferred to the cryo-electron microscope where the positions of interest can be recovered and the labelled structures imaged in 2D and/or in 3D with cryo-ET. Cryo-ET is a very powerful technique to perform 3D structural studies of large pleiomorphic objects, such as organelles or cells with a resolution of a few nanometers (Medalia et al. 2002). The main limitations of this technique are the low contrast of unstained biological material in vitreous ice and the high sensitivity to radiation damage that results in a poor signal-to-noise ratio in the electron micrographs. The implementation of a cryo-CLEM approach ensures a guided search for the labelled features of interest and minimises radiation damage during search (Fig. 14.1). The key feature and novelty in this approach consists of imaging the frozen sample with cryo-FLM, a new technique introduced independently by two research groups (Sartori et al. 2007; Schwartz et al. 2007). It is based on the design of modular stages that fit common light microscope platforms. These dedicated stages keep the frozen sample constantly at liquid nitrogen temperatures and preserve its vitreous state by protecting it from ice crystal contamination. Although the use of immersion objectives with higher numerical aperture and magnification is precluded by contact with a cold surface, imaging with FLM at low temperatures has the advantage of reducing the photobleaching by a factor of 10–15 times and thus has the potential to provide enhanced detection of weak or sensitive fluorescent signals (Fig. 14.2). Additionally, the use of ultrathin frozen cryosection provides further advantages for high-resolution fluorescence microscopy in cryogenic conditions (Gruska et al. 2008) if compared with cells grown on grids and plunge frozen (Sartori et al. 2007; Lucic et al. 2007). It offers excellent axial resolution and image clarity because the out-of-focus signal common to wide-field epifluorescence imaging and the background noise generated by autofluorescent molecules in ice are limited by the relatively uniform and thin specimen thickness set by the sectioning device, typically less than 200 nm (Robinson et al. 2001).

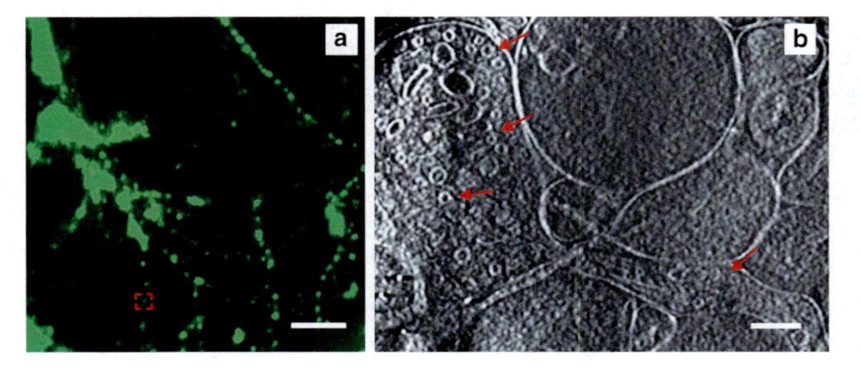

Fig. 14.2 Correlative microscopy at cryo-temperatures. **a**. Cryo-fluorescence microscopy of rat hippocampal neurons, cultivated on gold EM grids: live labelled for pre-synaptic vesicles with the dye FM1-43. Scale bar: 14 m. **b**. Cryo-electron tomography of the fluorescent spot (bouton) boxed in **a**, representing a pool of pre-synaptic vesicles (red arrows) on an axonal varicosity. Scale bar: 100 nm

Finally, Verkade (2008) designed a "rapid transfer system" allowing within 5 s the transfer of a sample from the LM to the HPF device to be cryo-fixed. This way, when we observe with the fluorescent light microscope an interesting event happening in the living cells, we can immediately transfer the sample for cryo-fixation and further cryo-sectioning and image the same event with cryo-electron microscopy of vitreous samples (CEMOVIS) with excellent ultrastructural preservation.

These approaches ensure the highest structural preservation of the biological material but, unlike the classic CLEM approach, they are generally not compatible with immunocytochemistry.

At this point, it still remains a major challenge to capture with CLEM events that happen very quickly, like synaptic vesicle fusion.

References

Baschong W, Stierhof YD (1998) Preparation, use, and enlargement of ultrasmall gold particles in immunoelectron microscopy. Microsc Res Tech 42:66–79

Betzig E, Patterson GH, Sougrat R, Lindwasser OW, Olenych S, Bonifacino JS, Davidson MW et al (2006) Imaging intracellular fluorescent proteins at nanometer resolution. Science 313:1642–1645

Diaspro A (2001) Confocal and two-photon microscopy: foundations, application and advances. Wiley-Liss, New York

Gaietta G, Deerinck TJ, Adams SR, Bouwer J, Tour O, Laird DW, Sosinsky GE, Tsien RY, Ellisman M (2002) Multicolor and electron microscopic imaging of connexin trafficking. Science 296:503–507

Giepmans BNG, Deerinck TJ, Smarr BL, Jones YZ, Ellisman MH (2005) Correlated light and electron microscopic imaging of multiple endogenous proteins using quantum dots. Nat Methods 2:743–749

Grabenbauer M, Geerts WJ, Fernandez-Rodriguez J, Hoenger A, Koster A, Nilsson T (2005) Correlative microscopy and electron tomography of GFP through photooxidation. Nat Methods 2:857–862

Gruska M, Medalia O, Baumeister W, Leis A (2008) Electron tomography of vitreous sections from cultured mammalian cells. J Struct Biol 161:384–392

Gustafsson MGL (2005) Nonlinear structured-illumination microscopy: wide-field fluorescence imaging with theoretically unlimited resolution. Proc Natl Acad Sci U S A 102:13081–13086

Lucic V, Kossel A, Yang T, Bonhoeffer T, Baumeister W, Sartori A (2007) Multiscale imaging of neurons grown in culture: from light microscopy to cryo-electron tomography. J Struct Biol 160:146–156

Medalia O, Weber I, Frangakis AS, Nicastro D, Gerisch G, Baumeister W (2002) Macromolecular architecture in eukaryotic cells visualized by cryoelectron tomography. Science 298: 1209–1213

Mironov AA, Polishchuk RS, Beznoussenko GV (2008) Combined video fluorescence and 3D electron microscopy. Methods Cell Biol 88:83–95

Murk J, Posthuma G, Koster A, Geuze H, Verkleij A, Kleijmeer M, Humbel B (2003) Influence of aldehyde fixation on the morphology of endosomes and lysosomes: quantitative analysis and electron tomography. J Microsc 212:81–90

Nickell S, Förster F, Linaroudis A, Net WD, Beck F, Hegerl R, Baumeister W, Plitzko JM (2005) TOM software toolbox: acquisition and analysis for electron tomography. J Struct Biol 149:227–234

Nickell S, Beck F, Korineck A, Michalache O, Baumeister W, Plitzko JM (2007) Automated cryoelectron microscopy of "single particles" applied to the 26 S proteasome. FEBS Lett 581:2751–2756

Robinson MJ, Takizawa T, Pombo A, Cook PR (2001) Correlative fluorescence and electron microscopy on ultrathin cryosections: bridging the resolution gap. J Histochem Cytochem 49:803–808

Sartori A, Gatz R, Beck F, Rigort A, Baumeister W, Plitzko JM (2007) Correlative microscopy: bridging the gap between fluorescence light microscopy and cryo-electron tomography. J Struct Biol 160:135–145

Schwartz CL, Sarbash VI, Ataullakhanov FI, McIntosh JR, Nicastro D (2007) Cryo-fluorescence microscopy facilitates correlations between light and cryo-electron microscopy and reduces the rate of photobleaching. J Microsc 227:98–109

Schwarz H, Humbel BM (2007) Correlative light and electron microscopy using immunolabeled resin sections. Methods Mol Biol 369:229–256

Studer D, Michel M, Müller M (1989) High pressure freezing comes of age. Scanning Microsc Suppl 3:253–268, discussion 268–269

Suzuki T, Matsuzaki T, Hagiwara H, Aoki T, Takata K (2007) Recent advances in fluorescent labeling techniques for fluorescence microscopy. Acta Histochem Cytochem 40:131–137

Takizawa T, Robinson JM (2003) Ultrathin cryosections: an important tool for immunofluorescence and correlative microscopy. J Histochem Cytochem 51:707–714

Tokuyasu KT (1980) Immunochemistry on ultrathin frozen sections. Histochem J 12:381–403

Verkade P (2008) Moving EM: the Rapid Transfer System as a new tool for correlative light and electron microscopy and high throughput for high-pressure freezing. J Microsc 230:317–328

Vicidomini G, Gagliani MC, Canfora M, Cortese K, Frosi F, Santangelo C, Di Fiore PP, Boccacci P, Diaspro A, Tacchetti C (2008) High data output and automated 3D correlative light-electron microscopy methods. Traffic 9:1828–1838

Willig KI, Rizzoli SO, Westphal V, Jahn R, Hell SW (2006) STED-microscopy reveals that synaptotagmin remains clustered after synaptic vesicle exocytosis. Nature 440:935–939

Index

A. Diaspro (ed.), *Optical Fluorescence Microscopy*,
DOI 10.1007/978-3-642-15175-0,

T

V

W

X